Lecture Notes in Computer Science　　　1605

Edited by G. Goos, J. Hartmanis and J. van Leeuwen

Springer

Berlin
Heidelberg
New York
Barcelona
Hong Kong
London
Milan
Paris
Singapore
Tokyo

Jonathan Billington   Michel Diaz
Grzegorz Rozenberg (Eds.)

Application
of Petri Nets
to Communication
Networks

Advances in Petri Nets

 Springer

Series Editors

Gerhard Goos, Karlsruhe University, Germany
Juris Hartmanis, Cornell University, NY, USA
Jan van Leeuwen, Utrecht University, The Netherlands

Volume Editors

Jonathan Billington
University of South Australia, Computer Systems Engineering
Mawson Lakes Boulevard, Mawson Lakes, SA 5095, Australia
E-mail: j.billington@unisa.edu.au

Michel Diaz
LAAS-CNRS
7, avenue du Colonel Roche, F-31077 Toulouse Cedex, France
E-mail: diaz@laas.fr

Grzegorz Rozenberg
Leiden University, Leiden Institute of Advanced Computer Science (LIACS)
Niels Bohrweg 1, 2333 CA Leiden, The Netherlands
E-mail: rozenberg@wi.leidenuniv.nl

Cataloging-in-Publication data applied for

Die Deutsche Bibliothek - CIP-Einheitsaufnahme

Application of petri nets to communication networks : advances in
petri nets / Jonathan Billington ... (ed.). - Berlin ; Heidelberg ; New
York ; Barcelona ; Hong Kong ; London ; Milan ; Paris ; Singapore ;
Tokyo : Springer, 1999
 (Lecture notes in computer science ; 1605)
 ISBN 3-540-65870-X

CR Subject Classification (1998): C.2, D.2.2, C.4, H.3.4, H.4.3

ISSN 0302-9743
ISBN 3-540-65870-X Springer-Verlag Berlin Heidelberg New York

© Springer-Verlag Berlin Heidelberg 1999
Printed in Germany

Typesetting: Camera-ready by author
SPIN: 10704876   06/3142 – 5 4 3 2 1 0   Printed on acid-free paper

Preface

The development of telecommunication systems dates back to the last century with the development of the electric telegraph, telephone, and the transmission, switching and signalling systems to support them. The forerunner of the Internet, the computer communications network ARPAnet, originated in 1969, when the US Department of Defence Advanced Research Projects Agency (ARPA) initiated experiments in resource sharing. Convergence of the two technologies has now occurred with the development of integrated digital networks to support multimedia applications involving voice, data, images and video. This application area, which covers a vast range of systems embodying traditional telecommunication systems and computer networks, is of utmost importance in the development of new and advanced information systems and services. The next ten years will see enormous changes in the way the world operates. For example, the increase in the use of the Internet as a means of communication (via electronic mail and bulletin boards) and for information discovery and advertising, using the World Wide Web, has been breathtaking. These systems will continue to expand and to influence most economic activity. Examples of the latter include the development of electronic commerce, advanced intelligent manufacturing systems comprising multiple co-operating agents, flexible delivery of global education products, on-line banking and trading, complex military command and control systems, space systems and transport control systems. This is an area of intense research worldwide, as is demonstrated by the many international conferences and countless journals that deal specifically with telecommunications, computer communications, networks, and advanced information services.

These systems will become vitally important to the operation of the global economy. The consequences of the failure of this new information infrastructure will range from minor annoyance to major disruption, costing millions of dollars (ECUs, Yen, etc.), such as in failures in networks supporting financial systems, or to loss of life in safety critical applications (such as air traffic control, the nuclear power industry, military operations, transportation, space and medicine). It is therefore vitally important that the communications systems that are built to support the information economy are engineered to rigorous standards of functionality and performance. To do this requires the use of rigorous methods for the specification, design, verification, performance evaluation, implementation, testing and maintenance of networks. These methods depend on the use of mathematical techniques, computer aided tools and methodologies.

Networks are inherently distributed and comprise systems with concurrently operating components. Petri nets offer a mathematically defined technique for the specification, design, analysis, verification and performance evaluation of concurrent distributed systems. They not only offer precise semantics and a theoretical underpinning, but also a graphical form that assists in understanding information and control flow within the same formalism. This is important for network specification and design. Petri net tool development during the last 15 years has also been invaluable in the management of large industrial applications. It is not surprising then, that Petri nets have been used in this domain for the last 25 years. The earliest work, which dates back to Merlin's work in 1974, seems to have been in the area of modelling and analysis of communication protocols. Communication protocols are the procedures used by software and hardware entities within networks to transfer not only user information (in several forms) between network customers, but also control information for managing connections and the network. They are the life blood and nervous system of the network.

This volume assembles a selection of the latest advances in the use of Petri nets for the modelling and analysis of communication networks and systems. The work on protocols is still very significant, and in this volume, is addressed directly by 4 of the 10 papers. However, Petri nets are now being used more generally in the modelling of communication systems, including high speed (asynchronous transfer mode (ATM)) packet switches and multiplexers, optical network receivers, intelligent network architectures (including call state models), feature interactions (conflicts) between different services (such as call forwarding and call waiting), bandwidth allocation policies, network management and local area networks. The papers in this volume cover most of these topics.

The book is structured into three sections.

The first section is concerned with functional modelling of communication systems and comprises three papers. The first, by Capellmann et al., shows how high-level nets and their associated tools can be applied to three aspects of the Intelligent Network. Lakos and Lamp provide an approach, using high-level nets with object oriented extensions, to the incremental modelling of an information retrieval protocol (Z39.50), illustrating how the evolution of standards can be handled. In the last paper of this section, Wheeler provides an extensive coloured Petri net model and a partial analysis of the IEEE 802.6 Metropolitan Area Network configuration control protocol.

The second section of the volume uses time extensions of Petri nets to examine performance. The first of these, by Mnaouer et al., uses Design/CPN's timed hierarchical coloured Petri nets to model and analyse the Fieldbus protocol used for factory automation. Moon et al. use Merlin's time Petri nets to analyse the logical link control procedures of IEEE 802.2 to establish appropriate parameter values. The final paper of this section, by Reid and Zuberek, provides a timed Petri net model for the analysis of layered protocols, using their university's ATM local area network and its higher layer protocols as an example.

The third section comprises a set of four papers that are concerned with the use of variants of stochastic Petri nets (SPNs) to analyse system performance. Haverkort provides an overview of polling mechanisms that are relevant to communication systems and the SPN models that are useful for the evaluation of their performance. These include token-bus and token-ring local area networks, and also an ATM multiplexer. In a second paper, SPNs are also used by Haverkort and Idzenga to investigate the cell-loss ratio of an ATM switch, using a structural decomposition approach. An extension to SPNs, called COSTPN (controlled SPNs) is introduced by de Meer et al. to allow for dynamic optimization. Their technique is applied to a multimedia server. The final paper of this volume, by Franceschinis et al., provides a stochastic well-formed net model of a multi-receiver system for an optical network. Optical networks are at the forefront of research for providing very high speed services. This paper examines packet loss due to contention for optical receivers, in an all-optical network, using state space reduction techniques.

This volume illustrates the extent of the use of Petri nets to the specification, modelling, analysis and performance evaluation of a range of communication systems, where some very useful results have been obtained. The volume has not been exhaustive in its presentation of the use of Petri nets for communication network applications (some notable exceptions being rigorous testing methods and network management), but it does provide a very good starting point for those interested in this rapidly evolving area.

All the papers selected for this volume have undergone an extensive refereeing process, involving at least three reviews and then a revision cycle if needed. For this, we are indebted to the referees who have given very generously of their time and expertise. Finally, we would like to thank the authors for their contributions and their patience with the reviewing and production processes. This volume is testimony to their insights, originality, creativity, analytical ability and communication skills.

January 1999   Jonathan Billington, Adelaide, Australia
                                Michel Diaz, Toulouse, France
                    Grzegorz Rozenberg, Leiden, The Netherlands

CONTENTS

Using High-Level Petri Nets in the Field of Intelligent Networks

Carla Capellmann, Heinz Dibold, Uwe Herzog

Deutsche Telekom, Technologiezentrum Darmstadt
P.O.B. 100003, D-64276 Darmstadt
E-MAIL: {capellmann|dibold|herzog}@tzd.telekom.de

Abstract. Telecommunication systems in general, and hence Intelligent Networks too, are large distributed systems with a high degree of concurrency. To develop and investigate specific aspects of such systems, adequate formal specification techniques are needed. Motivated by a short historical overview of Petri net usage at our premises, we introduce the requirements we have to pose on a formal specification language. We present three selected applications of high-level Petri nets in the area of Intelligent Networks. They cover modelling aspects under a functional-oriented modularisation as well as within an object-oriented approach. Formal verification techniques are addressed additionally. Our experiences prove, that, on a technical level, high-level Petri nets are well suited to specify telecommunication systems and services. However, to enable their broader usage in the telecommunication area, still some hurdles have to be taken. We discuss some of them.

1 Introduction

Today's telecommunication networks are facing the additional requirement to provide the ability to collect and compute information, i.e. to provide network intelligence, in addition to the classical transmission and switching functions. In this, the Intelligent Network (IN) is one of the most important evolution directions that are recognisable these days. The IN is understood as a concept for the fast and economical provision of new telecommunication services that go beyond the limits of conventional telephony. Starting from the assumption that each IN service can be assembled from a number of single, service-independent components, a logical and a physical architecture are defined for IN service provision. Details and an exemplary list of IN services can be found e.g. in [Dibo90] and [Thör94].

To investigate and develop specific aspects of such systems, formal specification techniques are needed that fulfil certain requirements. In the area of system and service development, the usage of SDL (Specification and Description Language, see [Z.100]), enhanced by MSCs (Message Sequence Charts, [Z.120]), clearly dominates. In the context of (pure) research projects, other formal description techniques are applied to meet specific needs and to overcome some drawbacks of SDL in this context. Among these techniques are (high-level) Petri nets (PNs), sometimes even used as the formal model underlying an SDL model.

This paper reports about motivation and experience of Deutsche Telekom Research on the usage of high-level PNs for the description of telecommunication systems and services. Section 2 starts with a short "historical" overview about our usage of PNs, thereby motivating important requirements on specification techniques in our application domain. Sections 3 to 5 present selected application cases from the IN area, considering first the modelling of a functionally structured system, the application of formal verification techniques to prove certain properties of the system, and eventually the modelling with PNs within an object-oriented paradigm. Finally, section 6 summarises our experiences with the application of high-level PNs in the field of IN and discusses the main factors which hamper their broader usage.

2 History of Petri Net Usage at Deutsche Telekom Research

2.1 Formal, Implementation-Independent Specification of Distributed Systems

A few decades ago, the development of telecommunication systems, i.e. switches, was mainly driven by technology, leading from pure mechanical switching networks with built-in hard-wired decentralised control towards processor controlled (but still mechanical) crossbars. After this approach became more and more common, two main streams were recognisable: Stored Program Controlled switches (SPCs) with a strict centralised control, using very high speed real-time processors - at least for that time -, versus decentralised control by means of several, less affording processor cores.

Like other public network operators, Deutsche Bundespost (as we called us at that time), had to decide which commercially available switch brand(s), based on one of the above control architectures, should be the future workhorse in its telephone network. In order to provide arguments from a technical viewpoint, we started to investigate whether a centralised or decentralised control architecture would be the "natural" approach with respect to a switch's tasks.

We soon found out that without a formal description of the switch's main tasks, being elaborated as implementation-independent as possible, this investigation would be fruitless. It also turned out that one important requirement for the selection of a feasible description language was the need to directly support concurrent behaviour on a fine grained basis, since telecommunication is concurrent by its very nature [Dibo88]. Looking at SDL which already was used and prolonged in-house at that time but still was very immature, these requirements were knock-out criteria. But one of the nice things with SDL was the offering of a graphical notation and its modelling approach being based on the concept of the finite state machine, a concept which seems to be easily understandable without a background in software engineering. This finally led us to discover Petri Nets ([Pete81]) and the work going on in the PN community.

 (We do not want to keep secret that the answer to the originally posed question of centralised vs. decentralised switch control became somewhat obsolete, since a) the two mainstreams converged in praxis by the introduction of additional (group) processors in the centralised approach and the crystallising of centralised structures in the decentralised approach, and b) it turned out that in procurements of such a dimension technical arguments are just one part of the overall decision matrix.)

2.2 The Need for High-Level Petri Nets with Structuring Means and Tool Support

Drawing up (literally taken) our first PNs during the above mentioned investigation, two more things became obvious: basic PNs were too elementary for the detailed modelling of concrete systems, and tool support, especially for "animation" of the described functionality and - due to the deeper insight into its dynamic behaviour - for the maintenance of the PN models, was urgently needed. The first requirement led us to shift to high-level PNs, adapting work on PNs with PROLOG inscriptions for data operations and declarations. PROLOG was favoured due to its declarative programming style. The second requirement was partially solved, i.e. not for graphics but for a textual representation of our so called Open Petri Nets (OPN), by writing our own OPN compiler and simulator based on a PROLOG solving engine [Dibo88].

Additional experience from our modelling tasks revealed that structuring means were required for practical work (e.g. the subnet concept as a possibility to refine a more complex transition). Hence we had an other requirement for our PN tool in use and we built in support for it in the OPN tool.

With the appearance of the PN tool Design/CPN [Design/CPN], we changed the PN class used for our modelling tasks towards Hierarchical Coloured Petri Nets (HCPNs, [Jens92]) in order to avoid the overhead to develop a graphical front end for our own OPN tools. Using Design/CPN also meant that we had to change to a functional programming language (SML) for data operations and declarations.

Having decided to use HCPNs as supported by Design/CPN, we re-focused on the actual application of PNs within research tasks, instead of working on the further development of PNs themselves. Our work on methodological aspects of using PNs [Dibo92b, CaDi93b, CaDi94] was continued though, as well as providing feedback to the tool suppliers.

2.3 Executable System Models to Investigate and Clarify Standards

Meanwhile, looking back at our proper research tasks, digital switches (with microprocessor control and a pure digital switching heart) spreaded all over the networks. While the classical task of telecommunication networks was (and still is) to provide resources for switching and transmission, a new demand arose: the ability to collect and compute information, i.e. to provide network intelligence - in a more and more service driven market, which led to the development of the concept of the Intelligent Network (IN, [Dibo92a]).

A key factor of ITU's (International Telecommunication Union) standardisation work on IN [Q.12xx] is the definition of the Basic Call State Model (BCSM). The BCSM represents an abstraction of the network's functionality as seen by (voice based) IN services. Being a core concept in the framework of IN, a precise and unambiguous description of the functionality of the BCSM is important. Furthermore, understanding the impact of its structure on service provisioning is essential for network operators aiming at introducing IN in its network. These questions motivated us to apply HCPNs in this context (see Section 3).

Working with the BCSM model, another important issue became apparent: The actual HCPN model deviated that much from the standard's BCSM shape, that communication with IN people about the model's behaviour (token flow) was hampered. Hence we investigated the possibility to additionally animate the token game in terms of the IN BCSM model. Section 3.5 reports in more detail about this work.

2.4 Making Use of Formal Validation

Our choice of HCPNs among the many PN classes around was guided by their ability to build compact, structured, and declarative system models, without compromising the mathematical power of PN theory too much. In most of our application cases, however, work focused on modelling issues. Validation in the sense of step by step investigations or simulation runs proved to be important in order to gain confidence in the model. Analysis of the system models by mathematical means was not possible or not important, though.

Considering the field of IN, however, there are certain problem areas where the application of formal validation techniques is feasible. One such area addresses the problem of service interaction: With the increasing number of IN and non IN service features, unexpected and/or undesired interactions between services or features are more and more likely to occur. Here, the usage of formal verification may prove to be helpful in order to detect such interactions and, in that way, to enable appropriate resolution activities to take place prior to the services' deployment in the network. Section 4 illustrates our approach towards interaction detection based on high-level PNs and temporal logic.

2.5 Going Object-Oriented

Currently, two trends can be distinguished in the area of IN: One covers the "classical" IN as defined in the ITU standards, the other addresses future developments of the IN. While the IN architecture of ITU is based on functional modularisation, most future developments propose to take advantage of object-oriented technology [P103, TINA].

Based on our contribution to this second trend, Section 5 presents in more detail an application of high-level PNs for object-oriented IN service modelling. Our main focus here was on the description of object behaviour, which was badly neglected in the "OO community" with respect to formality and hence unambiguity. The chosen PN constructs and their embedding in an overall object-oriented method are discussed as well as the problems that we faced.

2.6 Resulting Requirements

To sum up, this walk through the "Petri net history" at Deutsche Telekom Research should have illustrated our motivation to apply high-level PNs in the area of research

projects related to the Intelligent Network. Embedded in an overall methodological approach for system analysis and specification, high-level PNs essentially fulfil the following list of major requirements posed by our tasks:
- to describe distributed systems with their inherent concurrency,
- to cope with complex systems by powerful structurisation means,
- to model complex data and data operations in a compact way,
- to lead to declarative specifications which are independent of implementation-technology, and
- to gain formal, executable, analysable, but still comprehensible, graphical system models.

Almost needless to say, that appropriate tool support for modelling, simulation and analysis is strongly required.

3 Modelling the IN Basic Call State Model

3.1 Background

In section 2.3 we already mentioned one important concept in the IN architecture, the Basic Call State Model (BCSM). The BCSM represents the switching and message transport functions of the conventional telephone network, as perceived from an IN point of view. Services are realised in an IN in terms of service logics. Upon certain criteria the BCSM is able to recognise IN service calls and hands control over to the service logic. The service logic makes sure that the appropriate features of the service are executed and controls further basic call processing. Therefore, service logic and BCSM have to exchange information in order to provide the requested IN service.

Structure and granularity of the BCSM define when invocation of and interaction with the service logic is possible. Hence they are of great importance for the provisioning of IN services: If the structure of the BCSM is too coarse, only a limited amount of simple services will be feasible. A very fine structure would allow lots of different and sophisticated services, but had to pay off with tremendous costs for adapting the basic network infrastructure, leading to a big complexity and higher failure probability. Therefore it is essential to find an appropriate granularity that enables the production of most required IN services at reasonable costs.

To investigate this issue as well as to clarify open questions in the IN standards, we needed a formal model of the BCSM and its relation to the service logic. Although several PN models exist that regard aspects of the establishment of connection in a basic call, none of them models the basic call connection set-up in the context of IN. The standard itself includes an SDL model of the BCSM, but this model was extremely time-consuming to read and understand. Therefore we decided to develop our own model in order to investigate the BCSM and to gain a thorough understanding of IN service execution.

3.2 Introduction into the BCSM

Purpose of the BCSM is to describe the basic call processing, i.e. the switching and transmission functions of the conventional telephone network, as they should be perceivable for IN service logic. The BCSM, as given in the ITU standards, consists of two finite state machines: the Originating-BCSM (O-BCSM) modelling the processing at the calling party's local exchange, and the Terminating-BCSM (T-BCSM) modelling the called party's local exchange.

The states of the BCSM are referred to as "Point in Calls" (PICs). A PIC comprises a set of switching and transmission functions that are indivisible for IN service logic. An example of a PIC is e.g. PIC 2 Collect_Info (see Fig. 1). This PIC is responsible for collecting a sequence of valid digits representing e.g. a dialled number.

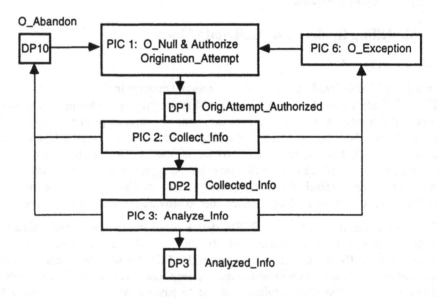

Fig. 1. Part of an O-BCSM [Q.12xx]

In order to be able to interrupt the basic call processing for an involvement of the IN service logic, the construct of "Detection Points" (DPs) was introduced. When call processing in the BCSM enters a DP which specific guard conditions are met, call processing is transferred from the switch to the IN service logic. DPs also are needed to return call processing from the IN service logic back to the switch, in general to proceed with another PIC and/or with changed parameters. The IN service logic itself is not covered by the BCSM.

Fig. 1 illustrates a part of an O-BCSM: PIC 1 comprises the idle state and takes care of authorisation of callers. It is followed by DP 1, where the first invocation of the IN service logic is possible, under certain conditions. Basic call processing proceeds with PIC 2, where number collection is accomplished. In case of termination of the call attempt by the user going on hook again, DP 10 is entered. Error cases, e.g. an invalid dialled number, are handled by transition to PIC 6, etc.

To sum up, O- and T-BCSM, with their PICs, DPs, and defined transitions, model the complete range of processing a basic call, from picking up the receiver (off_hook) until the receiver is put down again (on_hook). They also are the means to make the switches' functionality accessible to the IN service logic for IN service provisioning.

3.3 Design of the Model

Based on a graphical representation of O- and T-BCSM similar to the one shown in Fig. 1 and a textual description of the individual PICs and DPs, it was our objective to develop a formal, but still illustrative BCSM model based on HCPNs. This model should help us in gaining a deeper insight into the BCSM, validating the standards description and building a basis for discussion with IN experts. Especially the last point has some impact on the way the model should be designed, or as it turned out later, required additional work on a suitable animation.

Structural Aspects.

To define the general structure of the PN model, it had to be basically clarified how to represent the two BCSM constructs, PICs and DPs, in terms of HCPNs. Two important requirements had to be taken into account: Firstly, the model and its behaviour should be easily understandable in order to communicate the modelled IN service provision to IN people, who usually are not familiar with PNs. Secondly, to support this, the graphical appearance of the model should be similar to the IN BCSM representation.

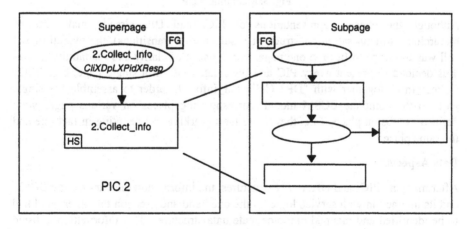

Fig. 2. Modelling a PIC

As PICs comprise states as well as ongoing processing until certain end conditions are reached, we decided to model each PIC by a place and a substitution transition. This allows us to keep the overall structure of the model clear and IN familiar, by hiding the details in a subnet on a separate page. The resulting structure for realising a PIC in HCPNs is shown in Fig. 2.

The second type of element in the BCSM, the DP, can be understood as a transition. The only action in each DP is to invoke the service logic when all criteria for invoking are met, otherwise it has to do nothing but to pass call processing on to the next PIC. This behaviour can be realised with two simple transitions: one passes processing to the next PIC, if the invoking criteria are not met, the other passes processing to the service logic, if the criteria are met. One common input place for both transitions completes our structure for DP modelling. DP2 is given as an example in Fig. 3.

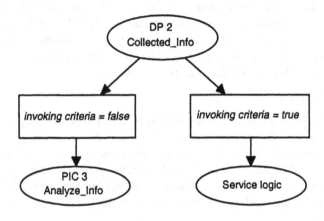

Fig. 3. Modelling a DP

Although the chosen representations of PICs and DPs already make use of hierarchical structurisation constructs (i.e. substitution transitions), the overall model still was too large to fit onto one page. For the sake of clarity and readability, it was thus decided to present every PIC and the related DP on a single page, e.g. "PIC2 Collect_Info" together with "DP2 Collected_Info". In order to assemble the single parts to the complete BCSM model, corresponding places on various pages were defined as "fusion places", i.e. they share their markings by denoting in fact one and the same place.

Data Aspects.

After mapping PICs and DPs to PN structures, the information relevant for the BCSM and its interaction with service logic on the one hand and users on the other hand had to be identified and mapped onto adequate data structures. This information is listed below:

Between User (calling/called person) and BCSM:
   - number of the connected line
   - all actions of the user (off_hook, dialled digits etc.)
   - responses to the user (tones etc.)

Between BCSM and Service logic:
- number of the calling line
- digits, user data
- position in the BCSM where the Service logic was invoked from
- responses to the user (content of announcements etc.)

Between O- and T-BCSM:
- number of the calling line
- number of the called line
- messages like 'request for connection' or 'terminating party busy' etc.

This data has to be defined in the functional language SML [HMT87, Wiks87]. SML offers a range of pre-defined basic data types, e.g. integer, real, string, boolean etc. and enables the construction of compound data types like lists, records etc. The basic data types together with the predefined type constructors were sufficient to model all the enumerated information in the BCSM model.

One type constructor not mentioned yet, the 'union', especially facilitates the modelling of data. According to the rules in HCPNs each place has its strict data type, and all input and output data must be of that type. This is inconvenient in cases where data of different types should be received via a given input arc. The following example illustrates this:

Data type 'Stimuli' is defined as union type, in order to enable the information transfer between user and BCSM with varying content:
- off_hook   (calling party lifted the receiver)
- on_hook   (calling party hanged up)
- digit (d)   (dialled digit d)

The declaration phrase in SML reads as follows:

color Stimuli = union off_hook + on_hook + digit:D;

A type definition starts with the keyword **color**. The single expressions *off_*hook and *on_hook* are type constants. *Digit* however indicates that it includes a variable of type D, which earlier was defined as type integer. Using the **union** construct thus enables us to define data in a problem oriented way by restricting data structures to the essential elements of the problem domain. This also contributes to a better comprehensibility of the model.

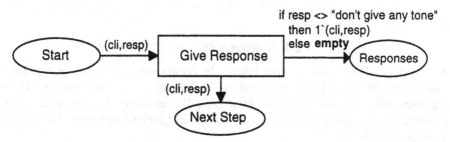

Fig. 4. Example for using the 'empty' construct

A second construct that facilitates the modelling of data in PNs concerns the production of output tokens when a transition fires. In the following example a transition has to present to an user different tones modelled by output tokens. When the input value is "dial_tone" or "busy_tone", then the token just has to be passed on as output token. However, if the input is "don't give any tone", we need no output token at all. This can be accomplished in Design/CPN by means of the 'empty' construct (see Fig. 4).

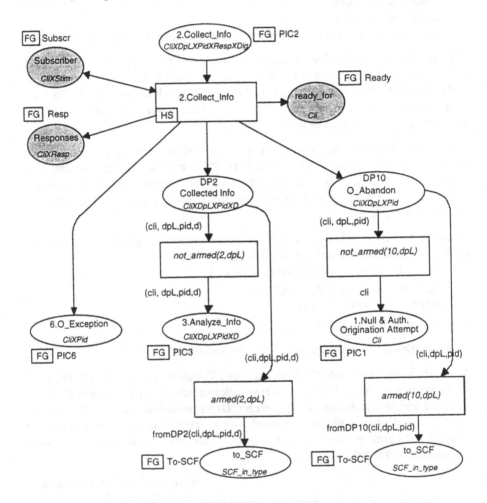

Fig. 5. PN model of PIC 2 and DP 2

A similar problem concerns input tokens. In certain cases, a particular transition should only fire when there are no tokens in a certain place. This could easily be expressed using the concept of inhibitor arcs. Unfortunately this construct is not supported by Design/CPN, though it has been defined for coloured Petri Nets (CPNs) in general [ChHa93].

3.4 The Resulting Petri Net Model

In this section part of the established PN model is shown. Since the complete model is too large to be included here, we exemplarily present the model of one PIC, namely PIC 2, and the related DP 2 (see Fig. 5).

Fusion place and substitution transition on top of Fig. 5 represent PIC 2 of the BCSM (compare Fig. 1). Depending on conditions in PIC 2, which are further specified in the subnet refining the substitution transition (see Fig. 6), the BSCM control flow branches either to PIC 6, DP 2 or DP 10. DP 2 is depicted in the middle part, DP 10 on the right hand side of Fig. 5. The three shaded places represent interfaces to the user.

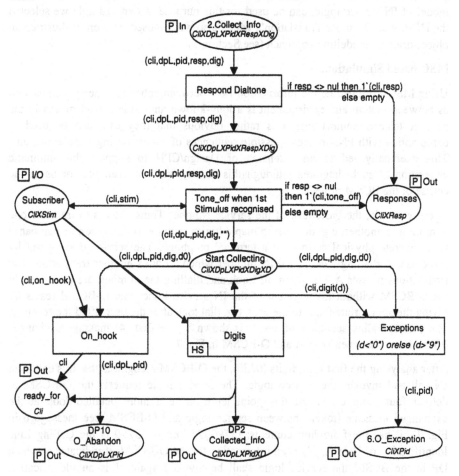

Fig. 6. Subnet refining the substitution transition of PIC 2

The refinement of the substitution transition of PIC 2 describes the collection of digits from the user and the control of tones to the user. Transition *Digits*, which checks the dialled number, again is realised as substitution transition, the corresponding subnet is not depicted in this paper. More details on the BCSM model can be found in [Herz95].

3.5 Simulation of the Model

Besides the HCPN model for O- and T-BCSM we had to specify the service logic, at least rudimentarily, in order to be able to show and investigate the BCSM behaviour in relation to IN services. Two typical IN services have been implemented and integrated in the model: Number Translation which is part of many commercial IN services, e.g. Freephone or Premium Rate Services, and VPN (Virtual Private Network), which is a service to build a private network by using public network resources. In this section we discuss how the presented BCSM model, extended by a model of IN service logic, can be used for this purpose. As an example we selected the IN service "Number Translation", which also will be used later on to illustrate an object-oriented modelling approach (see Section 5).

MSC based Simulation.

Using MSCs to illustrate the information flow between entities of the system as well as between system and environment is a long-known and wide-spread means in the area of telecommunications. It is rather obvious that they can also be used in combination with PNs to show possible sequences of events during simulation runs. This eventually led to an extension of Design/CPN to support the automatic generation of such diagrams - though this facility was not available for us in the course of the BCSM project.

In our example, the basic task of the service Number Translation is to translate one number into another, e.g. in order to map a logical number (of a person or company) to a concrete physical address of a terminal or phone. The service is triggered by some user A dialling a dedicated IN number which requires number translation. The first actions of user A, i.e. going off hook and dialling the number, are processed in the O-BCSM without involvement of the IN service logic. The O-BCSM reacts by giving the appropriate tones to the user, e.g. dial tone after the user lifted the receiver. The corresponding sequence of events is shown by the first (4) message exchanges (token flows) between User A and O-BCSM in Fig. 7.

After analysing the first four digits (*0130*), the O-BCSM recognises that this call is an IN call and invokes the service logic. The service logic requests the O-BCSM to deliver four more digits. At this point let us have a more detailed look at the exchanged message (token) between service logic and O-BCSM (see message 6 in Fig. 7). It has the following contents: (*3790,[2],1,nul,4*). *3790* is the Calling Line Identification of User A. *[2]* is an information for the O-BCSM; it indicates at which DP in the BCSM the service logic shall be invoked again. *1* is an identification number for the service instance started in the service logic. When the O-BCSM answers on the request, it includes this number in the message so that the message can be assigned to the correct service instance in the service logic. *nul* means that no message or tones shall be sent to User A, and *4* gives the requested number of digits.

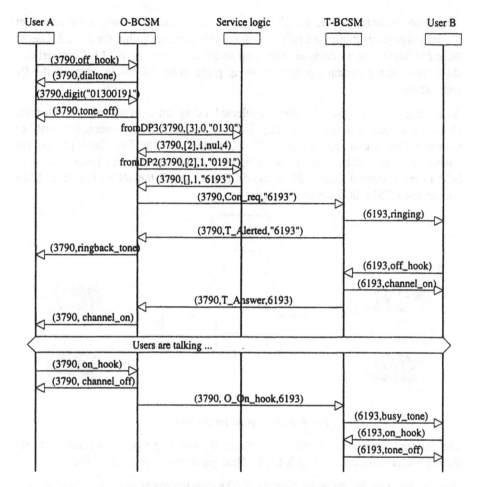

Fig. 7. Exchange of tokens during service execution

In the further processing, the requested digits are sent to the service logic which translates the dialled number (*0191*) into the real destination number (*6193*). This number is sent back to the O-BCSM together with the information that the call processing can continue as usually, i.e. no further involvement of the service logic is requested for this service. See Fig. 7 for the remaining message exchange until the end of the call is reached.

Animation of the BCSM Model.

In section 3 we mentioned that one of the objectives of the modelling work was to develop a formal and executable model of the BCSM that still reflects the BCSM structure. People from the field of IN should be able to understand and use the model without the need to learn (a lot) about PNs. As proven by the previous sections, this objective could not be reached yet! It was not possible to reflect the BCSM structure in the PN model, and even worse, the model became quite large and complex, mainly due to the inherent complexity of the BCSM functionality. Also the MSC based simulation presented above, though illustrating the information flow, is not sufficient for our purposes as it does not reflect what actually happens within the BCSM. Due to these considerations we decided to develop an animation of the BCSM PN model that shall visualise the dynamic processes taking place in the BCSM in terms of the IN architecture.

As starting point for this animation we selected a diagram representing relevant parts of an IN-structured network (see Fig. 8). This diagram shows several terminals or phones where a user can interact with the system, i.e. the IN. These phones are connected to two SSPs (service switching points), which again are connected to an SCP (service control point). SSPs and SCP are part of the IN architecture. SSPs contain the BCSM, SCP the service logic.

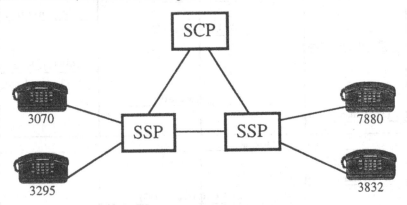

Fig. 8. Starting point for the animation

The phones represent the interface to the user. By selecting a phone, an user indicates that he wants to initiate a call (off-hook). Next the user may input a number.

By selecting a phone, the animation starts. The selected phone will be highlighted and the O-BCSM will pop up. Thereby, the actual state of the BCSM for this call will be highlighted as well. In that way, the call processing will be reflected by indicating in which PIC or DP the BCSM is at a certain step. Fig. 9 shows the case where an user is dialling a number, i.e. the O-BCSM is in PIC 2 "Collect_Info".

The implementation of this animation concept makes use of a library that was developed especially for the purpose of graphical animation in Design/CPN [Mimic]. In general, the animation is implemented by adding code segments to specific transitions.

3.6 Experiences

The suitability of HCPNs for modelling the BCSM and its interactions with service user and service logic was investigated. It was demonstrated that HCPNs and the language SML are well suited for modelling processes and messages in the telecommunication area. The hierarchical structurisation constructs of HCPNs help to clarify and overcome the often complex structures of real systems. This and the executability of HCPNs allow the validation of system specifications, e.g. of standards.

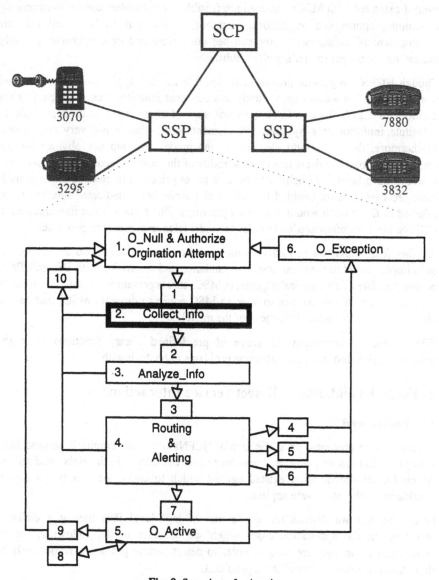

Fig. 9. Snapshot of animation

Besides a deeper insight in the processes of the BCSM that we gained during the course of modelling, we could identify some errors in the ITU-T standard on IN BCSM. This concerned mainly inconsistencies between graphical and textual representation. Another problem of the standards are sometimes vague descriptions that may lead to ambiguities.

Regarding our third objective, to use the HCPN model as communication tool with IN experts, we encountered more problems than expected. A direct use of the PN model, even if extended with MSCs, seemed not feasible. This problem can be overcome by developing appropriate animations. However, it also has to be stated that the development of animations is (or can be) quite time and cost intensive, possibly increasing the proper modelling effort substantially.

Though HCPNs in general proved to be feasible for this application case, there are some features that we miss(ed). Firstly this concerns inhibitor arcs. Though they are defined for CPNs, Design/CPN does not support them yet (?). It is possible to substitute inhibitor arcs by some PN elements - but this is not very convenient. Furthermore, this could add elements to the model, that are not obvious for the respective problem. It decreases the readability of the model. Secondly, in some cases it would have been very helpful to be able to set priorities to enabled transitions. If there are two or more enabled transitions at a time, but a dedicated one has to be selected to fire, then it would help to set priorities. This feature is not implemented in HCPNs, but very often needed in order to model telecommunication processes.

Another feature that we missed very much at the time we developed the model has been implemented in the meantime. For validation and simulation with Design/CPN it is now possible to automatically generate MSCs that represent the simulation runs. In addition, it would also be nice to have an MSC editor to draw up MSCs that can be checked against the actual behaviour of the model.

With respect to animation the usage of pre-defined library functions eases the implementation, but there are still some problems to be dealt with.

4 Formal Validation to Detect Service Interactions

4.1 Background

In section 3 we presented an application of HCPNs with main emphasis on modelling aspects. Validation was only used in order to gain confidence in the elaborated model. As no formal specification existed against which to check the model, no formal verification techniques were applied.

In this section we discuss the application of high-level PNs having a different objective. In the application case, which we consider in the following, formal verification techniques are used in order to detect certain problems, respectively to show that the model is free of these problems.

In contrast to our other application cases we used a different PN dialect throughout this project, called Product Nets [OcPr95]. Basically Product Nets are comparable to CPNs, though there are some differences. The language used for data declaration and manipulation is a different one. Some constructs, like the substitution transition, are missing whereas others, like inhibitor arcs, are included. The reason for using a different PN dialect and tool (the Product Net Machine) lies mainly in their strength with respect to formal verification techniques. In addition, the practical advantage of a cooperation with the Product Net and tool developers (GMD: German National Centre for Information Technology), situated close to our premises, should be exploited within this project.

4.2 Introduction to the Service Interaction Problem

As mentioned in the beginning, the IN aims at a rapid introduction of new services into the network. To reach this goal, services are developed independently of each other. One problem that may occur is that the operation of newly developed services or service features (unexpectedly) influences and alters the behaviour of the existing ones. Fig. 10 shows a well-known example where IN services CFU (Call Forwarding Unconditional) and SCR (Selective Call Rejection) interact.

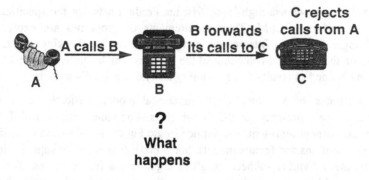

Fig. 10. Example to illustrate the problem of service interaction

In general there exist different ways how to handle the service interaction problem:

- Avoidance: Services (and networks) are developed in such a way that no interactions occur. However, it is unrealistic to assume that new architectures will be able to completely solve, i.e. avoid the service interaction problem.

- Detection and Resolution:

  Detection: (During service creation) potential service interactions are identified. Based on the service specifications, an analysis will take place that looks for conflicts, inconsistencies, etc. between the (combined) specifications.

  Resolution: A solution to detected interactions is provided. Such a solution may vary from a complete re-specification and re-design of a service, service prioritisation, run-time mechanisms to handle the interaction when it occurs, to the decision to do nothing.

Our objective was to apply formal analysis techniques defined for high-level PNs to detect interactions prior to the services' deployment in the network. More precisely we aim at developing a method for interaction detection that can be used in practice. The application case presented in the subsequent section should evaluate our approach on a realistic example.

4.3 Interaction Detection by Verification of High-level Petri Net Models

A lot of the research in the area of service interaction has addressed the problem of how to detect interactions on the basis of formal specification techniques and corresponding analysis methods. Many of them use the same basic idea, but differ in the selected formal techniques they apply. Our approach belongs to the category "satisfaction-on-a-model". The underlying idea of such approaches is as follows: Given a functional model of a service in some formal description technique and a formal specification of its requirements, it can be validated whether the service (model) fulfils its requirements. Considering now possible combinations of services (their models), an interaction can be detected if one of the services does no longer satisfy all its requirements.

Applying this idea, we use high-level PNs, i.e. Product nets, for the specification of IN services and the BCSM. Their requirements or properties are expressed by temporal logic formulae. Interactions can be detected by checking the temporal properties on the combined behaviour of the services which can be represented by an automaton obtained as result of a reachability analysis on the PN models.

Fig. 11 illustrates the structure of our functional model. Basically, we have two components. One represents the BCSM which is subdivided into O- and T-BCSM. The other component represents the Service Logic Function (SCF) and comprises the service specifications and feature models. Since Product nets do not support hierarchy constructs, the individual subnets are glued together via fusion places. The BCSM model contained here is similarly structured as the one presented in section 3, though it is modelled at a higher abstraction level.

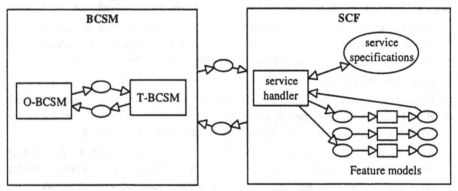

Fig. 11. Structure of the functional model

A service is modelled by stating the features it is composed of, its initial Detection Point, the dynamically armed Detection Points, if any, the point in the BCSM to which to return, and if the service has an access code, including this. E.g. for a service like the one presented in section 3 the service model would consist of feature "Number Translation", DP 3, nul, DP 3, and "0130". This information is defined by a corresponding data structure and put onto place *service specifications* in Fig. 11.

Each feature is specified by a separate subnet. The one for feature Number Translation is shown in Fig. 12. The place *Feature_queue* represents the interface to the service (logic) handler. A token on this place might trigger the feature's execution and will provide it with the needed input parameters. The result of the feature execution will again be put on this place.

The feature functionality itself is specified in the right hand part of the net. The feature translates a given number into another number. In its database *Database_NT* (see Fig. 12), number pairs are stored, representing for example UPT numbers together with their current destination numbers. When transition *NT* fires, the feature will be executed. It will look up the number *tli* (terminating line identification) that is stored for the input number *d* and return *tli* as result. Transition *No_NT* describes the case when the feature functionality can not be executed and no number translation will take place.

In addition to the functional model of services and BCSM, a formal specification of the services' requirements is needed. This is given in form of temporal logic formulae which can be checked against the model behaviour. However, in general, the behaviour of the specified services including the BCSM is too complex to efficiently check properties on the automata representation of the behaviour (compare [CMR93]). A complete analysis of a simple BCSM model already leads to a behaviour representation having several thousand states. Including services, an order of magnitude of the number of states will be reached where (efficient) analysis will no longer be possible.

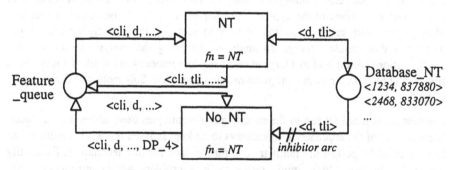

Fig. 12. Feature net for Number Translation.

To overcome this complexity problem, we applied an abstraction technique [Nits94a, Nits94b, Nits96, Ochs94]: Its idea is to first compute an abstraction of the behaviour that has a sufficiently small automata representation and contains sufficient information to check temporal properties of interest efficiently (see Fig. 13). When computing the abstract behaviour, it has to be made sure that behaviour important with respect to service interactions is not lost. One important class of properties that has to be addressed here is the class of lifeness properties (e.g. it should always be possible to make a call eventually), and it is especially this class of properties that is delicate to handle with respect to abstractions [NiOc95]. A small example illustrating the approach is given in [CDGNO96].

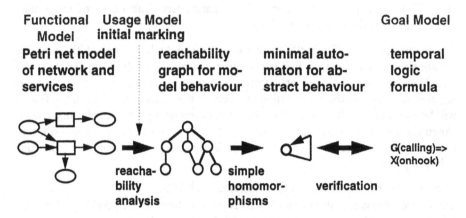

Fig. 13. Principles of our approach towards interaction detection

4.4 Experiences

The presented approach was applied to five services consisting of several features. It proved to be feasible, i.e. it is possible to detect interactions based on PN specifications and related analysis techniques. However it was also soon recognized that certain extensions of the approach are needed if it should be useful in practice. Mainly this addresses the question of how to identify those configurations (initial markings) that should undergo an analysis. Applying the analysis to all possible configurations would lead to a huge amount of often unnecessary work. To this end, a first model to compute such configurations was specified. Still more work in this area is needed.

Another point more related to the specification techniques used addresses the "user-friendliness" of the approach. It has always to be kept in mind that such a method has to be applied by people not familiar with formal specification techniques. Especially the use of temporal logic might prove to be a problem for an introduction into practice.

Finally regarding Product nets as compared to HCPNs our experience was that both net classes can be used in our application domain. HCPNs have their strength in modelling. Hierarchical constructs are extremely important when modelling realistic or real applications. Here Product Nets offer only fusion places as sole structurisation means. However they also have nice features, like the inhibitor arcs, that HCPNs, at least Design/CPN, do not support at the time being.

5 An Object-Oriented Model of an IN service

5.1 Background

As third application case of PNs in this paper we present the modelling of IN services in an object-oriented paradigm. The motivation for this shift in paradigm is the aim to overcome several shortcomings of service creation in a "classical" IN.

In the "classical" IN service creation is based on the notion of SIBs (service independent building blocks) which are functional modules providing basic service functionalities independent of specific services. Developing a new service with SIBs means to compose it on the basis of a predefined set of SIBs. With the advent of object-oriented techniques and their increasing spreading, it was quite natural to investigate whether and how object-oriented approaches might help to improve service creation (see for example [P103, SCORE, TINA]). The key concerns addressed by object-oriented methods
- "industrializing the software development process;
- making systems with open interfaces;
- ensuring the reusability and extensibility of modules;
- capturing more of the meaning of a specification"
as given in [Grah91, p 27/28] appear to address most of the major concerns of service creation in an IN.

When trying to adopt an object-oriented method for the specification of IN services, it became clear that, at least at that time, none of the existing methods offered a (formal) specification technique that was suitable for our purposes. Based on these considerations, we decided to develop the Object-Oriented Petri Net Method (OOPM), in order to integrate the benefits of an object-oriented method with those of high-level PNs as a formal specification technique proven in our application domain. At that time no other approaches aiming at integrating PNs with object orientation were available. Our method shall be suited for the development of IN services, taking into account the requirements specific to analysis and specification of IN systems or services. It is based on the Open Petri Net method [Dibo92b] and the object-oriented method OORASS [Reen92].

In the following, an object-oriented PN model of an IN service (or parts thereof) will be presented and discussed. Emphasis will be put on the integration of PNs in an object-oriented method and the problems that arose. The method OOPM itself will only be introduced as far as needed for the understanding of the service model; for more information, see [CaDi94] where also a more complete service specification is given.

5.2 Introduction into OOPM

The basic idea of OOPM is to regard a system (or service) as a system of interacting objects. At the specification level, a system will be represented as structure of collaborating classes, here called roles, where the structure defines the collaboration between objects: objects that play a certain role (belong to a certain class) need to interact with objects playing other roles by exchanging certain (defined) messages, in order to fulfil the task of the system. Such a model is called a role model. It consists of three elements: a role diagram, contract definitions and role specifications.

In Fig. 14, a simple *role diagram* is given in order to explain the notation (see also Annex B). The role diagram consists of three roles (big circles). The environment role (shaded circle) may only interact with role A. The messages it may send to role A are defined in contract 1 (small circle). Role A may interact with both other roles: contract 2 defines the messages it may send to the environment role and contract 3 those to role B. Interactions between environment and system are represented by bold lines, system internal interactions by normal lines. A single small circle indicates that the near role knows one object playing the remote role. A double circle means that the near role knows one or more objects playing the remote role.

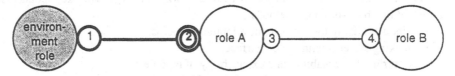

Fig. 14. A simple role diagram

Contracts define the messages that one role may send to another. They are defined in SML. The behaviour of *roles* is specified using CPNs. The basic idea underlying the use of CPNs for the specification of roles is as follows: Each role will be described by a place surrounded CPN, i.e. a subnet that is surrounded by a border of places. The representation of the internal state structure of the role is thereby modelled by one or more places. The methods that may change the internal state and/or invoke the sending of messages are modelled by one or more transitions.

An object playing a role may be seen as an instance of a role. Accordingly, an object is represented as an instantiated CPN, i.e. an CPN with a marking. The concrete state of an individual object is thus modelled by the set of tokens/data that is actually lying on the places comprising its state.

The methods that an object offers are invoked by sending a corresponding message to the object. These messages are represented as tokens that are lying on a special border place. These border places serve as interfaces between the objects. A specific type of places is used here that allows direct communication between the objects, although each object is represented by a separate CPN. This approach realises asynchronous object communication. Note, that only the border places of an CPN are accessible from the outside in accordance with the encapsulation principle holding for objects.

In Fig. 15 a simple CPN is shown that illustrates the PN concepts and introduces our notation (see also Annex B): Places are represented by ellipses, transitions by rectangles, and the connecting arcs by arrows. The border places serving as interfaces are graphically distinguished from internal places by a hatched surrounding. Names for places and transitions are given inside of the ellipses and rectangles respectively. The data types of the places are shown below the names in italic. The arrow annotations describe the data that may flow over the arc. Variable names are written in lower case letters whereas constants like message names start with an upper case letter.

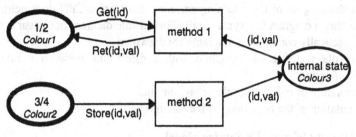

Fig. 15. A simple CPN

As a summary Table 1 relates the object-oriented concepts with the corresponding PN concepts:

Object-oriented concept	Petri net concept
role description	place surrounded CPN
internal structure	place(s)
method	transition(s)
object playing a role	instantiated CPN, i.e. CPN with marking
internal state	marking on place(s)
message	token on an interface place

Table 1. Relationship between object-oriented and PN concepts

Since a system or service is usually rather complex and may be quite large, a role model describes usually a coherent and self-contained sub task of the overall system's task setting. In a synthesis process, the individual role models specifying the different sub tasks of the system are integrated into one system role model. Hereby, related roles will be merged, or put differently, new synthesised roles will inherit from roles of the individual models.

For the development of real systems *tool support* is mandatory. A tool should support the method in each of its steps. For OOPM no tool exists that supports exactly this method taking its process model into account. Our idea was to (mis)use the editor of an available PN tool to bridge this gap as much as possible.

For the main part of OOPM, the role modelling and validation, we were able to use Design/CPN, although sometimes not as intended. Representing role diagrams in Design/CPN is a bit tricky: Roles are internally modelled as sub-transitions that are refined by an CPN. Note that this was only done to establish a link between a role and its specification. All other elements of the role diagram are represented as auxiliary nodes that have no semantics for the tool.

Thus the use of the tool allows

- the drawing up of the role diagram; note that Design/CPN does neither check whether a diagram is a syntactically correct role diagram nor ensures that only syntactically correct role diagrams are drawn up,
- the role specification by CPNs with a direct link between a role and its specification,
- simulation of individual role behaviour, and
- simulation of the behaviour of the role model.

5.3   An Object-Oriented IN Service Model

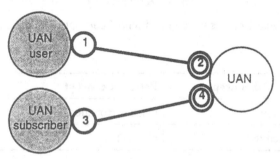

Fig. 16. UAN service and its environment

Stimulus	Response
1 Dialling of UAN by service user	2 Call connection if possible otherwise call rejection
3 Request from service subscriber for subscription to UAN service	4 Confirmation of UAN service subscription for the service subscriber if possible otherwise notification of rejection
Request from service subscriber for cancellation of his subscription	Confirmation with date of subscription end
...	...

Table 2. Stimuli and responses

To illustrate the application of OOPM, we present parts of a model for a fictitious IN service. This service is similar to the Number Translation service already used before. Its main task is to translate an Universal Access Number (UAN) dialled by a service user into a concrete terminating line number, depending on origin of call (feature ONE: one number) or origin of call and time and date of call origination (features ONE and TDR: time dependent routing). The assignment of a terminating line in the above service features ONE and TDR is modifiable on-line by the service subscriber. Furthermore, the UAN service can include other service features which are defined for incoming calls to terminating lines. Fig. 16 presents an abstract model of the UAN service and its environment. The contracts between UAN service and environment are summarised in Table 2.

The overall task of the UAN service can roughly be divided into two sub tasks: subscription, cancellation and modification on the one hand, and service usage on the other, each of which may be further divided. In this paper, we will focus on service usage and consider only the role model for the UAN translation.

The role model *UAN translation* describes the translation of an UAN into a destination line number: When an user dials an UAN, the service first has to identify the destination line that is defined in the translation scheme for this UAN, thereby taking origin of the call or origin and time of the call into account. In a second step, the service will try to connect to the computed destination line. This latter task is specified in *terminating line* role model which is not included here.

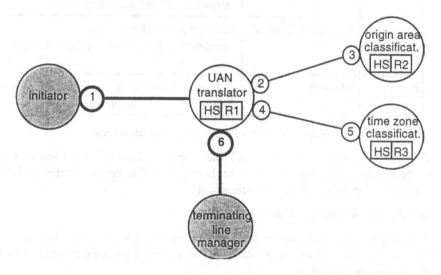

Fig. 17. Role diagram for UAN translation

The role diagram for *UAN translation* is given in Fig. 17. The following roles have been identified:

- Role *initiator* belongs to the environment of this role model. This role initiates the translation of an UAN.

- Role *UAN translator* is responsible for the computation of the call destination on the basis of the UAN data, i.e. the translation scheme.

- Roles *origin area classificator* and *time zone classificator* are responsible to deliver the origin area and the actual time zone of a given call.

Role *terminating line manager* receives the computed destination line and is responsible for the further processing. Since this will be specified in a separate role model, the role belongs here to the environment.

contract	SML expression	description
1	(cli,uan)	tuple of call line identification and universal access number (no message name)
2	ReqOrgZ(cli)	message: request for origin area parameter: call line identification
3	RetOrgZ(cli,orgZ)	message: return of origin area parameters: call line identification and origin area
4	ReqTimeZ(cli)	message: request for time zone parameter: call line identification
5	RetTimeZ(cli,timeZ)	message: return of time zone parameters: call line identification and time zone
6	(cli,dest)	tuple of call line identification and computed destination (no message name)

Table 3. Contracts for role model *UAN translation*

The contract definitions are given in Table 3. In the following the PN specifications of the (internal) roles will briefly be explained. Note that the environment roles are not specified in the context of this role model.

Specification of role *origin area classificator*.

Role *origin area classificator* maps a call origination (cli) onto its origin area and returns the origin area when requested (see Fig. 18). For a precise definition of the net inscriptions, see Annex A.

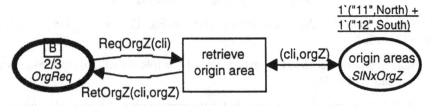

Fig. 18. Specification of role *origin area classificator*

Specification of role *time zone classificator*.

Role *time zone classificator* determines the actual time zone in which a call takes place and returns it on request (see Fig. 19).

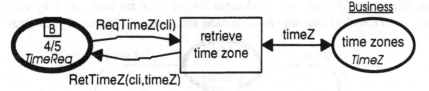

Fig. 19. Specification of role *time zone classificator*

Note that the role itself does not need to know the call line identification (*cli*) in order to determine the time zone. This parameter functions as process identification: If several calls are processed and thus the actual time zone is requested several times, it has to be indicated to which call the result refers.

Specification of role *UAN translator*.

Role *UAN translator* is responsible for the computation of the call destination for an incoming UAN call (see Fig. 20).

Triggered by an incoming call (*(cli,uan)* on place *1*) containing call line identification and dialled UAN, the translation scheme for the dialled UAN is looked up. Depending on the returned translation scheme, either transition *request data for one dim function* (feature ONE) or transition *request data for two dim function* (features ONE and TDR) will be enabled: In case of an one dimensional translation scheme, the call destination for the UAN under request depends only on the origin area of the call. Thus, only the origin area for the given call will be requested. In case of a two dimensional translation scheme, both, origin area and time of call, determine the call destination and have to be requested.

The computation of the call destination will be done by either transition *compute dest(org)* or transition *compute dest(org,time)*: Function *one_dim()* or *two_dim()* will extract the destination out of the corresponding translation tuple of UAN data. The result is put as tuple (cli, destination number) on place *6*.

Validation of Role Model.

When all roles of the role model are specified and validated by symbolic execution, the role model itself has to be checked: It has to be investigated whether the interactions between the roles are as defined and whether the role model fulfills its task. As for the validation of the individual role specifications, the role model will also be validated by simulation. The symbolic execution of the model helps a lot to avoid and/or detect specification errors, especially with respect to concurrent behaviour.

Note that although the role diagram has currently no formal semantic support by the tool used, we are able to do a simulation without extra work necessary, since the sub nets representing the role specifications communicate via interface (border) places defined according to the contract links of the role model.

Fig. 21 shows an MSC that represents a "test run" of the UAN translation role model. Each role instance is represented by a vertical line with a starting rectangle containing

the role name. The messages exchanged by the roles are represented by arrows
between the vertical lines with message name and parameters annotated to the arrow.

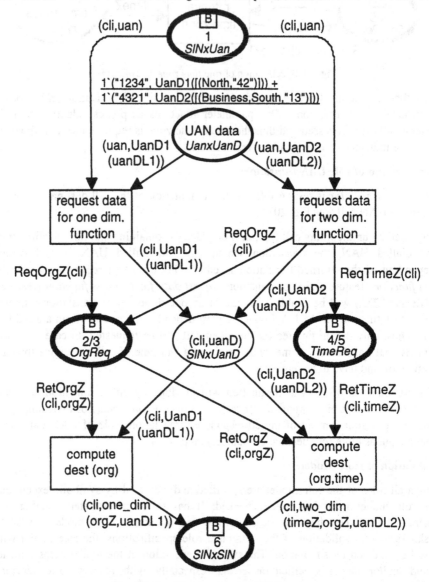

Fig. 20. Specification of role *UAN translator*

Note that this MSC shows only one possible sequence of the message exchange that
is allowed between the roles of the role model. All possible sequences are defined by
the PN specifications of the roles. In the case shown, the translation scheme is one
dimensional, i.e. the computation of the call destination depends only on the call
origin. Role *time zone classificator* has thus been omitted in the diagram.

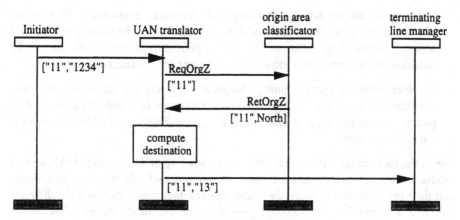

Fig. 21. MSC for a test run of UAN translation role model

5.4 Experiences

In this section, selected parts of an object-oriented PN model of an IN service have been presented. The original model (see [CaDi94]) consists of 4 role models specifying the different sub tasks of the service and one service role model. The 4 sub task models contain altogether 27 roles, the service model 17. The size of the CPNs defining the roles varies from very simple ones with one interface place, one internal place and one transition to nets with 6 interface places, 5 internal places and 8 transitions. In average, a role specification consists of about 3 interface places, 1.5 internal places and 3 transitions. Altogether, the UAN service model contains 61 places (41 interface places and 20 internal places) and 43 transitions.

The synthesis process is not included here. In this step, a new role model is built on the basis of existing role models by defining synthesised roles that inherit from different other roles. In this area further work is certainly needed. An automatic or (tool) supported synthesis of different role models is one important issue. Consistency checks, ensuring that there are no conflicts between the different role behaviours that a merged role inherit, refer to another important point.

Considering the integration of PNs into an object-oriented approach, CPNs offer structurisation mechanisms fitting to the concepts of role modelling. The behaviour of roles may be specified in a declarative way by giving the pre and post conditions of each of the role's methods. Thereby full concurrency is taken into account which is often neglected. The executability of CPNs allows the system description to be verified against one's own (informal) conceptions of the desired role behaviour, role model behaviour and system behaviour.

However a number of difficulties and problems that arose during the elaboration of the service model have to be mentioned as well. Quite a lot of decisions had to be made how to map the object-oriented concepts onto PN concepts. A comparison with possible alternatives and a thorough assessment can not be given at the moment, but some general problems have been recognised:

- There exists no mechanism that implicitly supports transactions. Due to the object-oriented structure of the model, some tasks have to be divided into several actions performed by different roles. As a consequence such tasks can not be modelled by only one transition that would ensure the transaction properties.

- In object-oriented (programming) languages, objects are addressed via their identifiers. In our approach, the recipient of a message is graphically addressed by putting the message on a specific (interface) place. This way of addressing roles is not always feasible.

Use of the tool Design/CPN enabled the fast drawing up of the complete UAN service model, containing role diagram, contract definitions and role specifications. Roles and their specifications are directly linked and communication between the different sub nets representing the roles is possible via "interface" places. This allows immediate validation of role specifications and role models by symbolic execution. The main drawbacks of the use of Design/CPN concerns the missing overall tool support. In order to cope with this problem, we (mis)used Design/CPN also for the role modelling. But it remains that the role modelling part of the method and the underlying process model is not supported by a tool.

Further problems related to Design/CPN are mainly the following:

- In Design/CPN, no local declarations are possible and names, e.g. of colours, must be unique. As a consequence, meaningful names have sometimes to be enhanced, just in order to make them distinguishable. This hampers reuse of message names for comparable purposes.

- Important object-oriented concepts like inheritance could not be completely integrated in the method due to missing tool support. It was quite often the case that different roles exhibit the same behaviour but on different data types. At the moment reuse can only be done by copy, paste and adapt. It would be nice to have tool support here, e.g. to allow for generic descriptions where the data types have to be refined.

With respect to a formal model of IN service provision, further work is certainly required. In order to support and improve this work, enhancements and improvements of OOPM including its tool support are especially needed in the following areas:

- refinement of the process model due to application feedback,

- better integration of object-oriented and PN concepts integrating also concepts like inheritance,

- better integration of role diagram support, either by adaptation of role diagram syntax to existing CPN constructs or by tool adaptation.

6 Conclusions

In this paper we presented a selected set of application cases of high-level PNs in the field of Intelligent Networks. The application cases differ in their objectives and their task settings: In the BCSM project, HCPNs are applied under a functional paradigm whereas in the UAN service modelling, they are used within an object-oriented approach. In both cases, the objective was to elaborate a formal and executable model of the functional behaviour of the service or system in order to gain a deeper understanding of the problem domain. Simulation of these (quite large) models certainly supports this and additionally gives some confidence in the "correctness" of the models. In addition to the observable PN token flow, animation of the models in terms of problem domain concepts supports the communication with the problem domain experts. While in these two tasks emphasis was laid on the elaboration of the model, the purpose of the service interaction project is to validate the PN models by applying formal analysis methods defined for high-level PN classes.

In all different application cases, high-level PNs have been successfully applied. Of course, problems were encountered in the course of the different projects (see the discussions in sections 3.6, 4.4 and 5.4), but mostly they could be solved without big impact on the projects' objectives and tasks. Based on these experiences, we conclude that high-level PNs are, in principle, well-suited for usage in the area of telecommunications.

Among our wishes related to HCPNs as supported by Design/CPN and the tool itself, the following points are most important in our application domain:

- As net inscription language, HCPNs use SML, a functional programming language that is more or less unknown in the telecommunication area, where ASN.1 very often is used. Using a new language might not be such a problem, but the use of a pure functional language requires a paradigm shift for many people in the telecommunication area. The differences to procedural programming languages like PASCAL or C, and hence the resulting difficulties, should not be underestimated.

- Although Design/CPN has proven to be a convenient tool, it has its drawbacks. We also were missing quite some important features for our work:
  - The turn around times between editor and simulator may be quite long. Especially for the specification of industrial applications, a long turn around time can be a real problem.
  - There is no automatic code generator that "translates" the PN model into C or C++ code, which currently are the most commonly used implementation languages in our area.
  - No editor for MSCs is integrated to define test cases against which the model can be validated.

Though we found the application of high-level PNs for telecommunication systems and services convenient, we believe that their broader usage in this area seems to be rather unlikely. Again based on our experiences (this time not with applying PNs, but with their promotion within the telecommunications world - searching, for a long time in vain, for support from the PN community), using PNs within the development domain seems still to be far away at the time being. The reasons have often been discussed (see for example [CaDi93a, CaDi94, Herz95]). The essential point is that PNs are (still) not standardised, neither by ITU-T nor by ISO. As a consequence (?), many different PN classes and dialects exist. This has several implications, the two most important are:

- PN models can not easily be exchanged.
- There exists no PN tool with professional support suitable for application on an industrial scale - at least at the moment and to our knowledge.

Both points address requirements that need to be fulfilled for an usage in the development process of telecommunications systems. Compare this with the situation of SDL, where an international standard exists and usage of SDL already spread substantially, from standardisation organisations to network operators and to telecommunication equipment suppliers. Also, at least two well-known commercially tools are available. These tools do not only support the specification language, i.e. SDL, but also object-oriented analysis and design, e.g. by offering a class diagram editor and an automatic translation of a class diagram to an SDL structure to give just one example. Embedding the specification language in a seamless development method where the different phases as well as the transitions between them are supported by a tool eases software development substantially. To our knowledge, there exists no PN tool that offers an equally complete support of software development.

However, the PN situation is changing. A first standardisation effort at ISO was recently launched, also supported by us. Discussions on a suitable format to exchange PN models between different tools started in the PN community. Still, in an area where already one specification language is broadly accepted, if any formal method is applied at all, it will be very hard to establish a second one. As for the question of centralised vs. decentralised switch control mentioned in the beginning, technical arguments are not the only factors in the decision matrix.

References

[CaDi93a]   Capellmann, C.; Dibold, H.: Petri Net based Specifications of Services in an Intelligent Network - Experiences gained from a Test Case Application. In Lecture Notes in Computer Science / Application and Theory of Petri Nets 1993, Vol. 691, pp. 542-551, Springer-Verlag, 1993. ISBN 3-540-56863-8.

[CaDi93b]   Capellmann, C.; Dibold, H.: The Object-Oriented Petri Net Method for the Specification of IN Services. Proceedings of the International Workshop on Intelligent Networks "Software Methods and Tools for IN Services", pp. 63-76, Lappeenranta/Finland, August 1993.

[CaDi94]   Capellmann, C.; Dibold, H.: Formal Specifications of Services in an Intelligent Network using High-Level Petri Nets. In Proceedings of the Case Studies Tutorial / Petri Nets'94, Zaragoza/Spain, June 1994.

[CDGNO96] Capellmann, C.; Demant, R; Galvez-Estrada, R.; Nitsche, U.; Ochsenschläger, P.: Case Study: Service Interaction Detection by Formal Verification under Behaviour Abstraction. In Proceedings of the Intelligent Networks 1996, Passau, March 1996.

[ChHa93] Christensen, S.; Hansen, N.D.: Coloured Petri Nets Extended with Place Capacities, Test Arcs and Inhibitor Arcs. Lecture Notes in Computer Science 691 - M.A.Marsan (Ed.) Application and Theory of Petri Nets 1993, 14th International Conference, Chicago, Illinois, USA, June 1993.

[CMR93] Combes, P.; Michel, M.; Renard, B.: Formal verification of telecommunication service interactions using SDL methods and tools. In: Faergemand, O.; Sarma, A. (eds.): SDL'93: Using Objects, pp 441-452. Elsevier, 1993.

[Design/CPN] Design/CPN Manual. Meta Software Corporation. Cambridge, USA, 1991.

[Dibo88] Dibold, H.: A Method for the Support of Specifying the Requirements of Telecommunication Systems. Proceedings of the International Zurich Seminar, Zurich, 1988, p. 115-122.

[Dibo90] Dibold, H.: Intelligente Netze - Einführung und Grundlagen. Der Fernmelde-Ingenieur, 44 (1990) 4.

[Dibo92a] Dibold, H.: Hierarchical Coloured Petri Nets for the Description of Services in an Intelligent Network. Proceedings of the International Zurich Seminar, Zurich, 1992, pp. 165-178.

[Dibo92b] Dibold, H.: Die Offene Petrinetz-Methode zur Analyse und Darstellung des funktional. Verhaltens verteilter Systeme. D. Hogrefe (Hrsg.): Formale Beschreibungstechniken für verteilte Systeme, Reihe: Informatik aktuell, Springer-Verlag, Berlin, 1992, 195-221.

[Grah91] Graham, I.: Object-oriented methods. Addison-Wesley, Wokingham, 1991.

[Herz95] Herzog, U.: Petri Net based Modelling of Interactions between Basic Call State Model and Service Logic in an Intelligent Network. Protocol workshop within the 16th International Conference on Application and Theory of Petri Nets, Torino, June 1995.

[HMT87] Harper, R.; Milner,R.; Tofte, M.: The Semantics of Standard ML, Version 1. Technical Reports ECS-LFCS-87-36, University of Edinburgh, LFCS, Department of Computer Science, University of Edinburgh, The King's Buildings, August 1987.

[Jens92] Jensen, K.: Coloured Petri Nets: Basic Concepts, Analysis Methods and Practical Use, Volume 1. EATCS Monographs on Theoretical Computer Science, Springer Verlag, Berlin, 1992.

[Mimic] Mimic/CPN. A Graphical Simulation Utility for Design/CPN. User's Manual, Version 1.5.

[Nits94a]   Nitsche, U.: Propositional linear temporal logic and language homomorphisms. In: Nerode, A.; Matiyasevich, Y.V. (eds.): Logical Foundations of Computer Science '94 - Logic at St. Petersburg, Vol 813 of LNCS, pp 265-277. Springer Verlag, 1994.

[Nits94b]   Nitsche, U.: A verification method based on homomorphic model abstraction. In: Proceedings of the 13th Annual ACM Symposium on Principles of Distributed Computing, p 393, Los Angeles, 1994. ACM Press.

[Nits96]   Nitsche, U.: Verification and Behavior Abstraction - Towards a Tractable Technique for Large Distributed Systems. To appear in the Journal of Systems and Software, Special Issue on Software Engineering for Distributed Computing, 1996.

[NiOc95]   Nitsche, U.; Ochsenschläger, P.: Approcimately Satisfied Properties of Systems and Simple Language Homomorphisms. Arbeitspapiere der GMD Nr. 965, December 1995.

[Ochs94]   Ochsenschläger, P.: Verification of cooperating systems by simple homorphisms using the Product Net Machine. In: Desel, J.; Oberweis, A.; Reisig, W. (eds.): Workshop: Algorithmen und Werkzeuge für Petrinetze, pp 48-53. Humboldt Universität Berlin, 1994.

[OcPr95]   Ochsenschläger, P., Prinoth, R.: Modellierung verteilter Systeme - Kozeption, formale Spezifikation und Verifikation mit Produktnetzen. Vieweg Verlag, Wiesbaden, 1995.

[P103]   EURESCOM Project P103 "Evolution of the Intelligent Network": Framework for Service Description with Supporting Architecture. Deliverable D6a, 1995.

[Pete81]   Peterson, J.L.: Petri Net Theory and the Modelling of Systems; Prentice Hall Inc., Engelwood Cliffs, N.J. 07632,1981.

[Q.12xx]   ITU (CCITT) Recommendations Q.12xx - Q series: Intelligent Network Recommendation, 1992.

[Reen92]   Reenskaug, T. et.al.: OORASS: Seamless support for the creation and maintenance of object-oriented systems. Journal of Object-Oriented Programming, October 1992.

[SCORE]   RACE Project SCORE: Report on Methods and Tools for Service Creation. Part I: Summary. R2017/SCOWP2/DS/P/027/b2. 1994.

[TINA]   TINA-C: Definition of Service Architecture. TB_B.MDC.012_1.0_93. 1993.

[Thör94]   Thörner, J.: Intelligent Networks, Artech House Inc., Boston, London, 1994.

[Wiks87]   Wikström, Â.: Functional Programming using Standard ML. Prentice-Hall Internat. Series in Computer Science, 1987. ISBN 0-13-33 1968-7, ISBN 0-13-33 1661-0 Pbk.

[Z.100]   ITU (CCITT) Recommendation Z.100: SDL, 1992.
[Z.120]   ITU (CCITT) Recommendation Z.120: MSC, 1992.

ANNEX A: Extract of the Global Declaration for the UAN Service Model

```
(* Role Model "UAN translation" *)
color OrgZ = with North|Middle|South;
color SlN  = string with "0".."9" and 2..2;
color TimeZ= with Business|Leisure;
color Uan  = string with "0".."9" and 4..4;

color OrgZxSlN = product OrgZ * SlN;
color TimeZxOrgZxSlN = product TimeZ * OrgZ * SlN;
color UanDL1 = list OrgZxSlN;
color UanDL2 = list TimeZxOrgZxSlN;
color UanD = union UanD1:UanDL1 + UanD2:UanDL2;
color UanxUanD = product Uan * UanD;

color SlNxSlN      = product SlN * SlN;
color SlNxOrgZ     = product SlN * OrgZ;
color SlNxUan      = product SlN * Uan ;
color SlNxUanD     = product SlN * UanD;
color SlNxTimeZ    = product SlN * TimeZ;

color OrgReq   = union ReqOrgZ:SlN + RetOrgZ:SlNxOrgZ;
color TimeReq  = union ReqTimeZ:SlN + RetTimeZ:SlNxTimeZ;

(* function declarations *)
exception one_dim and two_dim;
fun one_dim(_,nil)       = raise one_dim
  | one_dim(a,(b,x)::l) = if a=b then x
                                   else one_dim(a,l);
fun two_dim(_,_,nil)         = raise two_dim
  | two_dim(a,b,(c,d,x)::l) = if a=c andalso b=d then x
                                   else two_dim(a,b,l);

(*declaration of variables used in arc inscriptions *)
var cli   : SlN;
var orgZ  : OrgZ;
var timeZ : TimeZ;
var uan   : Uan;
var uanD  : UanD;
var uanDL1 : UanDL1;
var uanDL2 : UanDL2;
```

ANNEX B: Legend of Graphical Conventions used in OOPM

Role diagrams

 environment role with name

 role (with name) that is refined by subnet with page number No   (HS: hierarchical substitution)

 contract with reference number No to its definition

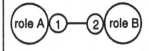 roles A and B may interact according to their contracts; role A may send messages defined in contract 1 to role B, role B may send messages defined in contract 2 to role A

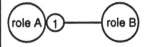 role A may send messages defined in contract 1 to role B, role B can not send messages to role A

 role A knows one or more objects playing role B to send messages to

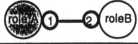 interaction crossing the border between system and environment roles

Role specifications

 transition with name, and optionally a guard

 place with name and colour set, and optionally an initial marking

 interface place with no1/no2 referring to the corresponding contracts of the role model (B: Border place)

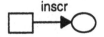 transition generates token "inscr" on place

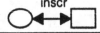 transition removes token "inscr" from place

 transition reads token "inscr" on place

The Incremental Modelling of the Z39.50 Protocol with Object Petri Nets

Charles Lakos[1] and John Lamp[2]
[1]Computer Science, University of Adelaide,
Adelaide, SA, 5005, Australia.
[2]Management Information Systems, Deakin University,
Waurn Ponds, Geelong, Victoria, 3217, Australia.
Charles.Lakos@cs.adelaide.edu.au
John.Lamp@deakin.edu.au

Abstract: This paper examines how object-oriented extensions to the Petri Net formalism provide flexible structuring primitives which can aid the modelling of network protocols. A key benefit is the support for incremental modelling. As a result, a protocol can be modelled as a collection of services, each of which can be expressed as the enhancement of a basic service, in which case, both the structure of the basic service and the nature of the enhancement can be clearly identified. More importantly, the evolution of a protocol through a sequence of standards can be expressed by progressive refinements. The object-oriented extensions are captured in the formalism of Object Petri Nets, with a textual language form referred to as LOOPN++, both of which are introduced in this paper. The incremental modelling capabilities and their benefits are demonstrated for the Z39.50 Protocol for Information Retrieval.

1 Introduction

For a long time, computer network protocols have motivated research in concurrent systems. The increasing complexity of concurrent systems has fuelled ever-increasing budgets and human resource demands for developing and maintaining the associated software. This has provided an impetus for flexible software development environments capable of handling concurrent systems and maximising software reuse; and for formal techniques which can ensure software reliability. Currently, for example, there is much interest in client-server applications and distributed object computing.

One formalism which has been applied to network protocols with beneficial results is that of Petri Nets [7, 8, 12, 21]. This has been facilitated by a number of attributes traditionally associated with Petri Nets – their formal definition, their graphical representation, the associated executable models, and their amenability to automated analysis.

This paper addresses one area which has previously been identified as a weakness in Petri Net formalisms: *the absence of compositionality has been one of the main critiques raised against Petri net models* [16]. This is primarily seen in the limited facilities for building complex systems out of simple components. Equally important there is a lack of facilities for building systems incrementally.

Network protocols, like software systems in general, have a number of aspects which are particularly suited to incremental modelling. They are typically composed of a number of protocol services, where each service is best understood (and modelled) as a basic service together with enhancements. Network protocols tend to have a number of configurable options. Finally, network protocols (like software systems in general) tend to evolve over time. It is desirable to be able to capture these enhancements, configurations and evolution directly with incremental specifications, rather than having to start with a fresh definition or having a number of specifications.

This paper considers the modelling of the Z39.50 Protocol for Information Retrieval [1, 2] which exhibits all of the above properties. It consists of a number of services which are best captured as basic services together with enhancements, it has a number of configurable options, and the protocol has evolved recently from the 1992 to the 1995 version.

The intent of the Z39.50 standard is to provide the kernel of a client/server system which allows computer-to-computer information search and retrieval. Z39.50 does not prescribe the way information is managed at the server, nor does it prescribe how information is presented at the client. This allows for a wide range of information sources to be accessed using Z39.50, with a client whose capabilities may range from the simplistic, to an intelligent software agent embedded in a larger information system.

The Z39.50 standard was initially proposed by the librarianship community to provide an open standard for networked access to bibliographic databases, and has now been extended to handle other forms of textual and non-textual databases such as graphical and geographic databases. It has also been used to support searches based on generalised pattern-matching techniques. These techniques will become increasingly important in the applications currently being developed for finding abstract information such as chemical structures, gene sequences, fingerprints, faces, video imagery, and numeric trend data. There are hundreds of information sources using Z39.50 or Wide Area Information Server (WAIS) protocols. (WAIS was based on an early version of Z39.50.)

The significance of Z39.50 has been recognised by the US Federal Government who have specified its use as part of the Government Information Locator Service (GILS). A number of US government agencies use Z39.50, as well as other government agencies such as the European Space Agency, many national libraries and universities, and commercial information sources such as Dialog, LEXIS/NEXIS, Online Computer Library Center (OCLC) and the Research Libraries Group. There are both freeware and commercial clients and servers available, as well as gateways for resources such as X.500, SQL and HTTP.

This paper is an extended version of an earlier one [31] which restricted its attention to Z39.50-1992 since Z39.50-1995 had not been standardised at that stage. By modelling the 1995 version as an extension of the 1992 version, this paper exhibits a realistic case study of incremental modelling.

The Petri Net formalism we use is called Object Petri Nets (OPNs) [24, 27], which is a modified Coloured Petri Net (CPN) formalism [17] incorporating object-oriented structuring. The goal of this formalism is to reap many of the benefits associated with object-oriented technology, such as more flexible and powerful structuring primitives and the practical support for extensibility and software reuse. An implementation of this formalism is the textual language LOOPN++.

The paper introduces both the graphical conventions for OPNs and the textual form of LOOPN++ in §2. The emphasis of this presentation is practical – the theoretical foundations for this work are found elsewhere [24, 27]. Z39.50-1992 is introduced in §3, its basic services are modelled in §4 and the enhanced services in §5. Z39.50-1995 is then introduced in §6, its new services in §7 and other significant extensions in §8. A number of other issues pertinent to protocol modelling are noted in §9, while §10 introduces the issues which are pertinent to the analysis of OPNs. Finally, the conclusions and proposals for further work are presented in §11.

2 Introduction to Object Petri Nets and LOOPN++

In this section we introduce Object Petri Nets (OPNs), their graphical conventions and their textual representation in the language LOOPN++. Each subsection begins with the relevant segment of the grammar for LOOPN++, and the following text discusses the implications of the grammar together with the associated graphical conventions. The grammar captures the possibilities of the language in a simple and concise manner, and provides a minimal set of contructs with orthogonal combinations. The graphical conventions extend the traditional Petri Net notation and are complemented by the textual representation, which supplies the annotations for the diagrams. Many of these annotations are only selectively displayed, so as to avoid cluttering the diagrams. Thus, the textual form can serve as an object-oriented language in its own right, as a graphics-independent interchange format for OPNs, and as a temporary test bed while sophisticated graphical tools are being developed. Currently, only the textual form has been implemented, but a graphical editor is nearing completion.

2.1 Nets, classes and instances

```
class   →   CLASS id [ : [ parent ] {, parent } ]    -- parents
            EXPORT ident {, ident }                   -- exported identifiers
            { field }                                 -- data fields
            { func }                                  -- functions
            { trans }                                 -- transitions
            { action }                                -- token + anonymous actions
            END id
```

An OPN specification or **net** consists of one or more **class** definitions. One class is designated the **root class**, and a single instantiation of this class constitutes the main program or net.

A **class** defines a set of **objects**, the **instances** of the class. The term **type** is used interchangeably with *class*. Types may be user-defined classes or basic, predefined types such as boolean, integer, real, char, string.

In general, a class consists of a number of components, each of which is a **field** (or data), **function, transition** or **action**. These are allocated and initialised on instantiation of the enclosing object.

Graphically, a class is drawn as a labelled frame enclosing the class components, as in fig 2.1. The label on the frame specifies the class name and the parent classes. This example shows a class with a number of constant definitions which are used in the definition of the Z39.50 protocol.

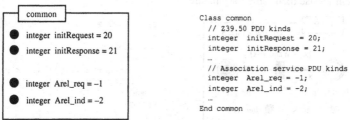

Fig 2.1: Graphical and textual representation of the class with common constants

Following Petri Net conventions, state components are drawn as ovals (or circles), and state change (or computation) components are drawn as rectangles (or squares). Fig 2.2 shows a class with a function component. While a function does not change the state, it does represent a computation which is performed at a certain point in time, and hence is drawn as a rectangle.

Fig 2.2: Graphical and textual representation of the class for the reference id

Classes may be arbitrarily instantiated as components or features of other classes. Thus, fig 2.3 shows a class with one field instantiating the class *RefId* (of fig 2.2).

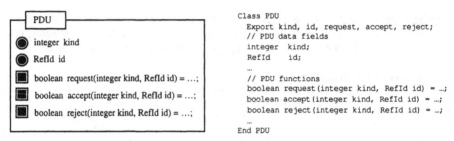

```
Class PDU
  Export kind, id, request, accept, reject;
  // PDU data fields
  integer  kind;
  RefId    id;
  ...
  // PDU functions
  boolean request(integer kind, RefId id) = ...;
  boolean accept(integer kind, RefId id) = ...;
  boolean reject(integer kind, RefId id) = ...;
  ...
End PDU
```

Fig 2.3: Graphical and textual representation of the class for Protocol Data Units (PDUs)

Each feature of a class may be local or exported, with only exported features externally visible and accessible. Textually, an **export** clause lists the identifiers which are exported. Graphically, an exported component is shown with a double outline. Thus, figs 2.2 and 2.3 show classes where all components are exported, while fig 2.1 shows a class where no components are exported.

A class may **inherit** the features of one or more **parents**, in which case all the features of the parents, together with the additional features declared within the class constitute the features of the new class. It is possible for a class to override a feature of the parent with a compatible one of the same kind – it is not possible to override a field by a function or an action. Graphically, inheritance is shown using two conventions – the parent classes are listed in the label of the class frame, and the components inherited without change are drawn with a grey shading, as in fig 2.4. In this example, class C inherits from class P. Class C has features n, p, q, t plus associated arcs. Components n, p are inherited without change, while q, t are introduced locally or override similarly named components of parent P. Only p and q are exported. Note that transition variables x, y can also be shown graphically, with a striped shading to emphasise their temporary nature.

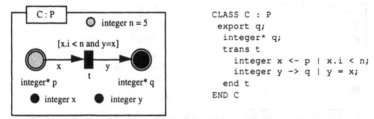

```
CLASS C : P
  export q;
  integer* q;
  trans t
    integer x <- p | x.i < n;
    integer y -> q | y = x;
  end t
END C
```

Fig 2.4: Graphical and textual representation of a class C which inherits from parent class P

OPNs support the usual **polymorphism** of object-oriented systems, which allows instances of a subclass to occur in superclass contexts.

Both OPNs and LOOPN++ maintain the bipartite nature of Petri Nets by distinguishing between state classes and transition classes. A **transition class** is one with arcs as (immediate) components, while a class without such arc components is a **state class**. (A state class may contain transitions but may *not* contain arcs.)

2.2 Types

type	$\rightarrow$	basic-type-ident	-- *int, bool, real, string, etc.*
	$\rightarrow$	class-ident	-- *predefined or user-defined*
	$\rightarrow$	type '*'	-- *multiset type for places*

Various net components may have a **type** which may be any built-in, predefined type or a user-defined class. The type may also be of the form `type*` which indicates a multiset (or bag) of objects each of type `type`. (The form `type*` is used to declare Petri Net places.)

2.3 Values and expressions

```
value   →  ident                                      -- copy of specified object
        →  '['ident:value {, ident:value }']'         -- new object with specified fields
        →  ident'['ident:value {,ident:value }']'     -- modify object by specified fields
        →  '['value {, value} [ | value ]']'          -- multiset value constructor
        →  expr                                        -- value given by expression
```

In certain contexts – in field and function definitions and in guards – it is possible to specify values. These values may indicate a copy of an existing object, the generation of a new object (by specifying values for some of the exported fields), a copy of an existing object with certain specified fields modified, a multiset of values, or a value computed by an expression. For a multiset of values, it is possible to specify a tail or remainder (following the vertical bar), which is helpful in generating multisets by recursive functions.

The format of expressions is left undefined, being determined by the underlying language (C++ in the case of LOOPN++).

2.4 Fields

```
field   →  type  ident [ = value ] [ | guard ]      -- data field + value + guard
                { , ident [ = value ] [ | guard ] } ;
```

A **field** is a class component of some type, optionally with a default initial value, as in:

    type ident = value;

If the field is local, then every instance of the containing class will associate this default value with the field on initialisation. If the field is exported, then the instantiation of the containing class may specify an alternative initial value for the field, using the notation:

    type object = [ident:value, ...];

in which case, each instance of the containing class may have a different value for this field.

The type determines the kind of field – a field of (predefined) type integer will simply be a constant; a field of (multiset) type integer* will be a place holding integer-valued tokens (the container doesn't change but its contents can vary); a field of (user-defined) type C could be a subnet instance. Where a field is an instance of a transition class its graphical representation is a rectangle. Where a field is an instance of a state class, its graphical representation is an oval.

Examples of field definitions were shown in figs 2.1, 2.3 and 2.4. Fig 2.4 showed one field n which was a constant, and two fields p and q which were places (since their type was *integer*).

A field specification may also specify a **guard** which is a boolean expression. The interpretation of the guard varies depending on whether the field occurs in a state class or a transition class. For a component of some state class, the guard specifies an **integrity constraint**. If the integrity constraint is ever violated, then the program will abort with an appropriate error message. For a component of a transition class, the guard specifies an **enabling condition**. If the guard is not satisfied, then the transition is not enabled and will not fire. The guard is like a class invariant of Eiffel [35].

2.5 Transitions and actions

```
trans    →   TRANS id [ : [ parent ] {, parent } ] ]   -- transition definition
             { field }                                  -- data fields
             { action }                                 -- token + anonymous actions
             END id

action   →   type  ident <- place  [ | guard ]         -- input action + select condition
         →   type  ident -> place  [ | guard ]         -- output action + output value
         →   type  ident -- place  [ | guard ]         -- test action + select condition
         →   procedure-call                            -- interact with environment
```

Traditional Petri Net formalisms include the fundamental concept of a transition. As already noted, OPNs and LOOPN++ build systems from typed components, where each type is predefined or defined by a class. A **transition** is an instance of a transition class. A **transition class** is distinguished from a state class by containing at least one arc (or named action) as an immediate component.

An **action** is the only construct for changing the state of an object. Actions may be input actions, output actions, test actions, or anonymous actions. Apart from anonymous actions (or procedure calls), each action has an associated identifier. A binding of a transition associates a value with each such action identifier. (Note that the grammar allows field declarations in transitions, in which case these fields are also bound to appropriate values in each binding of the enclosing transition.)

The anonymous actions (or procedure calls) of a state class are executed on instantiation of the class. The actions of a transition class are executed each time the transition fires. In order to support formal analysis, anonymous actions must have no effect on the firing of transitions in the net. Graphically, anonymous actions are drawn as annotations of transitions, while named actions are drawn as directed arcs, following the usual Petri net conventions. Textually, each named action may have a guard, while graphically there is only one guard annotating each transition as a whole.

Rather than declaring a class and instantiating it for each transition, syntactic sugar is provided for declaring a transition class with a singleton instance:

```
Trans ident
    type  x <- p | ...;          -- zero or more input actions
    type  y -> q | ...;          -- zero or more output actions
    procedure-calls              -- zero or more anonymous actions
End ident
```

An **input action** extracts a value from an object and is written in LOOPN++:

```
type  x <- p | guard;
```

where the constraint " | guard" is optional. For such an input action to occur, the value of x obtained from p must be of an appropriate type and satisfy the condition, if any. The value x is called a **token** (or tokens), while the object p is called an **input place**. The guard is a boolean expression which *may* contain terms of the form:

```
x = value
```

which is interpreted as testing that the components of x match those specified by the value. For example, the transition of fig 2.4 has an input action which selects an integer token provided its value is less than n.

A **test action** examines a value in an object and is written:

```
type  x -- p | guard;
```

where the constraint " | guard" is optional. For such a test action to occur, the value of x in p must be of an appropriate type and satisfy the condition, if any. The guard is a boolean expression which *may* contain terms of the form:

```
x = value
```

which is interpreted as testing that the components of x match those specified by the value.

An **output action** deposits a value into an object and is written:

```
type  x -> p | guard;
```

where "| guard" is optional. For such an output action to occur, the value of x must be acceptable to p. Again, the value x is called a **token** (or tokens), while the object p is called an **output place**. The guard is a boolean expression which *will be* of the form:

```
x = value
```

which is interpreted as generating an object with components matching that of the value. For example, the transition of fig 2.4 has an output action which generates a token *y*, a copy of *x*.

The current reference semantics for OPNs dictates that output tokens are always newly-generated objects or newly-generated copies of existing objects. In this way, it is not possible to carry a reference to a remote object (such as a place) around a Petri Net – it is only possible to refer to locally accessible objects. Alternative reference semantics may be possible, but have not yet been investigated.

2.6 Functions

```
func    →  type  ident  ( parms ) [ = fvalue ]      -- function definition
fvalue  →  expr                                      -- possible function values
        →  FORALL type ident -- place [ | guard ]; end FORALL
        →  EXISTS type ident -- place [ | guard ]; end EXISTS
        →  COUNT  type ident -- place [ | guard ]; end COUNT
```

A function defines a parameterised expression, which returns a value based on the other features (and hence state) of an object. Functions take parameters and return values of some type (which may be a predefined type, a class type, or a multiset type).

Functions can examine the state of an object, but cannot change that state. Quantification can be used to determine some value based on some or all of the tokens resident in a place. If the result of such functions are to be well-defined, their evaluation cannot be concurrent with transitions which modify the state under inspection. In other words, functions have some aspects of transition semantics and are hence graphically represented by rectangles.

2.7 Places and super places

In OPNs and LOOPN++ a **place**, in the simplest case, is an object of multiset type, declared as in:

```
type* ident
```

This type implies that a place may have incident arcs, and thus supports input, output and test actions. In general, any class which inherits from a multiset class will support input, output and test actions, and can be instantiated to form a place or, more precisely, a super place. The inherited multiset class determines the type of tokens which can be exchanged with the environment, while the components of the class determine what information is stored and how tokens are exchanged. Graphically, the possible exchange of tokens with the environment is indicated by arcs incident on the class frame. While these arcs may be inscribed by arbitrary multisets, efficient implementation is possible if these arcs exchange one token at a time (here called *token*). Note that this does not mean that a super place may only accept or offer one token at a time, but that each such token offer or acceptance involves one occurrence of an interface transition (here *get* or *put*). Experience seems to indicate that this restriction is acceptable in practice and merely reflects the notion that tokens in a place are independently accessible. Semantically, tokens can be deposited into a super place if the internal activity of the super place accepts exactly those tokens, and conversely for extracting tokens. An example of a super place which will buffer integer tokens is shown in fig 2.5.

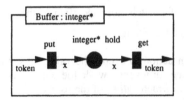

```
Class Buffer : integer*
    integer* hold = ...;
    trans put : put
        integer x -> hold | x = token;
    end put
    trans get : get
        integer token <- hold;
    end get
End Buffer
```

Fig 2.5: A super place for an integer buffer

The textual version assumes that the support of a multiset type for input, output and test actions is provided by (pseudo) transitions *get*, *put* and *see*, which serve to define *token* or use it for output. Defining how a super place supports these actions is a matter of inheriting the previous definitions and overriding them with extended definitions.

2.8 Substitution transitions and super transitions

In OPNs, it is possible to emulate the HCPN notion of substitution transitions [17] by defining classes with exported places. When these classes are instantiated, their exported places can be bound to places external to the instance (as in HCPN port assignments).

In this paper, we have followed the HCPN graphical conventions of drawing such substitution transitions as rectangles with incident arcs. However, such conventions are considered to be misleading since they suggest transition semantics, i.e. the synchronisation of the arc actions.

On the other hand, the formal definition of OPNs [27] allows for the notion of super transitions which do have transition semantics. However, the strict synchronisation of arc actions seems to be too restrictive in practice. We are currently investigating a compromise alternative, where the external environment perceives the arc actions to be synchronised, while internally they may be achieved by some sequence of actions [29].

3 Introduction to Z39.50-1992

In this section, we introduce the essential features of the ANSI Z39.50-1992 Standard for Information Retrieval [1], prior to specifying the protocol in the subsequent section. We concentrate on the transition table which formalises the protocol and the identification of the services.

The standard defines the operation of what is called the Z39.50 origin and target, which are those parts of the client and server respectively, which provide the facilities associated with networked information search and retrieval, as shown in fig 3.1.

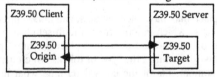

Fig 3.1: Basic structure of the Z39.50 protocol

A Z39.50 session consists of three phases: the establishment of the session, information transfer, and termination. Each phase uses a number of facilities. The establishment phase is handled by the Initialization Facility. The main facilities used during the information transfer phase are the Search Facility (for querying databases at a target), the Retrieve Facility (for retrieving copies of database records) and the Result-set-delete Facility (for deleting result sets known to the target). The termination phase is signalled either by the explicit use of the Termination Facility or the implicit termination by a communication failure or other external event. Other facilities overlay these and handle issues such as access control, accounting and resource control. The role of the origin and target cannot be reversed.

OSI terminology [38] defines protocols as a collection of services and thus the networking standards documents tend to reflect this terminology and this modularisation of protocols. Thus, the Z39.50-1992 Standard speaks of a number of facilities (as already noted), each of which consists of one or more services, as shown in fig 3.2.

Facility	Service
Initialization Facility	Init Service
Search Facility	Search Service
Retrieval Facility	Present Service
Result-set-delete Facility	Delete Service
Access Control Facility	Access-control Service
Accounting/Resource Control Facility	Resource-control Service Trigger-resource-control Service Resource-report Service
Termination Facility	IR-abort Service IR-Release Service

Fig 3.2: The facilities and services of Z39.50-1992

Unfortunately, the modularisation concepts and terminology are not always consistently applied. The Z39.50 standard identifies each service and describes it individually within the standard in unconstrained natural language. It also provides the format of Protocol Data Units (PDUs) in ASN.1 notation. However, the only other description of the protocol is the monolithic state transition table(s) of fig 3.3. It is difficult to separate out the individual services from this state transition table, let alone identifying the normal and abnormal activity of each service. This can only be done by reference to the informal, natural language description of the services. This style of transition table mitigates against modularity.

As well as being deficient with regard to modularisation, the transition table does not even constitute a formal definition. It includes loosely-defined entries such as *stkst* and *popst* for saving and restoring the state. In fact, the standard states: *The IRPM state table does not constitute a formal definition of the IRPM. It is included to provide a more precise specification of the protocol procedures.*

It is clearly desirable for the protocol to be formally specified and for the specification to match the OSI terminology of services. Such modularity will help to achieve an intellectually manageable model or specification. This is done below for the Z39.50-1992 standard (cf. [31, 32]).

Abbreviations							
A	Association control	Iab	IR abort	req	request
Aab	Association abort	ind	indication	resp	response
Acc	Access-control	Init	Initialise	Rsc	Resource-control
Arel	Association release	IR	Information Retrieval	Rsrp	Resource-report
conf	confirmation	Irel	IR release	Srch	Search
Dlte	Delete	Prsnt	Present	Trigrc	Trigger-resource-control

Table 10a: State Table for Origin – Part 1							
State Event	closed 1	Init sent 2	Open 3	Search sent 4	Prsnt sent 5	Delete sent 6	Rsrp Sent 7
Init req	Init PDU (2)						
Init resp PDU (ACCEPT)		Init conf + (3)					
Init resp PDU (REJECT)		Init conf –; Arel req (10)					
Srch req			Srch PDU (4)				
Srch resp PDU				Srch conf (3)			
Prsnt req			Prsnt PDU (5)				
Prsnt resp PDU					Prsnt conf (3)		
Dlte req			Dlte PDU (6)				
Dlte resp PDU						Dlte conf (3)	
Rsrp req			Rsrp PDU (7)				
Rsrp resp PDU							Rsrp conf (3)
Trigrc req		Trigrc PDU (2)		Trigrc PDU (4)	Trigrc PDU (5)	Trigrc PDU (6)	

Table 10a: State Table for Origin – Part 2									
State Event	Init sent 2	Open 3	Search sent 4	Prsnt sent 5	Delete sent 6	Rsrp Sent 7	Rsctrl recvd 8	Acctrl recvd 9	Rlease sent 10
Rsc PDU (Resp)	Rsc ind; stkst (8)		Rsc ind; stkst (8)	Rsc ind; stkst (8)	Rsc ind; stkst (8)	Rsc ind; stkst (8)			
Rsc PDU (Noresp)	Rsc ind (2)		Rsc ind (4)	Rsc ind (5)	Rsc ind (6)	Rsc ind (7)			
Rsc resp							Rsc resp PDU; popst		
Acc PDU	Acc ind; stkst (9)		Acc ind; stkst (9)	Acc ind; stkst (9)	Acc ind; stkst (9)	Acc ind; stkst (9)			
Acc resp								Acc resp PDU; popst	
Aab ind	Iab ind (1)	Iab ind (1)	Iab ind (1)	Iab ind (1)	Iab ind (1)	Iab ind (1)	Iab ind (1)	Iab ind (1)	Iab ind (1)
Apab ind	Iab ind (1)	Iab ind (1)	Iab ind (1)	Iab ind (1)	Iab ind (1)	Iab ind (1)	Iab ind (1)	Iab ind (1)	Iab ind (1)
Iab req	Aab req (1)	Aab req (1)	Aab req (1)	Aab req (1)	Aab req (1)	Aab req (1)	Aab req (1)	Aab req (1)	Aab req (1)
Irel req		Arel req (10)							
Arel conf									Irel conf (1)

Fig 3.3: Transition table and abbreviations for Z39.50-1992

The transition table specifies the allowable events depending on the current state of the Z39.50 origin. Thus, in the initial or closed state (1), the client may request the establishment of a connection by submitting an *Init req*. In response, the origin sends an *Init PDU* to the target and transfers into state 2. If the target accepts the connection by sending an *Init resp PDU (ACCEPT)*, then the origin notifies the client to this effect with an *Init conf +* and moves into state 3. In this state, the origin is prepared to accept service requests such as *Srch req*, *Prsnt req*, etc.

4 Basic Services of Z39.50-1992

We now use OPNs and LOOPN++ to model the Z39.50-1992 protocol. We have already noted that the Z39.50-1992 standard formalised the protocol in a monolithic transition table. It has long been recognised [13, 37] that the successful management of complexity requires support for abstraction with clean module boundaries. In this section, we demonstrate how this can be achieved in OPNs by modelling the basic services of Z39.50 (leaving the enhanced services to §5). We consider the passive data components, the generic services, and the composition of those services.

4.1 Passive data – the Z39.50-1992 message formats

We commence by considering a number of definitions which are commonly required in the protocol description. Firstly, there are a number of constant definitions which distinguish the different kinds of protocol data units (PDUs). These would be defined in a class as in fig 2.1, which could then be inherited by other classes so as to share these common definitions.

Secondly, all PDUs include an optional *Reference-id*. The format of this and its usage are not specified by the standard, except to require that a request PDU including a *Reference-id* must be matched by a response PDU with the same *Reference-id*, and to require that any intermediate Access control or Resource control request must specify the same *Reference-id*. Because of the vagueness of these requirements, our earlier definition of the Z39.50-1992 protocol omitted this optional *Reference-id* [31]. However, its usage is clarified in the Z39.50-1995 standard and is there required for concurrent operations. Accordingly, we define a class *RefId* (as in fig 2.2) with a function to determine if one *RefId* instance matches another. Note that the definition of the function is not supplied, although some simple definition could be given, such as the expression $id=self$. In reality, the function result depends on the way *Reference-id*s are stored, which is not defined by the standard nor in the above class. In a sense, the class serves as an abstract or deferred class, i.e. it will not be instantiated as is. Instead, we will instantiate a subclass which *will* include one or more data fields to store the *Reference-id*. Polymorphism means that these subclass instances can be used anywhere that a superclass instance is specified (see §9.1).

Thirdly, the class for the protocol data units (PDUs) can be given, as in fig 2.3. (The class for messages (MSGs) exchanged with the protocol user will have a similar format.) This class includes a field for the kind of PDU (which will be a value from the class *common* above); it includes the *Reference-id* by instantiating the class *RefId*; and it supplies functions to determine if the PDU is a request, an accepting response or a rejecting response for a certain kind of service with a given *Reference-id*. Note that when the class *PDU* is instantiated, the field *id* can be instantiated to be an instance of some subclass of *RefId*.

Once again, the functions defined for *PDU* are sufficient to determine the behaviour of the protocol entities, and it is a natural object-oriented solution to encapsulate the information in this way. Additional information (specified in the standard for each kind of PDU) will be included in subclasses of *PDU* which can then be used polymorphically in superclass contexts.

Fourthly, before turning to the modelling of active components of Z39.50, namely the protocol services, we note that those services will need to hold the state of the protocol, i.e. the service in progress together with its *Reference-id*. It is also appropriate to retain information about the *allowable* services of the protocol, which are determined by negotiation during initialisation. Accordingly, we define a configuration class to store the *Reference-id* and the set of allowable services as in fig 4.1. Once again it is not necessary to spell out the details of the fields which store the allowable services – it is sufficient to have a function which determines if a service is allowed.

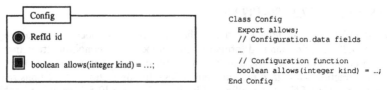

Fig 4.1: Graphical and textual representation of the configuration class

4.2 Modularity – the generic services of Z39.50

We have already noted in the introduction to §4 that the Z39.50-1992 standard identifies the various services of the protocol in the text of the standard, but not in the transition table. Careful study of the Z39.50 standard reveals that a number of the origin-initiated services (including *Initialize, Release, Search, Present, Delete, Resource-report*) can be captured by a particular style of subnet with an initial, intermediate, and final state (as shown in fig 4.2). This class can then be instantiated for each of the above services, possibly fusing the initial and final states.

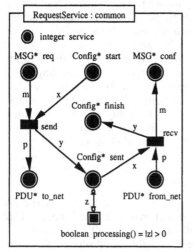

```
Class RequestService : common
    Export service, req, conf, to_net, from_net,
           start, sent, finish, processing;
    integer service;
    Config* start, sent, finish;
    MSG*    req, conf;
    PDU*    to_net, from_net;
    boolean processing() = exists z <- sent;

    trans send
        Config x <- start | x.allows(service);
        MSG    m <- req | m.request(service, x.id);
        Config y -> sent | y = x[id: …];
        PDU    p -> to_net | p = m[…];
    end
    trans recv
        Config x <- sent;
        PDU    p <- from_net | p.accept(service,x.id);
        Config y -> finish | y = x;
        MSG    m -> conf | m = p[…];
    end
End RequestService
```

Fig 4.2: Graphical and textual representation of a request service

Note that places are identified by their multiset type. In this class, all the places are exported and hence may be bound to external places when this class is instantiated. (This binding persists for the life of the class instance.) The arcs are annotated with token variables. The *send* transition can fire if there is a request message to transmit, and the protocol component is in its initial (*start*) state. The appropriate PDU is sent to the net. The *recv* transition can fire if the appropriate reponse PDU is received from the net and the protocol component is in its intermediate (*sent*) state. The appropriate indication message is sent to the protocol user and the protocol component enters its final (*finish*) state. The function *processing* is defined to return *true* if the place *sent* contains at least one token. Following the conventions of [22], this is shown graphically with the use of an equal arc, i.e. an arc which is enabled only if its inscription (here the variable z) is identical to the marking of the place (here *sent*).

A similar class is defined for target-initiated services and/or responses, as in fig 4.3. Note that *recv* transition stores a configuration x in place *recd* which is the same as y but with its *Reference-id* set to the *Reference-id* of the incoming request. This makes it possible to include it with the response.

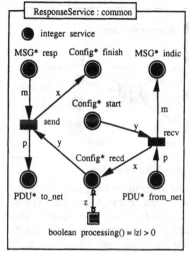

```
Class ResponseService : common
    Export service, resp, indic, to_net, from_net,
            start, recd, finish, processing;
    integer service;
    Config* start, sent, finish;
    MSG*    req, conf;
    PDU*    to_net, from_net;
    boolean processing() = exists z <- recd;

    trans recv
        Config y <- start | y.allows(service);
        PDU      p <- from_net | p.request(service,y.id);
        Config x -> recd | x = y[id:p.id];
        MSG      m -> indic | m = p[...];
    end
    trans send
        Config y <- recd;
        MSG      m <- resp | m.accept(service,y.id);
        Config x -> finish | x = y;
        PDU      p -> to_net | p = m[...];
    end
End ResponseService
```

Fig 4.3: Graphical and textual representation of a response service

4.3 Compositionality – the basic Z39.50 origin entity

The Z39.50 Search facility, like a number of other facilities, can be modelled by instantiating the request service of fig 4.2 as in fig 4.4. Since this is just one component of the Z39.50 origin entity, it has been drawn without a class frame. The textual form (which would appear as an annotation on the diagram) indicates that *search* is an instance of the *RequestService* class, that the exported field *service* is bound to the constant *searchRequest*, that the exported places *req, conf, to_net, from_net* are bound to similarly-named places in the context of the instance, that the exported places *start* and *finish* are both bound to the place *open* (which holds a *Config* token once a connection has been established), and that the exported place *sent* is not bound, and hence will be local to the instance.

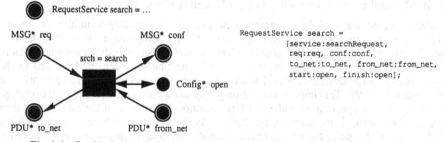

Fig 4.4: Graphical and textual representation of the instantiation of the Search request service

Specifying the data field *search* is sufficient to instantiate the service, complete with the binding of exported components. As observed in §2.8, this is semantically identical to the annotation used in the Design/CPN tool [19]. However, it is traditional for high-level Petri Nets to show the interaction of the service with the places using the graphical transition notation. This notation has been used in this paper, but we consider it somewhat misleading since it does not imply the usual transition semantics (see §2.8).

Similar instantiation of request services for the Z39.50 Present, Delete and Resource-report facilities will cover the entries in part 1 of the state transition table (of fig 3.3) for states 3, 4, 5, 6, 7. The same class is instantiated for the Initialize and Release services (states 1, 2, 10), but their interaction needs to be captured, as is done in the combined service of fig 4.5.

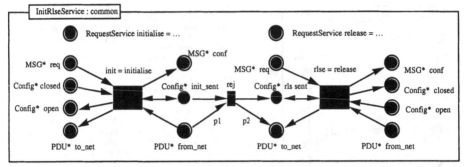

Fig 4.5: Graphical representation of the Initialze and Release services

Note that the various places have been duplicated or aliased to avoid crossing arcs. Note also that the *Initialize* facility normally takes you from the *closed* to the *open* state, via the *init_sent* state, while the *Release* facility normally takes you from the *open* to the *closed* state, via the *rls_sent* state. However, an *Initialize* request may be rejected, in which case a *Release* request is sent. This is achieved by the transition *rej* which would have a guard of the form *p1.reject(initRequest, x.id)*.

The above demonstrates how it is convenient to capture the various services of the Z39.50 protocol as subnets. Such modularity is readily supported by many Petri Net formalisms such as CPNs [17], though the ability to specify partially defined components like *RefId* (of fig 2.2) is not. In the following sections, further aspects of the protocol are discussed using OPN facilities which are not readily supported by other Petri Net formalisms.

5 Enhanced Services of Z39.50-1992

Having considered the basic services of Z39.50-1992, we now consider the enhancements. In the following sections we consider the *Trigger Resource Control* service, *Access Control*, *Termination Control*, and the origin entity including all these enhanced services.

5.1 Weak coupling of subnets – the Trigger Resource Control service

As already noted, the modular construction of a net by instantiating a number of subnets interacting with shared places is a common form of net decomposition. It is often the only one. For example, the original proposal for Hierarchical CPNs [14] advocated place fusion, transition fusion, and invocation transitions. Only the first has been implemented in the widely used package Design/CPN [19] while the second is common in other object-oriented net formalisms [6, 9]. This implies that the only way to interact with a subnet is to exchange tokens with it or to synchronise with one of its transitions. The logic of the subnet must then cater for every distinct interaction style. Even if the interaction simply involves examining some aspect of the state, the subnet needs to explicitly receive the request and return the result. This is contrary to the principle enunciated by Meyer [35] that objects should have high internal cohesion and weak external coupling.

A fundamental technique for achieving weak coupling between objects in object-oriented languages is to define exported functions which can evaluate selected aspects of an object's state without changing that state. It seems that Petri Net models have generally been slow to adopt this fundamental technique. LOOPN++ (like its predecessor LOOPN) supports the definition of access functions which may examine the state of a subnet. Elsewhere [22] it has been shown that this provision can be formally defined to support the usual step semantics of Petri Nets, and furthermore, nets with such extensions can be transformed into behaviourally equivalent CPNs. Another case study showing the value of this feature can be found in [30].

In modelling the Z39.50 protocol, the advantages of this facility in maintaining modularity can be demonstrated by considering the *Trigger-resource-control service* and the *Resource-control service* (where no response is required). Both of these require the origin entity to have sent an *Initialize*, *Search*, *Present* or *Delete* request, but not to have received a corresponding reply. In other words, they require one of the listed services to be in their intermediate state. While the place indicating this intermediate state has been declared as exported (fig 4.2), this has only been exploited in the *Initialize-Release* service (fig 4.5). This encapsulation should be maintained in the interests of the weak coupling of subnets. Accordingly, a function *processing* was defined for the various services to determine whether the subnet was in its intermediate state. This now makes it possible to define the Trigger-resource-control service as in fig 5.1.

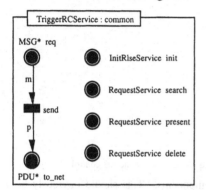

```
Class TriggerRCService : common
  Export req, to_net,
            init, search, present, delete;
  integer service = triggerResourceControlRequest;
  MSG* req;
  PDU* to_net;
  InitRlseService init;
  RequestService  search, present, delete;

  trans send
    MSG  m <- req | m.request(service) and
                          (init.processing() or
                          search.processing() or
                          present.processing() or
                          delete.processing());
    PDU  p -> to_net | ...;
  end
End TriggerRCService
```

Fig 5.1: Graphical and textual representation of the Trigger-resource-control service

Note that this subnet assumes that the various services will be bound to the local data fields so that they can be used to interrogate the state of those services. The simplicity of this interface is possible because of the weak coupling achieved by the use of access functions. What is true of this example is even more imperative for complex systems. As far as possible, the complexity of module interactions should be minimised.

5.2 Incrementality and Access Control

A further demonstration of software reuse enabled by inheritance and polymorphism can be made by considering the provision of access control in the the Z39.50 protocol. This was not examined in §4 where attention was focussed on what might be called the normal operation of each service. It will be noted from part 2 of the transition table of fig 3.3 that the *Access Control* facility temporarily interrupts a service (with a transfer to state 9).

It is possible to include *Access Control* as net components undifferentiated from the normal operation, but this can all too easily obscure the central function of the service (or facility), as is the case with the monolithic transition table. With OPNs, it is possible (and desirable) to specify these extensions as extensions of the normal service. Thus, the generic *RequestService* of fig 4.2 can be extended to a generic service with access control as shown in fig 5.2. This *RequestAccService* inherits from *RequestService*, with inherited components shown graphically by a grey shading. (The arcs which are inherited without change should also be shaded in grey, but the drawing tool used to prepare this paper did not have that capability.) *RequestService* is augmented with an instance of *ResponseService*, which responds to the Access Control request.

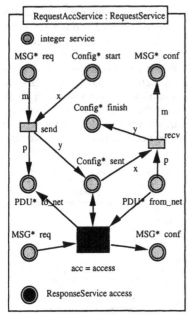

```
Class RequestAccService : RequestService
    Export access;
    ResponseService access =
                [service:accessControlRequest,
                 req:req,
                 conf:conf,
                 to_net:to_net,
                 from_net:from_net,
                 start:sent,
                 finish:sent];
End RequestAccService
```

Fig 5.2: Graphical and textual version of a request service with access control

Note that we have here displayed the inherited components. This should obviously be a configurable option of a graphical editor, but our experience indicates that where the graphical representation of a Petri Net conveys the pattern of interaction, this information is lost if only the incremental changes are shown (as would be the case in fig 5.2). On the other hand, if the inherited components do not reflect patterns of interaction, then their omission is not a problem (as is the case for *common* in figs 4.2 and 4.3).

An interesting question arises here concerning the *ResponseService* which provides the Access control for each *RequestService*. Should there be one such *ResponseService* for all *RequestService*s or one for each. As shown in fig 5.2, the field *access* is exported. If it is desirable to have one instance for all *RequestService*s, then the environment can bind this field to some global instance. Otherwise, the environment can omit a binding for this field, in which case there will be one instance of *ResponseService* for each *RequestService*. The choice will be constrained by whether the *ResponseService* can uniquely determine the *RequestService* in which it was initiated. That would be possible if *Reference-id*s were always used, but otherwise is problematical.

5.3 Incrementality and Termination Control

The Z39.50-1992 protocol specifies two services as part of the *Termination Facility* – an abrupt termination or abort service, and a graceful termination or release service. The release service is an origin-initiated, acknowledged service like any other. It has already been considered in fig 4.5. The abort service is different to the others since it is not acknowledged. It can be initiated at any time, and it can be initiated by any party – the origin, the target, or even the underlying Association Control. The abort service must leave the origin entity in the closed state.

The abort service breaks encapsulation – no matter how deeply nested in other services (e.g. within an access control request, within a retrieval request) the origin must transfer into the closed state after having received an abort request. This is apparent in the transition table (of fig 3.3) – the reception of any abort request or indication leads to an immediate transfer into the closed state (state 1). This is in marked contrast to the *Access control* service which is properly nested, since it causes the current state to be saved on a stack and later restored.

In modelling the abort service as an OPN, it is possible to reflect the breaking of encapsulation by including an additional exported abort state in each of the generic services of §4.2. When the services are instantiated, this abort state could be bound to the closed state. Each service could transfer to this abort state whenever an abort request or indication were received. Thus, no matter how deeply nested within services, every service could immediately return to the closed state in response to an abort request or indication.

Alternatively, the abort service could be modelled in a more encapsulated way. Each service could simply test for an abort request or indication and could then transfer to its final state. Since the service only tests for the abort, it is not consumed. In other words, each level of encapsulation would test the same abort request and transfer to its final state. Eventually, some global transition could consume the abort request or indication and transfer into the closed state. Because we have chosen to emphasise the modularity of our solution, this latter approach is adopted (as in fig 5.3).

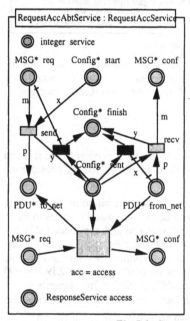

```
Class RequestAccAbtService : RequestAccService
      trans abort_request
         Config x <- sent;
         MSG      m -- req | m.request(abort,x.id);
         Config y -> finish | y = x;
      end abort_request
      trans abort_indication
         Config x <- sent;
         MSG      p -- from_net | p.request(abort,x.id);
         Config y -> finish | y = x;
      end abort_indication
End RequestAccAbtService
```

Fig 5.3: Request service with access control and abort

5.4 The enhanced Z39.50-1992 Origin Entity

The Z39.50 origin entity can now be defined as a collection of the above extended services, as shown in fig 5.4. The transitions in the Z39.50 origin, while not annotated, will deal with the abort requests and indications, which are simply examined by the more deeply nested services (see §5.3). In both cases, there is a transfer from the *open* to the *closed* state. In the case of an abort request, the abort is passed to the network. In the case of an abort indication, the indication is passed to the user of the origin entity.

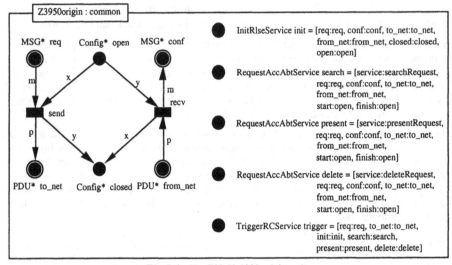

Fig 5.4: The Z39.50-1992 origin entity

Rather than clutter up the diagram of the *Z3950origin*, the super transition abstractions for the services have been omitted, although they could be supplied as in fig 4.4. Note that a number of the services are passed as parameters to the *Trigger-Resource-Control* service, as required (see §5.1).

6 Introduction to Z39.50-1995

Having modelled the Z39.50-1992 protocol, we now turn to the 1995 version to see if it can be modelled as an incremental extension of the 1992 version. In this section we consider the main features of the Z39.50-1995 standard.

The primary aim of the development of Z39.50-1992 was to achieve bit-compatibility with the ISO *Search and Retrieval* standard. Z39.50-1995 was developed to add features to those implemented in Z39.50-1992. Accordingly, Z39.50-1992 replaced and superseded Z39.50-1988, and specified version 2 of the protocol. Z39.50-1995 specifies both version 2 and version 3. In the words of the standard: *Z39.50-1995 is a compatible superset of the 1992 version. An implementor may obtain complete details of version 2 from the Z39.50-1995 document and build an implementation compatible with Z39.50-1992.* This suggests that it should be possible to develop a model of version 3 as an incremental extension of version 2.

The relationship between facilities and services for Z39.50-1995 is given in fig 6.1. The similarities with fig 3.2 reinforces that version 3 operates in much the same way as version 2. The Retrieval Facility has been enhanced to cope with the segmentation of records; the Termination Facility has been simplified; and three additional facilities have been added, Sort, Extended Services and Explain.

The Sort Facility allows for the presentation of records retrieved in a specified order. Scan allows for the interactive browsing of a list or index. Extended Services is used to initiate a specific operation which is executed outside of the Z39.50 session and whose progress may be monitored using Z39.50 services.

Facility	Service
Initialization Facility	Init Service
Search Facility	Search Service
Retrieval Facility	Present Service Segment Service
Result-set-delete Facility	Delete Service
Access Control Facility	Access-control Service
Accounting/Resource Control Facility	Resource-control Service Trigger-resource-control Service Resource-report Service
Sort Facility	Sort Service
Browse Facility	Scan Service
Extended Services Facility	Extended Services Service
Explain Facility	(see below)
Termination Facility	Close Service

Fig 6.1: The facilities and services of Z39.50-1995

The Explain Facility uses the normal Z39.50 facilities to access a special database on the server which then provides details of the server implementation – hours of operation, charges, contact information, databases available, attribute sets or extended services supported. Some of this information is intended for direct display to the user of the client, and others for the internal use of the client.

Significantly for an examination of this protocol, is the introduction of the capacity to handle multiple concurrent operations. There are also other relatively minor, but significant changes, such as support for different character sets, units, and for negotiation on the presence or absence of features.

As with Z39.50-1992, each service is individually described within the standard in unconstrained natural language. However, instead of one monolithic state transition table, there are several (see fig 6.2). These tables are loosely differentiated on function. Table 1 is in three parts, covering the initialisation, processing and termination phases. Table 2 specifies the Present Service (with its possibility of segmented responses), and Table 3 specifies *operations other than Present*. The standard says that these tables describe: *three protocol machines, one for the Z-Association (called the "Z-machine") and two for Z39.50 operations (called "O-machines"). ... There is one instance of the Z-machine (within a given application association) each for the origin and target; there may be multiple concurrent instances of the O-machines.*

Therefore, the modularity is not to enhance understanding, per se, but to show how multiple concurrent operations are achieved within the standard. In order to handle the complexity of concurrent operations, variables are used in the state tables and a syntax for testing these variables has been developed.

Abbreviations							
Acc	Access-control	Init	Initialise	resp	response	Srch	Search
conf	confirmation	Prsnt	Present	Rsc	Resource-control	Trigrc	Trigger-resource-control
Dlte	Delete	req	request	Seg	Segment	Z	Z-association

Table 1, part 1: State Table for Origin Z39.50 Association: Initialization Phase

Event \ State	Closed 0	Init sent 1	Acc recvd 2	Rsc recvd 3
Init req	Init PDU; (1)			
Init resp PDU+		Init conf +; setopCnt=0; :[conc] (5) else (4):		
Init resp PDU-		Init conf –; (0)		
Acc PDU		Acc ind; (2)		
Acc resp			Acc resp PDU; (1)	
Rsc PDU		Rsc ind; :[resp] (3) else (1):		
Rsc resp				Rsc resp PDU; (1)

Table 1, part 2: State Table for Origin Z39.50 Association: Processing Phase

Event \ State	Serial Idle 4	Concurrent Idle 5	Serial Active 6	Concurrent active 7	Z-Acc recvd 8	Z-Rsc recvd 9
<op> req	Initiate <op>; (6)	Initiate <op>; (7)		Initiate <op>; (7)	Initiate <op>; set RetSt=7; (8)	Initiate <op>; set RetSt=7; (9)
EndOp ind			(4)	Decr; :[noOps] (5) else (7):	Decr; :[noOps] set RetSt=5:; (8)	Decr; :[noOps] set RetSt=5:; (9)
Z-Acc PDU		Acc ind; set RetSt=5; (8)		Acc ind; set RetSt=7; (8)		
Z-Acc resp					Acc resp PDU; (RetSt)	
Z-Rsc PDU		Rsc ind; set Retst=5; :[resp] (9) else (5):		Rsc ind; set Retst=7; :[resp] (9) else (7):		
Z-Rsc resp						Rsc resp PDU; (RetSt)
Close req	Close PDU; (10)	Close PDU; (10)	Close PDU; KillOps; (10)	Close PDU; KillOps; (10)	Close PDU; KillOps; (10)	Close PDU; KillOps; (10)
Close PDU	Close ind; (11)	Close ind; (11)	Close ind; KillOps; (11)	Close ind; KillOps; (11)	Close ind; KillOps; (11)	Close ind; KillOps; (11)

Table 1, part 3: State Table for Origin Z39.50 Association: Termination Phase

Event \ State	Close sent 10	Close Recvd 11
AnyOpPDU	(10)	
Z-Rsc PDU	:[noResp] Rsc ind:; (10)	
Z-Acc PDU	(10)	
Close resp		Close PDU; (0)
Close PDU	Close conf; (0)	

Event \ State	Present sent 1	Rsc recvd 2	Acc Recvd 3
Rsc PDU	Rsc ind; :[resp] (2) else (1):		
Rsc resp		Rsc resp PDU; (1)	
Acc PDU	Acc ind; (3)		
Acc resp			Acc resp PDU; (1)
Trigrc req	Trigrc PDU; (1)		
Seg PDU	Seg ind; (1)		
Prsnt resp PDU	Prsnt conf; EndOp ind; exit		

Table 2: State Table for Origin Present Operation

Event \ State	<op> sent 1	Rsc recvd 2	Acc Recvd 3
Rsc PDU	Rsc ind; :[resp] (2) else (1):		
Rsc resp		Rsc resp PDU; (1)	
Acc PDU	Acc ind; (3)		
Acc resp			Acc resp PDU; (1)
Trigrc req	Trigrc PDU; (1)		
<op> resp PDU	<op> conf; EndOp ind; exit		

Table 3: State Table for Origin Operation other than Present

Fig 6.2: Transition table and abbreviations for Z39.50-1995

The transition table also uses the following miscellaneous actions:

- Initiate <op>:
  1. Initiate an O-machine for operation <op> – table 2 or 3
  2. send <op> PDU
  3. set initial state for operation to 1
  4. if concurrent operations in effect, increment opCnt by 1
- KillOps
  1. Immediately terminate all active operations
  2. Any PDUs pertaining to these operations are sent to the Z-machine
- Set <variable> = <x> Set the variable to the desired value
- Decr Decrement the variable opCnt by 1
- Exit Terminate the O-machine
- :[cond] act1 else act2: If *cond* is true then perform action *act1*, otherwise do *act2*
- conc True if concurrent operations in effect, otherwise false
- noOps True if no active operations, otherwise false

It is worth noting that the whole notion of concurrent execution of different tables for the Z-machine and the O-machine is not precisely defined, nor is the communication between them, including the ability of one (the Z-machine) to control another (the O-machine) with operations such as *Initiate <op>*, *EndOp ind*, and *KillOps*.

7 New services of Z39.50-1995

Like many protocols, Z39.50 has evolved over time with the addition of further functionality. In this section we consider that additional functionality and how it can be modelled as extensions of the 1992 protocol.

58

7.1 Additional services and facilities of Z39.50-1995

Z39.50-1995 introduces two new services – *Scan* and *Sort*. The former is used to scan terms in a list or index. It is currently the only service in the Z39.50 Browse facility. The latter is used to sort a result set. Both of these services follow the paradigm of the *Search* service discussed in §4.3. They therefore require no further attention beyond the observation that the functions to test for an appropriate response will need to be redefined.

Z39.50-1995 also introduces two new facilities – *Explain* and *Extended Services*. The former allows a client to retrieve details of the server implementation. The latter is used to initiate a specific extended service task, which is executed outside of the Z39.50 session and whose progress may be monitored using Z39.50 services. The beauty of polymorphism is that these new facilities can be encompassed in the existing model without further change.

7.2 Extended service of Z39.50-1995

A significant modification of Z39.50-1995 over the 1992 version is the support for segmentation of retrieval responses, a modification of the *Present* service. Once a search has taken place, the result set is established. The actual records can then be retrieved. Given the variability of the length of the records, it is possible to return multiple records in each response (if the records are relatively short), or only part of a record (if the records are relatively long). Provision for the first is referred to as level 1 segmentation, while additional provision for the second is referred to as level 2 segmentation. Both are accomplished by the target sending a number of *Segment* requests followed by the *Present* response. (These are referred to as *Segment* requests rather than responses in order to fit the request–response paradigm.)

The modelling of this extended service is a simple matter of defining a refined request service as shown in fig 7.3. Note that the refinement simply adds a new transition. The enabling of this transition depends on whether segmentation is supported, an issue which is resolved by negotiation.

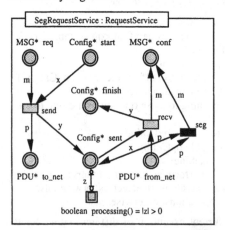

```
Class SegRequestService : RequestService
   trans seg
      Config x <- sent | x.allows(Segment);
      PDU    p <- from_net | p.request(Segment,x.id);
      Config y -> sent | y = x;
      MSG    m -> conf | m = p[...];
   end
End SegRequestService
```

Fig 7.3: Enhanced request service for segmented Present

Note that this enhancement has been shown on top of the *RequestService* (of fig 4.2) so as not to clutter the diagram. It could equally have been shown on top of the *RequestAccService* (of fig 5.2), or preferably the *RequestAccAbtService* (of fig 5.3), since it is an enhancement over what has been done before. This approach is quite adequate for our purposes. Another possibility would be to model each enhancement as a separate class and then form the enhanced services by multiple inheritance from the original service and the enhancements.

8 Other extensions to Z39.50-1995

Z39.50-1995 also introduced what it refers to as *Miscellaneous enhancements*. We consider two of these here, both of which do not so much add new functionality but modify the existing protocol scheme. Both of them can be seen as primarily impinging on the state of the connection.

8.1 Negotiated services

Z39.50-1995 modifies the functionality of the 1992 version by extending the negotiation of supported services. Depending on this negotiation, some services will never be used as part of a session. This is simply an extension of the negotiation included in the 1992 version, and the *Config* class (of fig 4.1) will need to be suitably extended.

8.2 Concurrent operations

Perhaps the most significant change introduced by Z39.50-1995 is the support for concurrent operations. This adds significant complexity to the transition table (as already noted in §6).

In considering how this extension can be addressed, we observe that in the first instance, the existing services are not modified. A *Search* service still has the same sequence of operations. It needs to have access to the the token in the *start* place (fig 4.2) which is bound to the *open* place (of fig 4.4). The difference is that the access to this token should not prohibit other services from similarly accessing this token. Clearly what has changed is the logic associated with accessing a token from this place.

The logic necessary for concurrent operations can be achieved by replacing the *open* place with a super place which has its own internal logic to control the access to tokens. For serial operations, only one token will be offered. For concurrent operations, multiple tokens will be available. This flexibility would not be available under an alternative definition of super places, which is currently under investigation [29]. Here, a super place has an associated abstract marking. Such a definition is more restrictive but does promise the possibility of more efficient analysis. (We return to this in §10.)

A further implication of the concurrent operations is that the *Reference-ids* specified in the 1992 standard now play a much more significant role. However, the support built into the earlier model is sufficient for our requirements here. In other words, the service definitions of §4 and §5 can be used here without change for concurrent operations. This is quite remarkable considering the magnitude of the changes required in the transition table to support concurrent operations. We believe that this supports the claim that the object-oriented structuring mechanisms supplied by OPNs and LOOPN++ are both powerful and flexible.

As a result of these arguments, the *open* place will be a super place which will offer and accept tokens of type *Config* (which indicate the supported services). Initially, the token deposited in the *open* place by the *Init* service needs to indicate the services which have been negotiated. Subsequently, it can be handled as shown in fig 8.1.

The intention is that place *avail* holds a single *Config* token which determines (see §4.1) the allowable services which have been negotiated. (The initial value will indicate that no services are allowed.) The place *active* will hold one token for each activated service. (The initial value will indicate that the only active service is initialisation.) The tokens in place *active* are copies of the tokens which have been offered to the environment. They are matched when the tokens are returned by the environment. They are therefore largely redundant, but in some situations, it will be helpful to have them collected together in this net. For example, the reception of an Access control request should be targetted at the active operation specified in the *Reference-id* in the PDU. If there is no such operation, the Access control is taken to relate to the Z-association. It will be necessary to be able to test this condition by comparison with the *Reference-ids* of all active operations.

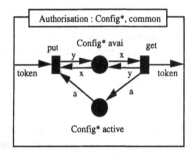

```
Class Authorisation : Config*, common
    Config* avail = ...;
    Config* active = ...;
    trans put : put
        Config a <- active | a = token;
        Config x <- avail | a.matches(x);
        Config y -> avail | y = x[...];
    end put
    trans get : get
        Config x <- avail;
        Config y -> avail | y = x[...];
        Config a -> active | a = x[...];
        Config token = a;
    end get
End Authorisation
```

Fig 8.1: Authorisation super place

Each time a *Config* token is removed from this super place, the token in place *avail* is modified to reflect the currently allowable services. Similarly, when such a *Config* token is deposited in this super place, the token in place *avail* is again suitably modified. A complex function can be used to compute the modified setting. For example, if only serial operations are allowed, then any removal of a *Config* token will result in the token in place *avail* indicating no allowable services. On the other hand, if concurrent operations are allowed, the removal of a *Config* token may result in no change to the token in place *avail*. A more subtle example is that the reception of an Access control PDU for the Z-association will prohibit the reception of further such Access control PDUs, until the current one has been dealt with.

8.3 Termination facility

A significant change in Z39.50-1995 concerns the *Termination facility*, so much so that it is difficult to know how the 1995 version can be described as a compatible superset of the 1992 version. The 1992 version supported both abrupt termination (the abort service) and graceful termination (the release service), while the 1995 version only supports a form of termination with both graceful and abrupt features. It is graceful because it is acknowledged. It is abrupt because all currently active operations are abandoned.

There are also more subtle differences. For example, if an initialisation request is refused in Z39.50-1992, a release request is sent to the association and eventually confirmed. In Z39.50-1995, no such release request is sent. Again, in Z39.50-1992 abort indications can be received both from the information retrieval service and from the association, while in Z39.50-1995, only a close PDU can be received.

This means that some functionality of the 1992 version needs to be removed. OPNs and LOOPN++ do not support the removal of functionality in a subclass. This is a significant problem and demonstrates that not all evolution of protocols fits appropriate patterns. To some extent, this problem can be hidden by extending the negotiation of services to ensure that the 1992 termination facility is never enabled.

Once the above problem has been identified and resolved, the handling of termination in Z39.50-1995 can be addressed, largely because of the modular approach adopted in §5.3. Firstly, the *InitRlseService* of fig 4.5 needs to be modified so that the rejection of the *Initialise* request results in a transfer to the closed state, and not the transmission of a *Release* request, in line with the transition table of fig 6.2. Secondly, the new transitions of fig 5.3 should be responses to *Close* rather than *Abort* requests and indications. Finally, the origin entity of §5.4 needs to be modified to ignore (possibly delayed) PDUs, other than the final *Close* request or indication (in line with the transition table of fig 6.2 – Table 1 part 3).

9 Other Object-Oriented Features

This section highlights some aspects of OPNs which are considered important for modelling network protocols, either in the example above of Z39.50, or in possible extensions of the example.

9.1 Support for inheritance and polymorphism

Languages which support the definition and instantiation of modules are classified as *object-based* [45]. Such languages encourage a certain amount of software reuse. Thus, in §4 and §5 the Z39.50 protocol was built up as a number of services, each of which instantiated one of two classes. Such multiple instantiation comes under the title of object-based and is commonly found in Petri Net formalisms such as Hierarchical Coloured Petri Nets (HCPNs) [17].

One of the primary motivations for the development of object-oriented technology was the quest for more effective mechanisms for software reuse [35]. This has primarily been achieved through inheritance and polymorphism, which qualify a language (or formalism) to be described as *object-oriented* [45]. Inheritance allows a class to derive its features from another (its parent class), and then to augment or modify them. Then, polymorphism means that an instance of the subclass may be used in a context specifying the parent. This facility has made possible the explosive growth in application frameworks, particularly in the realm of graphical user interfaces [3, 44, 46].

OPNs and LOOPN++ support inheritance and polymorphism, which can be used to structure the models and to benefit from this style of software reuse. This has been extensively used in the modelling of the Z39.50 protocol above. It made possible the minimal definition of the classes *RefId*, *PDU* and *Config* in §4.1. These were defined with sufficient information to cater for the required operations, but with the understanding that an actual implementation would attach additional information. This is particularly apparent for the class *PDU*. For the purposes of the protocol (figs 4.2 and 4.3), it is sufficient for this class to have functions which determine whether an instance is a request, an acceptance, or a rejection of a particular kind of service. In reality, a *PDU* will carry extensive information pertinent to the service request or response, but this is not relevant to the operation of the protocol.

9.2 Support for genericity

One of the primary motivations for the development of the OSI Reference Model [15] was the appropriate modularisation of the complexities associated with networking. Each layer of the Model addressed certain issues and delegated others to lower layers. A consequence of this is that a layer should not be concerned with higher layer issues, but simply provide a set of services to these higher layers through a number of primitives. As a result, any data submitted for transmission by a higher layer via one of these primitives should remain uninterpreted. Thus a network layer receiving a packet for transmission to some destination should perform the same way independent of the particular application requesting the transfer.

In other words, the layering of protocols demands the ability to define generic software components. One simple approach (and the one commonly implemented in networking software) is to treat the data simply as a (generic) sequence of bits or bytes (octets) and to supply appropriate encoding and decoding routines. In the abstract modelling of protocols, it is preferable to retain the type information, in which case the protocol layers will need to be able to transfer a variety of differently typed messages, and hence the need for genericity.

In modelling such generic protocol layers, OPNs and LOOPN++ can take advantage of the support for polymorphism already discussed in §9.1. A protocol layer defined to transfer tokens of a given class, can also be used to transfer tokens of any subclass. The simplest case in LOOPN++ is to define a protocol layer to transfer tokens of type *null*, which is a predefined class having built-in functions but no data. Since every LOOPN++ class inherits from *null*, the protocol layer will be able to transfer any token type.

There is a subtle twist to this strategy, since a protocol layer will normally add some header information on transmission, and remove it on reception. For example, a data link layer will typically add a header including the kind of frame, its sequence number and its acknowledgement number. Then the kind of token handled by the higher layer interface will differ from the kind of token handled by the lower layer interface. It is then important that the proposed polymorphic use of the layer will be consistently defined. Another example would be the *RequestService* of fig 4.2 which receives tokens of type *MSG* from the higher layer and submits tokens of type *PDU* to the lower layer. So far, we have assumed that these classes are the same, but it would be more common to find that the class *PDU* augments the class *MSG* with some control information.

The requirements in such situations can be highlighted by examining typical transitions which would add and delete such header information for a data link layer, as in fig 9.1.

```
TRANS send
    null x <- network_layer | can_accept_message;
    seq  y -> physical_layer | y = x [seq: nextseq(), …];
END
TRANS recv
    seq  x <- physical_layer | sequence_number_is_OK;
    null y -> network layer | y = x;
END
```

Fig 9.1: Generic send and receive transitions

Note that the *send* transition adds a sequence number to the incoming token, while the transition *recv* removes it. We assume the following notation:

(a) $class(x)$, $class(y)$ is the declared class of (tokens) x and y

(b) $class(x) <: class(y)$ means that the declared class of x is a subclass of that of y

(c) given $class(x) <: class(y)$, we write $class(x) = class(y) + C$, where C is the class containing those components which augment $class(y)$ to give $class(x)$

(d) $class(x')$, $class(y')$ is the actual class of the tokens bound to x and y at run-time

In order to use the *send* and *recv* transitions generically, the following conditions must hold:

(e) $class(x') = class(x) + C <: class(x)$

(f) $class(y') = class(y) + C <: class(y)$

(where the operator $+$ has higher priority than $=$ and $<:$) Both points (e) and (f) state that the actual token class is a subclass of the declared class. They also demand that both augment the declared class in the same way. Thus, the hidden information, given by the class C is transferred intact.

The above constraints can be (and have been) implemented with run-time type-checking. It is also desirable to be able to guarantee type safety at compile time. In order to support this, languages like Eiffel and C++ [35, 40] define generic classes with a type parameter which is then specified each time the class is instantiated. We prefer the more flexible approach of Palsberg and Schwartzbach [36], which allows *any* component class to be consistently renamed while still retaining type safety. An alternative approach, which is worth investigating, is that adopted by the BETA programming language, where the concept of virtual function components is extended to include the notion of virtual class components [34].

9.3 Support for mobile objects

Recent years have seen an increasing interest in distributed object systems and the use of mobile objects. The unified class hierarchy of OPNs readily supports the modelling of such systems – tokens may instantiate arbitrary classes, and class components may also instantiate arbitrary classes. Thus, in §4.1 the Z39.50 protocol data units were defined by classes encapsulating both data fields and functions. Elsewhere [27], we have used OPNs to model the documents of a hypothetical Electronic Data Interchange system [20], where the documents encapsulated transitions, in addition to fields and functions.

10 Analysis of OPNs

As noted in the introduction, the amenability of Petri Nets to automated analysis has been a key aspect which has facilitated their beneficial application to the development of reliable network protocols. The introduction into OPNs of powerful object-oriented structuring primitives has not yet been matched by appropriate analysis techniques. In fact, this is considered to be a significant area of research in which only the preliminary steps have been taken. For example, the design of the OPN formalism has been constrained so as to be able to transform OPNs into behaviourally equivalent CPNs [23]. It is anticipated that this will provide the foundation for adapting existing CPN analysis techniques for OPNs.

There are two forms of analysis commonly applied to high-level Petri Nets. Reachability analysis generates the graph of all possible reachable states and the transitions between those states. This form of analysis is normally plagued by the problem of state explosion where the number of possible states is exponential in the size of the net model. Techniques have been developed to reduce the size of the reachability graph without losing the essential properties of the system. Invariant analysis, as its name suggests, determines some property which is invariant over the execution of the net. A place invariant specifies a weighted marking which does not change over the firing of any (enabled) transition in the net, while a transition invariant specifies a weighted step which leaves the marking unchanged.

In this section, we highlight some of the issues which will need to be addressed in providing appropriate analysis techniques for OPNs. We consider the general issue of modular analysis and observe how this has lagged behind the development of modular specification techniques. We then consider the specific issues pertinent to reachability analysis and invariant analysis. Finally, we consider the ideal goal of developing incremental analysis techniques.

10.1 Modular analysis

Hierarchy constructs were first proposed for CPNs in 1990 [14]. Of these, only substitution transitions and place fusion were implemented in HCPNs [17, 19]. The behaviour of these nets was then defined in terms of the expanded CPN (after the substitution transitions were appropriately expanded and the relevant places fused). In other words, the modular definition of HCPNs was not matched by the notion of modular behaviour or modular analysis. It is our contention that the effective analysis of complex systems modelled by Petri Nets will increasingly demand the development of modular analysis techniques.

As far as CPNs are concerned, we note that symmetry analysis [18] can be most effective in reducing the size of the reachability graph by collapsing symmetrical states onto the one state. This technique seems most appropriate to CPNs where colours have been used to fold similar subnets onto the one subnet. Similarly, stubborn set analysis [41] has been used to reduce the number of possible interleavings of independent concurrent processes which are covered by the reachability graph. These two techniques are orthogonal and can be extremely effective in reducing the size of the reachability graph, but neither of them is particularly relevant to modular analysis.

The development of modular analysis techniques has lagged behind the development of hierarchy constructs for CPNs. However, proposals have been made recently for the modular analysis of Modular Coloured Petri Nets (MCPNs) [10, 11]. For place invariants [10], the weighted marking is specified on a module-by-module basis. Provided the weight functions are consistent (i.e. they agree on fused places), and provided that they constitute a place flow for each module, then the local weight functions determine a global weight function which constitutes a global place flow for the whole net. For reachability analysis [11], it turns out that modular analysis is much simpler where transition fusion is used as opposed to place fusion. Here, the computation of the state space (or reachability graph) for the individual modules is interleaved with the computation of a synchronisation graph, which captures the interaction points of the various modules, i.e. the firing of fused transitions.

While these modular analysis techniques are important contributions, we believe that their results will be limited because of the overly general form of modular nets that they address. A net is formed from a number of modules which are combined using the arbitrary mechanisms of place fusion and transition fusion. The modules are not constrained to have any particular properties, nor are the results constrained to place-bounded or transition-bounded modules. We believe that better results will be obtained by restricting attention to super places and super transitions, i.e. modules which have place or transition properties. Some preliminary steps in this direction have already been made [29].

A different approach to modular analysis has been developed using process algebra techniques [42]. Here, process algebra reduction techniques are applied to place-bounded subnets so that the same externally visible behaviour is obtained from simpler subnets. Clearly, this approach has great relevance to complex systems modelled by modular Petri Nets.

10.2 Reachability analysis

As already noted, reachability analysis involves the generation of every reachable state and every possible transition between these states, and consequently is severely affected by having the number of states being exponential in the size of the model. One simple implication of this, which is not unique to OPNs, is that predefined types need to be restricted to finite subranges. Thus, while *integer* type fields have been used in the specification of the Z39.50 protocol (e.g. for the PDU *kind* in fig 2.3), only a small number of values is actually used, and this should be made explicit.

OPNs introduce another potential problem area with the use of object identifiers to identify or address the various objects or class instances. The formal definition of OPNs specifies one set of object identifiers per class [27], and these sets are potentially infinite, because the number of instances of any class is potentially infinite. In order to make reachability analysis computable or tractable for OPNs, the sets of object identifiers will need to be finite (as in [5]). It will also be desirable (if not necessary) to include some form of symmetry analysis [18], so that states differing only with respect to their labelling by object identifiers are treated as identical.

10.3 Invariant analysis

As already noted, invariant analysis identifies some property which is invariant over the firing of the net. Modular invariant analysis [10] can be extended to OPNs [25], but the weight functions cannot be statically determined for each class instance, since the number of instances varies over the life of the net. Instead, the weight function for a class instance is formed from the composition of a weight function for the class and a weight function determined by the context of the module instance (relative to the root class).

Simpler invariant results can be obtained if classes are constrained to define super places and super transitions, i.e. subnets with place and transition properties [29]. This follows since those place and transition properties are captured by weight functions and local invariant properties. For example, the abstract marking of a super place is defined by a weighted marking of the associated subnet. This should be invariant over the internal activity of the subnet. Consequently, a place invariant over the abstract net (where super places are treated as places) can be extended to a place invariant over the expanded net.

10.4 Incremental analysis

Just as it is important to develop modular analysis techniques for modular nets, so the ultimate goal in analysing OPNs is to develop incremental analysis techniques to match the incremental modelling capabilities. The ideal is that when one net component is replaced by a more refined component, then the analysis results can be suitably modified rather than recomputed from scratch.

There are many issues to be addressed here, not the least of which is the specification of what behaviour is preserved from a parent class in the subclass. Some proposals have required a form of bisimilarity, while other have required a state preorder [5]. Our experience with some practical case studies indicates that both of these are too restrictive [26].

In some cases, we anticipate that incremental analysis will be relatively straightforward. For example, the additional support for *Segment* requests (in fig 7.3) and even the addition of *Access Control* (in fig 5.2) do not detract from the previous behaviour but only extend it – the added components only respond to different kinds of PDU. The original state sequences are retained, though possibly augmented (as in the state preorder property of [5]).

In other cases, we anticipate that incremental analysis will not be possible. For example, the modified support for termination control (as discussed in §8.3) is so different to the earlier version that it is difficult to see it as an incremental change.

Finally, there are some cases where it is problematic whether incremental analysis will be possible. For example, the addition of an abort capability (in fig 5.3) means that the original sequences of states are still possible, but parts may now be bypassed.

11 Conclusions

This paper has introduced Object Petri Nets, both in their graphical form and in their textual form (in the language LOOPN++). There is an economy of notions in OPNs, particularly with respect to the unified class hierarchy. Thus, classes encompass simple predefined types, user-defined classes without actions, and user-defined classes with actions. Consequently, the notion of a field encompasses a number of different notions from more traditional Petri Net formalisms – simple constants, Petri Net places, and subnet instances. The same flexible type system applied to functions and tokens means that tokens may be associated with subnets and functions may return subnet instances. The unified class hierarchy therefore caters directly for the arbitrary nesting of objects, which sets OPNs and LOOPN++ apart from other object-oriented net formalisms [4, 6, 9, 39, 43].

A class may be declared to inherit the features of one or more parents, in which case all the features of the parents, together with the additional features declared within the class constitute the features of the new class. As is standard in object-oriented languages, subclass instances can be used polymorphically in superclass contexts. This makes it possible to introduce minimal definitions into classes (as in §4.1) knowing that more complex definitions can be substituted later without affecting the logic of the net. In other words, inheritance can be used to structure the model of the protocol, and also to capture the different configurations and the evolution of the protocol.

OPNs support both super places and super transitions. A super place can accept or offer tokens to its environment under the control of its internal logic. Since the token type of the interface is determined by the multiset class that it inherits, polymorphism means that such super places can be substituted for simple places in a net. Similarly, a super transition has transition characteristics, but the extent to which this is the case varies between different Petri Net formalisms and is the subject of further investigation (see §2.8).

In summary, OPNs incorporate a uniform and flexible type system which provides good support for modularity, inheritance, polymorphism, and mobile objects. This paper has demonstrated how these attributes can be of significant benefit in the incremental modelling of network protocols. The possibility of incremental modelling is seen as one of the chief benefits of OPNs. It allows the modeller to capture the basic services of a protocol and clearly identify the enhancements. It allows the identification of different protocol configurations, and it allows the evolution of protocols to be captured in the model.

Perhaps the most dramatic demonstration of this was the way that the support for concurrent operations, introduced in Z39.50-1995, could be captured with minimal change to the previously developed net. The only thing required was to override a simple place with a super place, the super place encapsulating the logic of when multiple concurrent operations could be allowed.

While this paper has not presented the formal foundations for OPNs, other papers have been referenced which give the formal definition and prove that OPNs can be transformed into behaviourally-equivalent CPNs. As noted in §10, this is considered to be a first step in adapting CPN analysis techniques for use with OPNs. We also noted that the development of better modular analysis techniques will also be of great importance to OPNs. Clearly, the ideal will be to develop incremental analysis techniques to match the incremental development possibilities of OPNs. A related issue is the kind of behaviour that should be preserved from a parent class in a subclass – it is unclear whether existing proposals are adequate for practical applications.

Currently, a preliminary version of LOOPN++ has been implemented [28]. By translating LOOPN++ programs into C++, it is intended that it will be easy to integrate LOOPN++ with other software packages, either to provide analysis tools, or to provide a prototyping environment. For example, we have observed that such an open environment can lead to a significant part of a protocol model being reused as the foundation of a prototype implementation [30]. Another variant of this compiler produces Java code, which could therefore be used in the development of web applications [33].

It is therefore anticipated that OPNs will reap the practical benefits of object-orientation including clean interfaces, reusable software components, and extensible component libraries.

Acknowledgements The authors gratefully acknowledge the contributions of the reviewers, which have served to improve the quality of this paper.

References

[1]   ANSI *Z39.50: Information Retrieval Service and Protocol* ANSI/NISO Z39.50-1992 (version 2), American National Standards Institute (1992).

[2]   ANSI *Information Retrieval (Z39.50): Application Service Definition and Protocol Specification* ANSI/NISO Z39.50-1995 (version 3), American National Standards Institute (1995).

[3]   Apple Corporation *MacApp Reference Manual* (1992).

[4]   M. Baldassari and G. Bruno *An Environment for Object-Oriented Conceptual Programming Based on PROT Nets* Advances in Petri Nets 1988, G. Rozenberg (ed.), Lecture Notes in Computer Science 340, pp 1–19, Springer Verlag (1988).

[5]   E. Battiston, A. Chizzoni, and F. de Cindio *Inheritance and Concurrency in CLOWN* Proceedings of Workshop on Object-Oriented Programming and Models of Concurrency, Torino, Italy (1995).

[6]   E. Battiston, F. de Cindio, and G. Mauri *OBJSA Nets: A Class of High-level Nets having Objects as Domains* Advances in Petri Nets 1988, G. Rozenberg (ed.), Lecture Notes in Computer Science 340, pp 20–43, Springer-Verlag (1988).

[7]   G. Berthelot and R. Terrat *Petri Nets Theory for the Correctness of Protocols* Proceedings of Protocol Specification, Testing and Verification II, pp 325-341, Los Angeles, North-Holland (1982).

[8]   J. Billington, G.R. Wheeler, and M.C. Wilbur-Ham *PROTEAN: A High-Level Petri Net Tool for the Specification and Verification of Communication Protocols.* IEEE Transactions on Software Engineering, **14**, 3, pp 301–316 (1988).

[9]   D. Buchs and N. Guelfi *CO-OPN: A Concurrent Object Oriented Petri Net Approach* Proceedings of 12th International Conference on the Application and Theory of Petri Nets, Gjern, Denmark (1991).

[10] S. Christensen and L. Petrucci *Towards a Modular Analysis of Coloured Petri Nets* Application and Theory of Petri Nets, K. Jensen (ed.), Lecture Notes in Computer Science 616, pp 113-133, Springer-Verlag (1992).

[11] S. Christensen and L. Petrucci *Modular State Space Analysis of Coloured Petri Nets* Application and Theory of Petri Nets, G.D. Michelis and M. Diaz (eds.), Lecture Notes in Computer Science 935, pp 201-217, Springer-Verlag (1995).

[12] M. Diaz *Modelling and Analysis of Communication and Cooperation Protocols Using Petri Net Based Models.* Proceedings of Protocol Specification, Testing and Verification II, pp 419–441, Los Angeles, North-Holland (1982).

[13] E.W. Dijkstra *Notes on Structured Programming* Structured Programming, O.J. Dahl, E.W. Dijkstra, and C.A.R. Hoare (eds.), pp 1-82, Academic Press (1972).

[14] P. Huber, K. Jensen, and R.M. Shapiro *Hierarchies of Coloured Petri Nets* Proceedings of 10th International Conference on Application and Theory of Petri Nets, Lecture Notes in Computer Science 483, pp 313-341, Springer-Verlag (1990).

[15] ISO *Information Processing Systems – Open Systems Interconnection: Basic Reference Model* International Organisation for Standardization and International Electrotechnical Committee (1984).

[16] K. Jensen *Coloured Petri Nets: A High Level Language for System Design and Analysis* Advances in Petri Nets 1990, G. Rozenberg (ed.), Lecture Notes in Computer Science 483, Springer-Verlag (1990).

[17] K. Jensen *Coloured Petri Nets: Basic Concepts, Analysis Methods and Practical Use – Volume 1: Basic Concepts* EATCS Monographs in Computer Science, Vol. 26, Springer-Verlag (1992).

[18] K. Jensen *Coloured Petri Nets: Basic Concepts, Analysis Methods and Practical Use – Volume 2: Analysis Methods* EATCS Monographs on Theoretical Computer Science, Springer-Verlag (1994).

[19] K. Jensen, S. Christensen, P. Huber, and M. Holla *Design/CPN™: A Reference Manual* MetaSoftware Corporation (1992).

[20] P. Kimberley *Electronic Data Interchange* McGraw-Hill (1991).

[21] R. Lai, T.S. Dillon, and K.R. Parker *Verification Results for ISO FTAM Basic Protocol* Proceedings of Protocol Specification, Testing and Verification IX, pp 223-234, North-Holland (1990).

[22] C. Lakos and S. Christensen *A General Systematic Approach to Arc Extensions for Coloured Petri Nets* Proceedings of 15th International Conference on the Application and Theory of Petri Nets, Lecture Notes in Computer Science 815, pp 338-357, Zaragoza, Springer-Verlag (1994).

[23] C.A. Lakos *Object Petri Nets – Definition and Relationship to Coloured Nets* Technical Report TR94-3, Computer Science Department, University of Tasmania (1994).

[24] C.A. Lakos *From Coloured Petri Nets to Object Petri Nets* Proceedings of 16th International Conference on the Application and Theory of Petri Nets, Lecture Notes in Computer Science 935, pp 278-297, Torino, Italy, Springer-Verlag (1995).

[25] C.A. Lakos *The Object Orientation of Object Petri Nets* Proceedings of Workshop on Object Oriented Programming and Models of Concurrency, Torino, Italy (1995).

[26] C.A. Lakos *Pragmatic Inheritance Issues for Object Petri Nets* Proceedings of TOOLS Pacific 1995, pp 309-321, Melbourne, Australia, Prentice-Hall (1995).

[27] C.A. Lakos *The Consistent Use of Names and Polymorphism in the Definition of Object Petri Nets* Proceedings of 17th International Conference on the Application and Theory of Petri Nets, Lecture Notes in Computer Science 1091, pp 380-399, Osaka, Japan, Springer-Verlag (1996).

[28] C.A. Lakos *The LOOPN++ User Manual* Technical Report R96-1, Department of Computer Science, University of Tasmania (1996).

[29] C.A. Lakos *On the Abstraction of Coloured Petri Nets* Proceedings of 18th International Conference on the Application and Theory of Petri Nets, Lecture Notes in Computer Science 1248, Springer-Verlag (1997).

[30] C.A. Lakos and C.D. Keen *Modelling a Door Controller Protocol in LOOPN* Proceedings of 10th European Conference on the Technology of Object-oriented Languages and Systems, Versailles, Prentice-Hall (1993).

[31] C.A. Lakos, J.W. Lamp, C.D. Keen, and B.W. Marriott *Modelling Network Protocols with Object Petri Nets* Proceedings of Workshop on Petri Nets Applied to Protocols, pp 31-42, Torino, Italy (1995).

[32] J.W. Lamp *Encoding the ANSI Z39.50 Search and Retrieval Protocol using LOOPN* Honours Thesis, Department of Computer Science, University of Tasmania (1994).

[33] G.A. Lewis *Producing Network Applications Using Object-Oriented Petri Nets* Honours Thesis, Department of Computer Science, University of Tasmania (1996).

[34] O.L. Madsen and B. Møller-Pedersen *Virtual Classes: A Powerful Mechanims in Object-Oriented Programming* Proceedings of OOPSLA 89, SIGPLAN Notices 24, pp 397-406, New Orleans, Louisiana, ACM (1989).

[35] B. Meyer *Object-Oriented Software Construction* Prentice Hall (1988).

[36] J. Palsberg and M.I. Schwartzbach *Object-Oriented Type Systems* Wiley Professional Computing, Wiley (1994).

[37] D.L. Parnas *On the Criteria to be Used in Decomposing Systems into Modules* CACM, **15**, 12, pp 1053-1058 (1972).

[38] M.T. Rose *The Open Book: A Practical Perspective on OSI* Prentice-Hall (1990).

[39] C. Sibertin-Blanc *Cooperative Nets* Proceedings of 15th International Conference on the Application and Theory of Petri Nets, Lecture Notes in Computer Science 815, pp 471-490, Zaragoza, Spain, Springer-Verlag (1994).

[40] B. Stroustrup *The C++ Programming Language (Second Edition)* Addison-Wesley (1991).

[41] A. Valmari *Stubborn Sets for Coloured Petri Nets* Proceedings of 12th International Conference on the Application and Theory of Petri Nets, Aarhus (1991).

[42] A. Valmari *Compositional analysis with place-bordered subnets* Proceedings of 15th International Conference on the Application and Theory of Petri Nets, Lecture Notes in Computer Science 815, pp 531-547, Zaragoza (1994).

[43] P.A.C. Verkoulen *Integrated Information Systems Design: An Approach Based on Object-Oriented Concepts and Petri Nets* PhD Thesis, Technical University of Eindhoven, the Netherlands (1993).

[44] J.M. Vlissides *Generalized Graphical Object Editing* Technical Report CSL-TR-90-427, Stanford University (1990).

[45] P. Wegner *Dimensions of Object-Based Language Design* Proceedings of OOPSLA 87, pp 168-182, Orlando, Florida, ACM (1987).

[46] A. Weinand, E. Gamma, and R. Marty *ET++ – An Object-Oriented Application Framework in C++* Proceedings of OOPSLA 88 Conference, ACM (1988).

The Modelling and Analysis of IEEE 802.6's Configuration Control Protocol with Coloured Petri Nets

Geoffrey Wheeler

Telstra Research Laboratories
770 Blackburn Road, Clayton North, Victoria, 3168, Australia

Abstract — The IEEE Standard 802.6 (*Distributed Queue Dual Bus (DQDB) Subnetwork of a Metropolitan Area Network (MAN)*) permits subnetwork reconfiguration, usually without loss of communication ability, whenever there are bus faults. The Configuration Control Protocol (CCP) is the protocol which enables this to occur.

In this paper, we report on the modelling of CCP using Coloured Petri Nets. The complete model is too lengthy to include, so we have illustrated the modelling using a number of carefully selected examples. The examples illustrate how CPNs can replace a number of commonly used informal specification practices. They also show how CPNs cope with defining the purpose of the protocol and the intended environment in which it is meant to operate — important features often missing from specifications. Finally, some analysis of CCP which revealed a problem is described.

1 Introduction

1.1 The Standard

The IEEE Standard 802.6 – *Distributed Queue Dual Bus (DQDB) Subnetwork of a Metropolitan Area Network (MAN)* [1] – enables equipment manufacturers to build components which offer integrated telecommunication services such as data, voice and video over a metropolitan area. It specifies the Physical and DQDB layers of subnetworks required to support the Logical Link Control (LLC) sublayer of the Data Link Layer. The DQDB layer includes the functionality of a medium access control (MAC) sublayer that provides services to the LLC sublayer in the same manner as other members of the IEEE 802 LAN/MAN family. DQDB subnetworks can provide switching, routing and concentration of certain high-speed data, voice and video services. They can also be used to interconnect LANs, hosts, workstations and PBXs.

1.2 Configuration Control

IEEE 802.6 defines a high speed shared medium access protocol for use over a subnetwork consisting of two counter-flowing unidirectional buses. The topology of the dual bus network can be an open topology in which the end points of the buses are distinct or it can be a looped topology in which the end points of the buses are co-located. Note, however, that in the looped topology, data does not flow from the end of one bus through to the head of the other bus.

The looped topology is important in that it permits a subnetwork reconfiguration, usually without loss of communication ability, whenever there are bus faults. With a bus fault, the fault is isolated and a healing mechanism invoked in which the heads of bus are shifted to the nodes adjacent to the break or as close as possible to the break (some nodes may not be able to act as head of bus). If there are further bus faults, the healing mechanism cannot maintain full connectivity, but by splitting the subnetwork into islands of communicating nodes it provides as much connectivity as possible given the dual bus architecture.

The Configuration Control Protocol (CCP) is used to pass information between DQDB Layer Management Protocol Entities of nodes. It supports the operation of the healing mechanism.

1.3 Modelling and Analysing the Configuration Control Protocol

The model of CCP is described by Coloured Petri Nets (CPNs) [3], which like other Petri net varieties [2] have a graphical representation. The CPNs were created using Design/CPN version 1.7 [10], a software tool from Meta Software Corporation. The expressions used in the CPNs are written in CPN ML [10]. CPN ML is an extension of the functional programming language Standard ML [7, 8], which provides an extremely rich language for this task. Extensions to ML are needed so that declarations of color sets and variables in CPNs can be presented in the usual way.

CPNs are described briefly in section 2. In section 3 we introduce the five elements of a protocol which are used to structure the modelling, and we describe the style of the CPN modelling. The complete specification [4] is too lengthy for inclusion (it has 21 pages of CPN subnets and 5 pages of CPN ML declarations), so seven examples of modelling are included in section 4 to show its salient features. There are parts of the modelling of each of the five protocol elements, including 3 examples of the procedures. The examples of procedures show how CPNs can model a relatively complex pair of state machines, a pair of large and complex tables, and some specification constraints.

The analysis of the model was performed using TORAS [5], a reachability analysis tool developed in Telecom Australia's Research Laboratories. Section 5 describes one of the more interesting problems discovered during the analysis.

Some conclusions are provided in section 6.

2 Coloured Petri Nets

A CPN model of a system describes the states a system can get into, and for each state, shows events which can occur and the states which will result if an event occurs.

A CPN state is broken into a number of component states, each component being determined by *tokens* in a *place*. *Tokens* can have arbitrary values determined by their type or *colour*. Each distinct *token* value can be thought of as a different coloured or shaped piece in a board game like chinese checkers or monopoly. The *places* are like the parts of a game board where you can put pieces.

Events are represented by *transitions*. They are connected to some of the *places* by *arcs* next to which are expressions that determine the redistribution of tokens that occurs when the event occurs.

Figure 1 is a simplified part of the CCP specification that illustrates the elements of a CPN. Transition *initialize_12* is the box in the middle of the diagram. It represents the initialization of two resource status indicators (LINK_STATUS_1 and LINK_STATUS_2) and a flag (CC_12_CONTROL) when a node powers up. This event can occur if the node has yet to power up, reflected by the arc labelled (OFF, xn), which requires that there is a token in place POWER (the circle labelled POWER) which has a first field which has the value OFF and a second field xn representing a node. Declarations (not shown) would indicate that xn is a variable that takes the value of (almost) any node value. I say "almost" because there is a constraint on the values that xn can take which is expressed by a *guard* on the transition which is written as $[xn <> Nd(d)]$ (see the interior of the box). The constraint says that the node identified by $Nd(d)$ is an unacceptable value for xn. The result of the event occurring is that the token (OFF, xn) is removed from place POWER, a token (ON, xn) is added to place POWER, a token $(CC\_12, DISABLED, xn)$ is added to place CC_z2_CONTROL, and two tokens — a $(CC\_1, DOWN, xn)$ and a $(CC\_2, DOWN, xn)$ — are added to place LINK_STATUS. This means that after the event *initialize_12*, the node is powered up, the link status indicators associated with that node (they have the same node identifier, xn) both have the value $DOWN$, and the CC_12_CONTROL flag has the value $DISABLED$.

Note that all the places have two labels. The one in upper case and mostly inside the circle is the place name. The other, slightly italicized and outside the circle, is the colour set. This is the set of allowable values that the tokens in the place can take (token types are strongly typed).

Variables (xn is an example) have a scope which is local to a transition. Consequently, in our example because there is only one transition, the value of xn is the same in its 6 appearances. However, in CPNs when there is more than one transition and the same variable name appears in arcs or guards with different associated transitions, then those values are unrelated.

An initial distribution of tokens called an *initial marking* (not shown) is associated with a CPN. The places which have tokens in the initial distribution have another label in the proximity of a place indicating the initial values.

3 Modelling Style

3.1 Structure of the Specification

Something that would improve the quality of protocol specifications such as those in IEEE 802.6 is the explicit recognition that protocols have five distinct elements. These five elements are: the service it provides (purpose), the assumptions about the environment in which it is intended to work, the vocabulary of

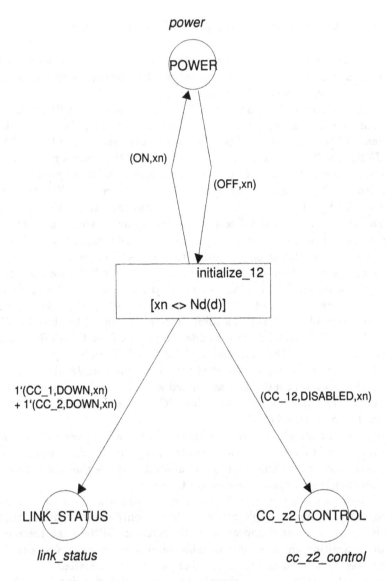

Fig. 1. Part of a CPN

messages, the format (encoding) of each message, and the procedures for message exchanges. Holzmann [6] has recently emphasized these elements. Usually when a protocol is designed, its vocabulary, format and procedures are given, but its purpose and the environment in which it is meant to operate are either not stated or indicated in a very indirect way. This may lead to various misunderstandings and inappropriate applications of the protocol. Often various relevant aspects are dispersed throughout a document. Cross-referencing is often used to deal with the problem but is seldom sufficient. The CCP specification suffers from these weaknesses.

If the 5 elements are not only addressed, but used to structure a specification, the specification becomes easier to understand. The CPN specification of CCP was developed in this spirit.

3.2 The Process of Modelling

Our first step in modelling is to identify the places. With CCP there were relatively few places involved, although they are repeated, often many times, throughout the collection of pages which constitute the specification. Places which are repeated have a nearby box marked FG, indicating global fusion [12], a term that means that the multiple appearances of a place are to be treated as if there was only one place. In Design/CPN the various instances of global fusion are distinguished by identifiers, this is not shown in the figures.

The next phase in the modelling is to add transitions, whose occurrences will represent events. These events are identified at a level of abstraction that corresponds to the informal description. The arcs and transition guards are then added to relate events and states (representing objects or data). To retain this desirable level of abstraction, variables are needed so that transitions refer to potentially many primitive events.

Iteration of the above steps continues until we obtained a model. The places and transitions are meant to be easily related to the IEEE specification. They were given suggestive names to indicate this. To retain places and transitions which correspond to some clear notion of state or event, which are the graphical aspects of the CPN, quite often we have to use somewhat complex text inscriptions. This seems to be preferable to allowing the number of graphical elements (places and transitions) to grow and keeping the inscriptions simple, because the number of graphical elements grows exponentially and the resulting nets are reminiscent of the Minotaur's labyrinth, and as difficult to follow.

We have also used the idea of transition fusion [12] quite extensively to structure the specification. Transition fusion allows the specifier to separate functionally separate procedures which occur at the same time and hence require the events involved to be synchronised. To do this a number of separate transitions are drawn that belong to the same fusion set. They are equivalent to a single transition, whose guard is the conjunction of the individual guards, and which has all the input and output arcs that the individual transitions have.

Other hierarchical constructs such as substitution [12] could have been used to provide a coarse-grained structure for the CPN. Instead related pages of the subnet were grouped in an informal way.

Most of the CPN modelling has been to try and create as clear a specification as we could. If we were thinking more about verification then the CPN may look somewhat different. For instance, the description of timers in section 4.5, can be simplified by removing place *TIMER_H* describing the state of the timers — the effect of the timers is still captured by the remaining transitions.

4 Modelling

4.1 The Purpose of CCP

CCP is responsible for configuring system resources into a correct dual bus topology. The resources it manages are the head of bus functions, bus identification functions and the timing reference functions. It is used to automatically start-up a network, that is, ensure a correct initial allocation of resources. It is also used when there are bus faults, to reconfigure the subnetwork into a correct topology. It is expected to return the subnetwork to its original configuration once the bus faults are repaired.

A correct dual bus topology satisfies the so-called *fundamental subnetwork requirements* of IEEE 802.6. These requirements use the idea of a stable network, which (presumably) is a network which is neither in the process of establishing the initial configuration at start-up nor re-configuring. They are reproduced (with some editorial licence) below:

FR1 There shall be one Head of Bus A function and one Head of Bus B function operating in a stable subnetwork.

FR2 There shall be one 125 microsecond timing reference for all nodes of a stable subnetwork.

FR3 All subnetworks shall contain exactly one node (D-node[1]) with an active Default Slot Generator (SG_D) function when started up. The mechanism by which this node is selected is outside the scope of this standard. The SG_D function shall provide the function of defining the bus identity for a subnetwork.

There are three varieties of correct dual bus topology: looped, open and island topologies.

A looped topology has co-located head of bus functions. The D-node's SG_D function provides the head of bus functions for a stable looped configuration. If a single bus fault occurs in a looped topology, a healing mechanism is invoked in which the subnetwork should be reconfigured into an open topology.

In an open topology the heads of bus A and B are located at different nodes. The nodes which act as head of bus are those which are capable of doing so and which are most upstream on their respective bus. For instance, when an open topology arises from a bus fault in a looped topology, the heads of bus will be located at the nodes which are capable and as close to the fault as possible. There may be nodes further upstream on the buses which are incapable of acting as head of bus — they will be isolated from all other nodes. This maximizing of the number of nodes in the resultant open topology is a feature of any open topology.

If a single bus fault occurs on an open topology (or two faults occur on the looped topology) the subnetwork should split into two parts. The component

---

[1] This is not IEEE 802.6 terminology, but having to use 8 words every time to identify this node surely justifies some additional terminology. Any other node will be called a 1-node.

with the D-node will have an open topology. The other component which is isolated from the D-node, is still operable. It is said to have an island topology. As before, a node will be isolated if it cannot act as head of bus and it is upstream of the first node capable of being head of bus.

Bus faults on an island topology should result in the subnetwork breaking into a number of island networks. If there are n bus faults on an island topology (or $n+1$ faults on an open topology or $n+2$ faults on the looped topology) then $n+1$ island topologies should arise.

In the reconfigurations we have described there is a possibility that the potential island or open topologies do not have the minimum of two distinct nodes which can act as head of bus. In such cases, the component nodes are isolated.

CPN Specification of Stable Configuration Although I have gone to some pains in the previous section to describe reconfiguration in natural language, there are still aspects not covered. For example, what happens if nodes are added or subtracted from the network? The formal definition will try to cover any remaining uncertainties (or perhaps more accurately — will propose a purpose for CCP that the designers can reflect upon).

Figure 2 is a specification of a correct stable configuration. It is a precise statement which encompasses the requirements on head of bus location, correct topologies and timing reference uniqueness. The modelling is explained in the next section.

The notion of a (correct) stable configuration is captured by defining a transition called *stable_config* which can occur if every subnetwork is in a stable configuration. Note that all places have double-headed arrows, which means that when the transition occurs there will be no change in the net's marking. The transition's only purpose is to test for the existence of stable configurations.

The representation of a stable configuration is in a sense an abstraction of other parts of the specification, yet is part of the specification. It is abstract in that the *partition* colour set, associated with place 'PARTITION', contains just enough information for describing stable configurations. It is redundant because other colour sets, already in the specification, could have been used. If we had avoided using the *partition* colour set we would have the disadvantage of a long and obscure specification.

This net is complemented by another (not shown) which has transitions *link_up* and *link_down* that reflect the way that topologies change as links go up and down. Other parts of the CPN reflect the changes in the location of the head of bus and in the timing source being used by the various nodes.

The specification is largely captured by the definitions of the ML functions used in the guards. The definition of the function *has_correct_configuration*, used in the guard of figure 2, is lengthy and has been omitted to save space. The other ML definitions are given in figure 3. They are included to exemplify CPN ML definitions. The reader should not be concerned with the details of CPN ML which would require lengthy explanations that are beyond the scope of this paper. Note, however, the use of keywords *color* and *var*, that have been added to describe CPNs. Also note the two styles in which functions are defined in ML — one using the keyword *fun* followed by the function name, then as many parameters referring to the functions arguments as needed, then "=" followed by the definition; the other style avoids naming the function and uses the keyword *fn* followed by argument parameters, then "=>" followed by the computation rule.

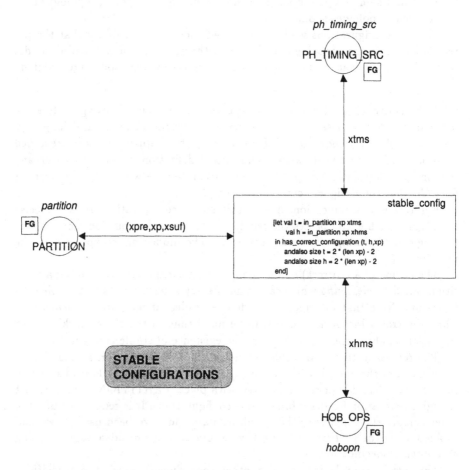

Fig. 2. CPN of Stable Configuration Definition

If the protocol is correct the abstract representation of the configurations given by the markings of place PARTITION will be consistent with the more concrete representation given by the protocol's data structures (not shown).

Explanation of Modelling of Stable Configuration Place PH_TIMING_SRC has tokens representing the timing reference resources used by each node. Af-

```
(* constant definitions *)
val d = 1;  (* position of node with DSG function *)
val n = 4;  (* number of nodes *)

(* preliminary colour set definitions *)
color nodes = index Nd with 1..n declare ms,rot;
color bif = with BusA | BusB | BusUNK;
color bus = subset bif with 1'BusA + 1'BusB;
color timing = with EXTERNAL | NODE | BUSB | BUSA;
color ccfn = with CC_D | CC_1 | CC_2;
color sgelems = with D | A | B;
color sgvalues = list sgelems with 0..3;
color node_list = list nodes with 1..n;
fun succ_nodes x :: y :: xs = y = rot'nodes 1 x andalso succ_nodes y :: xs
    | succ_nodes x :: nil = true
    | succ_nodes nil = true;
color succ_node_list = subset node_list by fn x => succ_nodes x;

(* the required colour set definitions *)
color timing = subset timing_req with 1'EXTERNAL + 1'NODE + 1'BUSA + 1'BUSB;
color ph_timing_src = product bus *  timing * nodes;
color hobopn = product ccfn * sgvalues * nodes declare ms;
color partition = product succ_node_list * succ_node_list * succ_node_list;

(* variable declarations *)
var xpre, xp, xsuf : node_list;
var xtms : ph_timing_src ms;
var xhms : hobopn ms;

(* definition of function len *)
local fun len' n nil = n
       | len' n (_ :: xs) = len' (n+1) xs
in fun len xs = len' 0 xs end;

(* definition of function in_partition *)
fun mem nil a = false
    | mem (x :: xs) a = a = x orelse mem xs a;
fun in_partition xp z = filter (fn (w,x,y) => mem xp y, z);
```

Fig. 3. Declarations for Stable Configurations CPN

ter initialization there will be 2 tokens for each interior node[2], representing the timing references used for bus A and bus B. The nodes which are heads of bus in open or island topologies have only one associated token. Thus the tokens in this place are of color *ph_timing_src* which is a triple identifying the node, the bus and the timing reference being used. Place HOB_OPS similarly identifies the current distribution of heads of bus.

[2] interior nodes are nodes that have 2 neighbours in the topology

Place PARTITION has tokens, each of which represents a subnetwork. At start-up, a single token representing the initial configuration is deposited into this place. Thereafter, if the subnetwork is split, any open topology, island topology or isolated node is represented by a distinct token. Each of these tokens has 3 lists of nodes. The tokens are of colour *partition*.

The second list of nodes in a *partition*-coloured token represents the members of the subnetwork which can communicate. If this list has only one node, it is a physically isolated node i.e. there are bus faults on both sides of the node. Otherwise, the node which is head of bus A is first in the list, its downstream neighbour is listed next, and so on until the head of bus B is listed. Note that in the case of a looped topology where the heads of bus are co-located, the first and last member of the list will be the same node.

The first list of nodes in a *partition*-coloured token represents any isolated nodes which are upstream of the head of bus A on bus A. These nodes have been isolated by bus faults because they do not have head of bus capability. Similarly, the third list of nodes represents any isolated nodes upstream of head of bus B on bus B.

Now transition *stable_config* ensures that each partition (subnetwork) in place PARTITION is checked for correctness. Each one's correctness depends on the timing references and head of bus locations of the communicating nodes i.e. the nodes belonging to the token's second list. The definitions of t and h in the transition's guard extract the appropriate tokens from the PH_TIMING_SRC and HOB_OPS places. The 'size' conditions in the guard ensure that the above extraction considers all the tokens in the respective places. The function application *has_correct_configuration* (t,h,xp) ascertains the correctness of the configuration.

To keep the net relatively simple it was necessary to allow the use of variables (*xtms* and *xhms*) whose type is a multiset of a defined colour set. This required a minor syntactic extension of Design/CPN's *var* definitions. Some definitions of CPNs explicitly allow this use of multiset-typed variables [11].

4.2 Assumptions About the Environment

In order to perform their function, protocols must communicate with their environment. For CCP, this environment includes the Physical Layer which allows remote communication; the internal communication within nodes so that Configuration Control functions in a node can communicate; and the DQDB Layer Management Interface. Assumptions about the environment affect the design of the protocol. We need to explicitly model the environment to complete our specification. This together with a specification of the service and of the protocol's procedures provides a basis for undertaking any proof obligations. As an example of this, part of the interface with the Physical Layer is shown below.

Links Going Up and Down Ph-STATUS indication primitives are generated to indicate a change in the status of a duplex transmission link associated with the Physical Service Access Point (Ph-SAP). They contain a status parameter which is UP if the Physical Layer considers the duplex link to be active and there exists an active head of bus upstream on the bus which enters the node at the Ph-SAP. A status value of DOWN indicates the Physical Layer thinks that one or both directions of the duplex transmission link are inactive or that there is no active head of bus upstream on the incoming bus. In fact, rather than actually detecting an active head of bus function upstream, what happens is that a node which doesn't support head of bus functions, and which detects an incoming transmission link failure, will signal the fact to the next node downstream on that bus. A node that receives the signal behaves as if the incoming link has failed.

Figure 4 is the CPN representing this behaviour.

The Link Status Indicators are set as a result of the reception of Ph-STATUS indication primitives. The CPN models the change in the Link Status Indicators, thus indirectly modelling the receipt of the primitives.

Transition *ls_up_immed*, models a Link Status Indicator going from DOWN to UP at node xnn when link xl becomes active (i.e. place LINK_CUT has a (false,xl) token) and when the node immediately upstream on the restored link is able to act as head of bus. As the immediately upstream node is able to act as head of bus, it will not signal link failure to node xnn, and a Link Status of UP will result. The guard ensures that all 3 varieties of Link Status Indicator are dealt with, and that node ynn is the node immediately upstream on the restored link.

Transition *ls_down_immed*, models a Link Status Indicator going from UP to DOWN at node xnn when link xl fails (i.e. place LINK_CUT has a (true,xl)-token). Note that as well as the Link Status Indicator corresponding to the failed link, being set to DOWN, the Link Status Indicator corresponding to the other Ph-SAP is also set to DOWN if the node cannot act as head of bus.

Transition *ls_down_upstream* models one element of the chain of signalling that may occur due to nodes being downstream of a failed link and having no nodes with head of bus functionality upstream. Note that in the case that the node itself does not have head of bus functionality, the guarded terms in the arc expressions to and from place LINK_STATUS come into effect. This ensures that the node behaves as if the incoming transmission link has failed. The change in link status at the node (token (xcc,UP,xnn) being replaced by token (xcc,DOWN,xnn)) occurs when the immediate upstream node ynn, believes that its incoming link on the same bus has failed (token (zcc,DOWN,ynn) is present). Of course, node ynn has no head of bus functionality. This is indicated by the token (NO,ynn) on the arc (HOB_CAPABILITY,ls_down_upstream).

Transition *ls_up_upstream*, models one element of the chain of signalling that may occur due to nodes being downstream of a recently restored link, and having no nodes with head of bus functionality upstream. The change in link status at the node (token (xcc,DOWN,xnn) being replaced by token (xcc,UP,xnn)) occurs when the immediate upstream node ynn, believes that it's incoming link on the same bus has been restored (token (ycc,UP,ynn) is present).

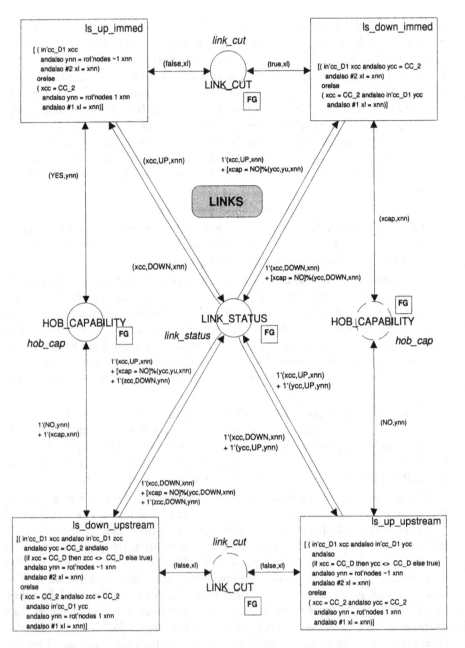

Fig. 4. CPN of Links Going Up and Down

4.3 The Vocabulary of DQDB Layer Management Information Octets

Management Information Octets pass downstream on a bus, going from node to node, conveying management information. They are transferred between peer

DQDB Layer Management Protocol Entities as octets. They are generated as octets of type DQDB_MANAGEMENT in Ph-DATA request primitives, and are received as octets of the same type in Ph-DATA indication primitives.

A Management Information Octet (MIO) is either for Configuration Control (CC) or MID Page Allocation:

$$color\ mio\ =\ union\ cc:\ cc\ +\ midpa\ :\ midpa;$$

The MID Page Allocation protocol is outside of the scope of this paper.

The Configuration Control MIO has 3 fields. The first field is a constant indicating that the MIO is for configuration control; the second field is the Bus Identification Field (BIF) — this is used to distinguish the two unidirectional buses, bus A and bus B; and the third field is the Subnetwork Configuration Field (SNCF) which contains the Configuration Control information. The Configuration Control MIO is defined as:

$$color\ cc\ =\ product\ type\_zero\ *\ bif\ *\ sncf;$$

One of these fields, sncf, has 3 sub-fields:

$$color\ sncf\ =\ product\ sncf\_dsgs\ *\ sncf\_hobs\ *\ sncf\_etss;$$

namely, the Default Slot Generator Subfield (DSGS) which has 2 possible values, PRESENT and NOT_PRESENT:

$$color\ sncf\_dsgs\ =\ with\ DPRES\ |\ DNOT;$$

the Head of Bus Subfield (HOBS) which has 3 possible values, STABLE, WAITING and NO_ACTIVE_HOB:

$$color\ sncf\_hobs\ =\ with\ HSTAB\ |\ HWAIT\ |\ HNOACT;$$

and the External Timing Source Subfield (ETSS) which has value PRESENT or NOT_PRESENT:

$$color\ sncf\_etss\ =\ with\ EPRES\ |\ ENOT;$$

The definition of the other fields is similar in form.

4.4 The Format of DQDB Layer Management Information Octets

Section 4.3 described MIOs in terms of their information content. This is used in the description of the protocol procedures that form the bulk of the specification. It is also usually sufficient for verification.

In order to implement CCP, we need to also know the encoding of the information as bit sequences. This is the subject of the current section. Upper case equivalents of the names used above are used to describe the format of the bit sequences.

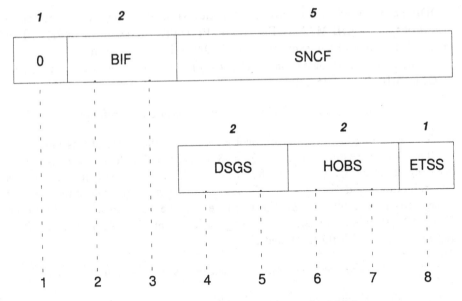

Fig. 5. Format of a Configuration Control MIO

An MIO is formatted as follows:

$$color\ MIO\ =\ union\ CC:\ CC\ +\ MIDPA:\ MIDPA;$$

Figure 5 informally presents the format of a CC MIO. The rest of this section uses ML to make the definition more precise (although the picture is much easier to digest).

Now, we will need to deal with sequences of bits of length 1, 2, 5 and 8. An octet is defined as:

$$color\ octet\ =\ string\ with\ "0".."1"\ and\ 8..8;$$

that is, an octet is a string, comprised of 0s and 1s, and is always of length 8. We can similarly define bit, two_bits and five_bits.

This enables us to say that the CC MIO is an octet whose first bit is a zero:

$$color\ CC\ =\ subset\ octet\ by\ fn\ s\ =>\ hd(explode\ s)\ =\ "0";$$

Note that *explode* is a standard ML function that transforms a string into a list of strings of size one (*implode* does the reverse).

Now CC is expected (from the previous section) to be subdivided into 3 fields. It has a type field which is the mandatory leading zero bit, a BIF field which is encoded by two bits, and a SNCF field which is encoded by 5 bits. We can rename one of the bit definitions to indicate this, for instance:

$$color\ SNCF\ =\ five\_bits;$$

The 5 bit SNCF field is subdivided into 3 subfields. It has the Default Slot Generator Subfield which is encoded by 2 bits, the Head of Bus Subfield which is encoded as 2 bits, and the External Timing Source Subfield which is encoded by 1 bit. They are defined in the style of SNCF.

The field and subfields values for the MIO are not expected to have just any value. The valid values are determined by the function valid_CC_MIO. They are the values for which it evaluates to true:

fun $valid\_CC\_MIO$ $(x :$ $octet)$ $=$
    $substring(1, 1, x)$ $=$ "0" $andalso$
    $substring(2, 2, x)$ $<>$ "11" $andalso$
    $(substring(2, 4, x)$ $=$ "00" $orelse$ $substring(2, 4, x)$ $=$ "11")$ andalso$
    $substring(2, 6, x)$ $<>$ "11";

```
The first field of substring is its size, the second its
position and the third the string.
```

Finally, we need to specify the order in which the fields are given and the correspondence between the abstract values of section 4.3 and the encoded values used. This information is given in a function CC_MIO_encoding which returns a pair whose first component is the encoding and the second component is the position within the octet at which the encoding appears. The details are not provided.

4.5 The Procedures

The procedures of the Configuration Control Protocol (CCP) describe the dissemination of information needed by the DQDB Layer Management Protocol Entities of nodes to provide the Configuration Control function (see section 4.1). There are various resources which are manipulated so that an appropriate dual bus topology results. They are the default slot generator function, the head of bus A function, the head of bus B function, the bus identification function and the primary timing reference for the subnetwork.

The information that supports CCP is carried in the SubNetwork Configuration Field (SNCF) and Bus Identification Field (BIF).

In the standard, each node's Configuration Control functionality is defined in terms of functionality associated with its two slot generation functions. For the D-node, we talk of the Default Configuration Control function, CC_D, which is associated with the Default Slot Generator function; and the Type 2 Configuration Control function, CC_2, which is associated with the Type 2 Slot Generator function. Each 1-node, similarly, has a Type 1 Configuration Control function, CC_1, which is associated with the Type 1 Slot Generator function; and a Type 2 Configuration Control function, CC_2, which is associated with the Type 2 Slot Generator function.

The Default Slot Generator function provides the bus identity for a subnetwork and if needed the Head of Bus A and Head of Bus B functions, whereas CC_D controls the activation and de-activation of the Default Slot Generator function, the Head of Bus A and Head of Bus B functions. CC_D also provides the 125 microsecond timing for Buses A and B, using either the external timing reference or the node clock.

The Type 1 Slot Generator function provides the Head of Bus A function if needed, otherwise it has null functionality. CC_1 controls the activation and de-activation of the Head of Bus A function. CC_1 also provides the 125 microsecond timing for Bus A, using either the external timing reference or the node clock.

The Type 2 Slot Generator function provides similar functionality to the Type 1 Slot Generator function but for Bus B rather than for Bus A.

The following three subsections give examples from the procedures that illustrate the way CPNs are used to model constraints, state machines, and tables, respectively.

CPN of Errors in Received SNCFs SNCFs received by CC_D which have a DSGS of PRES and a HOBS of WAIT are in error (see section 4.3 for the acronyms). This is reflected in the CPN of figure 6. If either transition f_pw1 or f_pw2 can occur then the CPN says there is a fault in the protocol which needs to be rectified. These transitions are special ones called **facts**. Facts are not expected to occur. They are treated differently during analysis.

Head of Bus State Machines for Configuration Control Type 1 and 2 Functions IEEE 802.6 provides two state machine diagrams for the Heads of Bus resources, one for each bus. The similarity of these state machines is significant, and is reflected in the CPN representation which uses common places and transitions, but differentiates between the two state machines by using distinct tokens to represent their state.

Figure 7 is the CPN of the Head of Bus State Machines.

Place HOB_STATE has tokens which indicate the state of the Head of Bus State Machine. The Head of Bus State Machine can be in one of the following states: Initialize, HOB_Dflt, HOB_Wait_Dflt, HOB_Active, HOB_Wait_Active. The tokens are triples, the first field of which indicates the bus being controlled (Bus A for the Configuration Control Type 1 function, and Bus B for the Configuration Control Type 2 function); the second field is the state of the state machine; whereas the third field is the node identifier. Note that it appears twice on the page to make the CPN more readable.

Place TIMER_H has tokens which if present indicate that a timer is running. The tokens are pairs, the first field of which is either CC_1 or CC_2, identifying the configuration control function involved, the second field identifying the node. If we are dealing with CC_1, then this timer is called Timer_H_1; for CC_2 we deal with timer Timer_H_2. In fact we do not need to include place TIMER_H in our net as its presence or absence does not affect the occurrence of transitions at

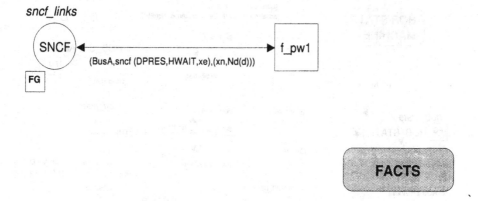

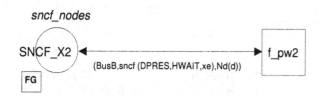

Fig. 6. CPN of Errors in Received SNCFs

all. So it is a redundant place. On the other hand, its inclusion makes it easier to read the specification.

Place LINK_STATUS represents the three status indicators that record the status of the duplex transmission link. Place CC_x2_CONTROL represents the two flags that control the operation of the Configuration Control Functions. Place HOB_CAPABILITY records the capability of a each node to act as head of bus or not. Place BIF records the BIF values received at a node.

Transition *hbx0* captures the transitions of IEEE 802.6 state machines which are labelled (10), (20), (30) and (40). The guard requires that the bif token (xb,bif yb,xl), describes a BIF value of yb that is coming in to node xnn on bus xb from link xl. The condition *inlink xl xnn* ensures that we deal with the node xnn that corresponds to link xl and bus xb. Condition $xb <> yb$, means that the received BIF value is not what is expected on that bus. The transition to state Initialize occurs in any state except Initialize $(xhob <> INIT)$. Finally, this transition only occurs if this combination of values is received at a node with a CC_1 function $(xnn <> Nd(d))$.

Transitions *hb01, hb12, hb23, hb34* and *hb43* model state machine transitions (01), (12), (23), (34) and (43) in a straightforward way.

Transition *hb41_timeout* occurs whenever (41) occurs due to a a timeout.

Transition *hb21_LS_UP* models (21) when they occur because the appropriate Link Status Indicator (LINK_STATUS_1 or LINK_STATUS_2) is UP.

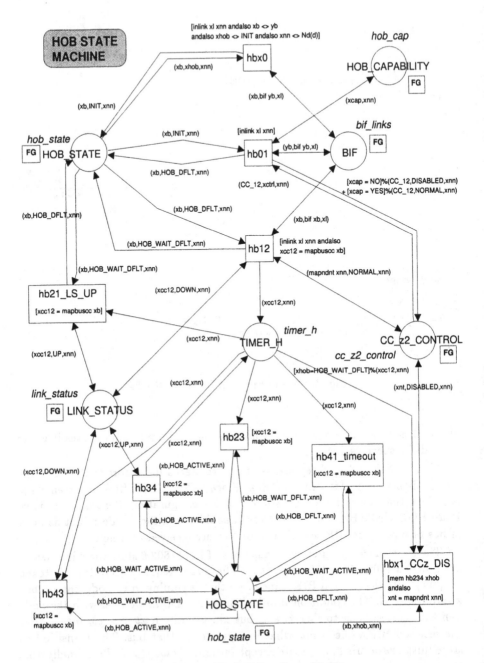

Fig. 7. CPN of Head of Bus State Machine

Transition *hbx1_CCz_DIS* models (21) and (41) when they occur because the appropriate Configuration Control Flag (CC_12_CONTROL or CC_D2_CONTROL) is DISABLED. It also models (31).

Operation of Configuration Control Type 1 and 2 Functions The operations tables in IEEE 802.6 which record the status of resources being controlled by Configuration Control Type 1 and 2 Functions are shown in figures 8 and 9. (These tables were unfortunately swapped in the final draft of the standard [1], but this error is corrected here.)

Figure 10 models the tables except for the first two rows.

Transition *ccz_normal* has a complex guard that includes all distinctions necessary to indicate the values in rows 3–13 of figure 8 and rows 3–10 of figure 9.

The guard has 7 conditions separated by commas which need to be fulfilled to enable the transition to occur. The two most interesting conditions are the penultimate and the ultimate.

The penultimate condition is a case condition preceded by the local declaration contained between the key words 'let' and 'in'. The case statement depends on the value of variable yhob, which is the state of the HOB state machine. For instance, if the HOB state machine is in HOB_Dflt state then the appropriate Configuration Control Status Indicator and Head of Bus Operation Indicator are set to INACTIVE and NULL, respectively. The values of DSGS and HOBS on bus xb are relayed to the other Configuration Control function at the node.

The ultimate condition is also a case condition. It determines the value of the ETSS field on both buses, and the timing source that the node will use. For instance, if the External Timing Source state machine is in state NO_ETS, then the ETSS value on bus yb is relayed from the incoming value on bus yb. The value of ETSS on bus xb and the other timing source values depend on the state of the HOB state machine. The nested case condition deals with this.

5 Analysis

The specification was analysed using the TORAS reachability analysis tool. This revealed a number of errors, some incompleteness and a number of ambiguities in the text. The start up procedures which were effectively distributed over a number of chapters of the specification, illustrate one sort of difficulty, but perhaps the most troublesome problem we discovered was an ambiguity.

There are a number of ambiguities in IEEE 802.6. One ambiguity has had a history of 3 different interpretations of the relevant text, none of which can easily be discounted, except for the fact that the later interpretations lead to fewer behavioural problems. The history of the interpretation of that text is given below.

First attempt: On page 264 of IEEE 802.6, referring to tables 10.10(a) and 10.10(b), it is stated that "The complete set of rows describes all possible input conditions to the CC_D function for that resource". I foolishly believed it and adopted a simple interpretation of input condition which corresponded to the arrival of MIOs. With this interpretation if one gets an input combination which is not in the table and no action is specified, nothing will happen. This leads to deadlocks because some analysis shows that such input combinations can occur.

INPUT					OUTPUT					
HBA St	ETS St	Bus B SNCF			STAT_1	HOB_1	Tim So	Bus A SNCF		
		DSGS	HOBS	ETSS				DSGS	HOBS	ETSS
Init	No_ETS	X	X	X	INACT	NULL	NODE	Relay	Relay	Relay
Init	ETS_Ac	X	X	X	INACT	NULL	EXT	Relay	Relay	Relay
Df	No_ETS	X	X	X	INACT	NULL	EITHER	Relay	Relay	Relay
W_Df	No_ETS	X	X	X	ACT	A	NODE	NOT	WAIT	NOT
W_Ac	No_ETS	X	X	X	ACT	A	NODE	NOT	WAIT	NOT
W_Df	ETS_Ac	X	X	X	ACT	A	EXT	NOT	WAIT	PRES
W_Ac	ETS_Ac	X	X	X	ACT	A	EXT	NOT	WAIT	PRES
Df	ETS_Ac	X	X	X	INACT	NULL	EXT	Relay	Relay	PRES
Ac	No_ETS	PRES	X	PRES	ACT	A	BUS_B[1]	NOT	STAB	NOT
Ac	No_ETS	PRES	X	NOT	ACT	A	BUS_B[1]	NOT	STAB	NOT
Ac	No_ETS	NOT	X	PRES	ACT	A	BUS_B[1]	NOT	STAB	NOT
Ac	No_ETS	NOT	X	NOT	ACT	A	NODE	NOT	STAB	NOT
Ac	ETS_Ac	X	X	X	ACT	A	EXT	NOT	STAB	PRES

KEY:

HBA St = Head of Bus state for CC_1
ETS St = ETS State
Tim So = Timing Source
STAT_1 = CC_STATUS_1
HOB_1 = HOB_OPERATION_1
SNCF = SubNetwork Configuration Field
DSGS = Default Slot Generator Subfield
HOBS = Head Of Bus Subfield
ETSS = External Timing Source Subfield
Init = Initiialise
Df = HOB_Dflt
W_Df = HOB_Wait_Dflt
W_Ac = HOB_Wait_Active
Ac = HOB_Active
X = Don't care
INACT = INACTIVE
ACT = ACTIVE
A = HOB_A
PRES = PRESENT
NOT = NOT_PRESENT
STAB = STABLE
WAIT = WAITING

NOTE:
(1) Timing source will be NODE if LINK_STATUS_2 is DOWN.

Fig. 8. Table 10.12 from IEEE 802.6 (CC_1 Function Operations)

| INPUT | | | | | OUTPUT | | | | | |
| | | Bus A SNCF | | | | | | Bus B SNCF | | |
HBB St	ETS St	DSGS	HOBS	ETSS	STAT_2	HOB_2	Tim So	DSGS	HOBS	ETSS
Init	No_ETS	X	X	X	INACT	NULL	NODE	Relay	Relay	Relay
Init	ETS_Ac	X	X	X	INACT	NULL	EXT	Relay	Relay	Relay
Df	No_ETS	X	X	X	INACT	NULL	EITHER	Relay	Relay	Relay
W_Df	No_ETS	X	X	X	ACT	B	NODE	NOT	WAIT	NOT
W_Ac	No_ETS	X	X	X	ACT	B	NODE	NOT	WAIT	NOT
W_Df	ETS_Ac	X	X	X	ACT	B	EXT	NOT	WAIT	PRES
W_Ac	ETS_Ac	X	X	X	ACT	B	EXT	NOT	WAIT	PRES
Df	ETS_Ac	X	X	X	INACT	NULL	EXT	Relay	Relay	PRES
Ac	No_ETS	X	X	X	ACT	B	BUS_A[1]	NOT	STAB	NOT
Ac	ETS_Ac	X	X	X	ACT	B	EXT	NOT	STAB	PRES

KEY:

HBB St = Head of Bus state for CC_2
ETS St = ETS State
STAT_2 = CC_STATUS_2
HOB_2 = HOB_OPERATION_2
SNCF = SubNetwork Configuration Field
DSGS = Default Slot Generator Subfield
HOBS = Head Of Bus Subfield
ETSS = External Timing Source Subfield
Init = Initialise
Df = HOB_Dflt
W_Df = HOB_Wait_Dflt
W_Ac = HOB_Wait_Active
Ac = HOB_Active
X = Don't care
INACT = INACTIVE
ACT = ACTIVE
B = HOB_B
PRES = PRESENT
NOT = NOT_PRESENT
STAB = STABLE
WAIT = WAITING

NOTE:
(1) Timing source will be NODE if LINK_STATUS_1/LINK_STATUS_D is DOWN.

Fig. 9. Table 10.11 from IEEE 802.6 (CC_2 Function Operations)

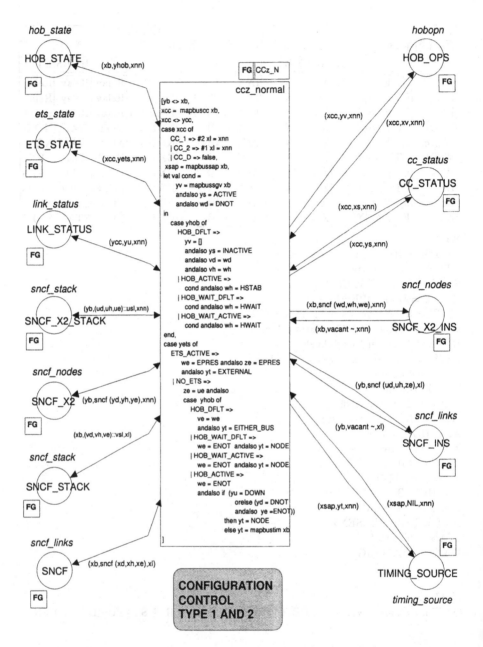

Fig. 10. CPN of Bus and Timing Source Operations for Configuration Control Type 1 and 2 Functions

Second attempt: Interpretation 1 would stop us from ever starting, so it was weakened to say if an input combination isn't in the table then ignore it and wait until we get a condition which is in the table. This solved most of the problems but didn't prevent the possibility of a startup deadlock. If LINK_STATUS_D was set to UP before we had received sufficient MIOs via Ph-SAP_A to have defined an input value as given in the table, then the startup deadlock would occur. One could add a few extra conditions to overcome this problem but it is a little messy.

Third attempt: My third interpretation was to relay rather than ignore an input which is not in the table either because it is not explicitly named or because it hasn't been received twice as required by section 10.2. of the text.

I think it is reasonable to say that the third interpretation is not the most obvious one, yet it is the simplest one that I could think of that avoids deadlocks.

6 Conclusion

This paper has reported on a complete specification of the Configuration Control Protocol in CPNs. The experience provides some experimental evidence to suggest that CPNs can provide elegant and complete specifications of protocols like the Configuration Control Protocol. The consistency in style and precision of the formal specification has created a specification that is arguably clearer and better than the original. Unlike informal specifications there is also the possibility of analysis. A brief indication of some of the analysis undertaken was included.

Another thread of the paper has been the suggestion that the explicit recognition of the 5 elements of a protocol also helps. It obliges the specifier to clearly state the purpose of the protocol and the environment in which it is designed to operate, features commonly left unstated. The experience of trying to use CPNs to express the purpose and environment demonstrated the difficulty of ascertaining what the designers' intentions were in this case study.

Acknowledgements

The permission of the Director, Telstra Research Laboratories, to publish the above paper is hereby acknowledged.

References

1. Distributed Queue Dual Bus (DQDB) Subnetwork of a Metropolitan Area Network (MAN), IEEE P802.6/D15, October 1 1990. Final draft, approved by IEEE standards board on December 6, 1990.
2. W. Reisig. *Petri Nets – An Introduction*, Springer-Verlag, 1985.
3. K. Jensen. *Coloured Petri Nets – Basic Concepts, Analysis Methods and Practical Use, Volume 1: Basic Concepts*, EATCS Monographs on Theoretical Computer Science, Springer-Verlag, 1992.
4. G. Wheeler. *The Configuration Control Protocol of IEEE 802.6 in Coloured Petri Nets*, Telecom Australia Research Laboratories Report 8134, May 1992.

5. G. Wheeler, A. Valmari and J. Billington. *Baby TORAS Eats Philosophers but Thinks about Solitaire*, Proceedings of the Fifth Australian Software Engineering Conference, Sydney, 22–25 May 1990, p. 283–288, ISBN 0 909394 21 0.
6. G.J. Holzmann. *Design and Validation of Computer Protocols*, Prentice-Hall, 1991.
7. Å. Wikström. *Functional Programming Using Standard ML*, Prentice-Hall, 1987.
8. R. Milner, M. Tofte and R. Harper. *The Definition of Standard ML*, MIT Press, 1990.
9. R. Harper. *Introduction to Standard ML*, School of Computer Science, Carnegie-Mellon University, Pittsburgh, PA 15213, September 1990.
10. *Design/CPN: A Reference Manual*, version 1.75, Meta Software Corporation, June 1991.
11. J. Billington. *Many-sorted High-level Nets*, Proceedings of the Third International Workshop on Petri Nets and Performance Models, Kyoto, 11–13 December 1989, pp. 166–179, IEEE CS Press.
12. P. Huber, K. Jensen and R. Shapiro. *Hierarchies in Coloured Petri Nets*, Advances in Petri Nets 1990, LNCS 483, pp. 313–341, Springer-Verlag, 1991.

Colored Petri Nets Based Modeling and Simulation of the Static and Dynamic Allocation Policies of the Asynchronous Bandwidth in the Fieldbus Protocol

Adel Ben Mnaouer[1], Takashi Sekiguchi[1], Yasumasa Fujii[1]
Toru Ito[2], Haruki Tanaka[2]

[1] Div. of Elec. and Comp. Eng., Yokohama National University
Hodogaya-ku, Tokiwadai 79-5, Yokohama 240, Japan.
Email: {adel,prof,yas}@sekiguchilab.dnj.ynu.ac.jp
[2] Development Dept, FUJIFACOM Corporation
1, Fuji-machi, Hino-shi, Tokyo 191, Japan

Abstract. This paper presents a Colored Petri net based modeling, simulation and analysis of the centralized architecture of the timed-token Fieldbus protocol. In such an architecture, the asynchronous traffic classes are served by using an emulation of a "circulated" timed-token mechanism, while the synchronous one is served by a centralized scheduling mechanism.

The performance evaluation is concerned with two different policies of asynchronous bandwidth allocations. The first is based on a Static Allocation Scheme (SAS), and the second is based on a Dynamic one (DAS), that uses the leftover bandwidth allocated to a specific station and adds it to the share of the following station dynamically.

Considering a symmetric and an asymmetric system, four Hierarchical Timed Colored Petri Nets (HTCPN) were build, each corresponds to a specific system using either the SAS or the DAS.

The simulation results show the superiority of the DAS over the SAS in improving the performance of the normal priority and time-available priority traffic classes in terms of increasing channel utilization, decreasing transfer delays and token delegation overheads.

1   Introduction

In recent years, the world of Factory Automation (FA) and Process control has witnessed new trends, calling for migration from centralized control architectures to distributed ones. These are triggered by the recent technological advances which gave birth to highly intelligent floor level devices (e.g., smart sensors, intelligent actuators, etc), that can perform parts of the control process, hence alleviating the burden of the controlling devices. Also, a need is being felt worldwide, for renovation of the communication systems on the floor, or in higher levels of the FA hierarchy, aimed at reducing cabling problems. Fig.1., depicts the future FA communication hierarchy.

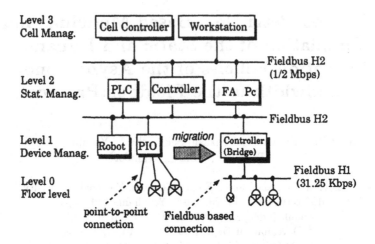

Fig. 1. The Factory Automation communication hierarchy

Industrial applications are, usually, characterized by two types of traffic: A synchronous time-critical one, and an asynchronous one, with no time-pressing constraints, except for alarm related data that require bounded delivery delays.

To meet the above challenges, new Local Area Network architectures, called Fieldbus Networks, are under intensive investigation and standardization efforts worldwide. The proposed Fieldbus architectures are intended to link floor level devices, at a low speed, to their controlling devices of the lowest level of the FA hierarchy (see Fig. 1). Furthermore, they can be designed to link controlling devices among themselves, at higher speeds, in upper hierarchical levels (i.e., levels 1 and 2) [1], [2].

The ISA/IEC[3] DLL protocol standard draft [3], which is in its last stage before final approval, is the subject of our study.

The Fieldbus protocol stack is characterized by a reduced architecture, comprising only three layers (i.e., the Physical layer, the Data Link Layer (DLL) and the application layer), in contrast to the OSI seven-layers ISO reference model.

The DLL's media access mechanism is based on a centralized control scheme: A central station assuming an elective role of master-ship of the local link, called Link Active Scheduler (LAS), controls and schedules the access to the link by the other connected stations. It has an initial pre-run schedule of synchronous (i.e., periodic) activities [7], and uses a timed token to allow accessibility to the common link.

To execute the synchronous phases the token is delegated, at predefined instants, to the station that needs to execute its synchronous transactions. To serve the asynchronous phases, the token is circulated among all the connected stations, with an associated duration, thus, emulating a "circulated" token mechanism (see section 3, for more detail). The difference between this mechanism

---

[3] Instrumentation Society of America, and the International Electro-technical commission

and the conventional token passing scheme [4], is that the token has to be returned after each delegation to the station that is ensuring the central control, so that it can meet the time requirements of the synchronous traffic.

In a previous study of this protocol [5], the problem of guaranteeing a bounded token circulation time and the upper bound delivery delay of urgent data has been addressed.

The idea was based on setting a theoretical maximum token circulation time as being less than the permissible upper bound delay of urgent data, then derive the allocable asynchronous bandwidth, after deducing the eventual synchronous phases, token delegation overheads, and total safety-related idle time. Finally, the asynchronous bandwidth is divided among all the connected stations, using either an Equal Partition Allocation Scheme (EPAS), in case of a symmetrically loaded system, or a modified version of the Normalized Proportional Allocation Scheme (NPAS) [6], considering an asymmetrically loaded one.

The problem is that the safety-related idle times, generated due to scheduling constraints, and the total token overheads that can be generated within one token circulation cycle, cannot be determined with certainty by analytical means. In [5], a parameter calibration of the idle time parameter, has been performed as a kind of heuristic, to determine the suitable asynchronous bandwidth that can be allocated, and that gives the best (or close to best) performance while guaranteeing the delay bounds of urgent data.

In that study, the Static Allocation Scheme (SAS) defined in the protocol's draft was used. Under the SAS the maximum token holding time (MTHT) of each station is set to a specific value that won't be altered during operation. However, there exists another approach based on a dynamic scheme, which utilizes the left-over bandwidth of the previously delegated duration, to a specific station, in order to increase the bandwidth that can be allocated to the following station in sequence. We shall call it the Dynamic Allocation Scheme (DAS) in the sequel.

This approach has been proposed and investigated for centralized architectures of the Fieldbus network in [8], [9], and for earlier versions of the ISA/IEC Fieldbus protocol in [11], [10].

As the DLL protocol evolved through time, those analyses became less accurate. The changes made concern mainly the management of the asynchronous traffic by the LAS, which evolved from a polling based collection of aperiodic requests, followed by a first opportunity token delegation, to a time-based, circulated token, limited to a circulation list (with a maximum of eight stations), that is returned to the LAS after a full cycle around that list. In the current version, the token has to be returned to the LAS immediately after each delegation to a single station, and the circulation list includes all the connected stations.

Thus, what we propose in this paper, is to reevaluate both allocations schemes through more detailed Colored Petri Net based simulations that take into account, most of the scheduling features of the most recent standard draft, that could not be included in the previous studies of the protocol.

The application of High level Petri Nets, such as CPNs and Stochastic Petri Nets (SPN) to the modeling and simulations of communication protocol has increased in recent years [10], [13], [14],[15], [16], [17], [18], [19], [22]. This because they have the particular feature of presenting concise and easy to understand graphical models that visualize the interactions between the different communicating and cooperating entities of the system. In addition they are backed by strong mathematical foundations that allow us to verify behavioral properties (such as liveness, boundedness, etc.) and check correctness.

CPNs and SPNs are two solutions to two conflicting requirements of system modeling and analysis [22]:

- The model is required to be detailed enough to include all system features that have a significant impact on performance.
- The model is required to be simple enough to be tractable

CPNs, and especially Hierarchical ones (HCPN) [23], are the response to the first requirement, as they have means for modeling and specifying very-large scale systems, with their colored tokens and hierarchy constructs, folding the system description into very compact forms.

While SPNs (with its extensions, Generalized SPNs and Deterministic SPNs) constitute an answer to the second requirement, as they can be useful in modeling complex system with a very high level of abstraction, that can be tractable when meeting certain conditions (e.g., In case of GSPNs the system should be cyclic so that an analytical solution can be found for it).

Besides, in recent years, the CPN formalism is being presented as a kind of Formal Description Technique (FDT), which can serve for the specification, modeling and verification of distributed systems, in general, and communication protocols in particular. Indeed, in [14] the HCPN formalism was proposed as a substitute of the Specification and Description Language (SDL), for the specification and correctness verification of communication protocols.

Also in [15], a HCPN model for the modeling and simulation of a network management system is presented. In a more recent work, in [20], a HCPN model was proposed for the modeling, and correctness checking of a communication gateway. Moreover, in [21], the HCPN formalism was shown to be useful for the modeling and analysis of a distributed program executions.

In most of the above works, there was a strong need for tool support. Therefore, the Design/CPN simulator package, together with the Occurrence Graph Analyzer (OGA) tool [25], and the Invariant tool [26] were used for the modeling and correctness verification of the above mentioned systems.

In addition to the above facilities, Design/CPN allows the use of the time concept by having the timed tokens carry a time stamp, indicating the time at which the token is ready to be used, i.e., consumed by a transition. Thus, making it possible to use the Timed CPN (or HCPN) formalism for performance evaluation purposes.

As an example, in [22], a HTCPN simulation model was used to verify the correctness of the results of an approximate Stochastic Petri Net (SPN) model of a large-scale concurrent system.

Exploiting the above feature of the HTCPN formalism, we used it to construct detailed HTCPN simulation models of the Fieldbus DLL protocol, each corresponding to a specific system (i.e., symmetric or asymmetric), using either the SAS or the DAS. These models capture most of the protocol's characteristics, that couldn't be included in the previous studies of this protocol presented in [10] and [11], like the Target Token Rotation Time (TTRT) and the MTHT parameters, and the timing-related explicit and implicit scheduling paradigms. The model was constructed using the Design/CPN modeling and simulation tool.

The paper is organized as follows: In the following section, a brief description of the project at hand is given, followed by a description of the Fieldbus protocol in section 3. Section 4, describes the modeling process, with an informal introduction to HTCPN formalism, and a detailed description of the models used in the simulation. In section 5, the simulation process and the analysis are presented. We conclude in section 6.

2 Project description

This project has for main goal the analysis of the Fieldbus protocol as proposed by ISA/IEC standard draft [3]. It involves three teams:

- Yokohama National University team comprised of three research staff and three master students. The project supervisor is the IEC/TC65 Japanese committee chairman.
- Fuji Facom company team comprised of two engineers, experts on Process Control and FA applications. One of them is in the Japanese technical team of the Fieldbus standardization committee. The company is engaged in manufacturing and testing the future Fieldbus Protocol chips.
- Kanagawa industrial technology research institute team comprised of two engineers engaged in experiments and emulation of the Fieldbus protocol applications using programmable Single Board Computer units.

The first part of the project is concerned with the scheduling of multiple synchronous phases on Fieldbus based control systems, with the aim of increasing the asynchronous bandwidth, while ensuring schedulability. Various scheduling techniques including Linear programming and heuristics were used.

The second part is concerned with the analysis of asynchronous phases' scheduling, asynchronous bandwidth allocation and parameter setting problems using analytical means and Colored Petri Net based simulations.

Concerning the part presented in this paper, it took almost three months to get the final and acceptable Colored Petri Net models. The CPN modeling and simulations was achieved by the first author (who has been using and working on Petri Nets and CPNs for 6 years) as a part of his Ph.D thesis.

A top-down modeling technique was adopted and used. Once the basic CPN model of the Fieldbus protocol was built, it was easy later to alter it, in order to consider other algorithms (e.g., for the token delegation process), and other system structures (e.g., symmetric or asymmetric, etc.).

The main benefit of the CPN approach is that it helped increase the understanding of the protocol's operation and algorithms, because of the graphical nature of the models. It has also provided flexible and detailed models that are useful for an in-depth study of complex protocols.

3 Fieldbus protocol description

The Fieldbus Medium Access Control (MAC) protocol is based on a centralized control scheme [3], and uses a timed-token mechanism to organize the access to the common link. A set of stations, with different functional classes, are connected to the bus. These classes are (with increasing complexity order) *Basic* class, *Link Master (LM)* class, and *Bridge* class. The two latter classes enable a station to assume the Link Active Scheduler (LAS) master-ship role.

Every station is considered as a Data Link Entity (DLE), and the DLE serving as the LAS, is called the LAS-DLE in the sequel. The LAS-DLE has an initial schedule, set by the management system, in which periodic activities are scheduled, and can be updated during operation.

The LAS-DLE uses the notion of Delegated Token (DeT), to transfer the right to transmit to another node DLE, to execute either its periodic or aperiodic services for a specified duration of time. The DLE which owns the DeT, becomes the only master of the link, and itself can delegate this right to another DLE station, creating a Reply Token (ReT), with which it can compel the sending of a data or an acknowledgment. When the delegated duration expires, the DeT holding station has to send back the token to the LAS.

The LAS-DLE uses the time between the end of the actual periodic delegation, and the next scheduled periodic activity, to allow for the execution of management related services as well as for the execution of asynchronous activities by all the active nodes. To do so, the LAS-DLE first executes the link-related management activities for a predefined duration, and then emulates a "circulated token" mechanism by sending a DeT to all active nodes, in turn, starting from itself (i.e., the LAS-DLE).

Prior to each delegation the LAS checks whether the time left till the next scheduled activity, allows either a previously requested additional delegation (received in a returned token), a full MTHT delegation, or in the worst case, at least the minimum permissible delegation of the token.

If none of the above conditions is met, then the LAS will keep the token until the next scheduled activity, generating an idle time. Each time a node receives the DeT, it has to give it back to the LAS after expiration of the delegation duration, or after completion of its asynchronous activities' execution.

The LAS uses an array of MTHTs, with an entry for each station, specifying its MTHT during one cycle of circulating the token. When, the time for the next scheduled activity comes the LAS interrupts the token circulation, allows for the execution of the scheduled activity (i.e., synchronous traffic), and then goes on, with the token circulation process. Thus, a station may see its allocable MTHT being split to two parts, and delegated to it in two separate delegations.

Furthermore, the protocol provides for the use of three levels of priorities, Urgent, Normal and Time-available. Urgent Priority (UP) is usually assigned to time critical data, such as alarms. Normal Priority (NP) is usually assigned to event based data such as collection of diagnosis data, collection of information about the configuration of the set of the devices operating in the system, etc. Time-available Priority (TaP) is usually assigned to the so called trend data (i.e., reporting the state of health of devices), program download data, etc. The permissible level of priority is set at the beginning of each cycle of circulating the token, and used in all the delegations.

At the beginning of a new cycle, the LAS uses the Target Token Rotation Time (TTRT) configuration parameter, and measures the Actual Token Rotation Time (ATRT) (i.e., the time elapsed between two successive token delegations to the LAS-DLE), to adjust the priority level. If ATRT is bigger than the TTRT, the priority level is increased, to restrict the traffic. On the contrary it is decreased, to relieve queued traffic.

Moreover, the protocol defines a default minimum delegation time (mindel), common to all the stations, below which a delegation does not take place.

4 The modeling process

4.1 Informal description of the HTCPN formalism

Colored Petri Nets represent a high level extension of ordinary Petri Nets, extending their modeling power and expressiveness [23]. A CPN is composed of places, transitions, arcs connecting them and colored tokens.

The coloring of tokens is realized by associating with each token a data value - called *token color*, which may be of arbitrarily complex type (e.g., integers, real, record etc.). Each place is associated with a *color set* which specifies the type of the tokens that can reside in the place. The *initial marking* of a place specifies a number of colored tokens.

Each transition is associated with a *guard* - i.e., a set of Boolean expressions, that constitute additional conditions for its enabling.

The current state of the CPN is represented as a distribution of tokens on the places. Each *arc expression*, may contain variables and functions using them. A transition is enabled if it is possible to bind the variables of the surrounding arc-expressions to values, in such a way that the arc-expressions of all input arcs evaluate to tokens which are present at the corresponding input places.

Furthermore, the Boolean conditions of the transition's guard (if present), should all evaluate to true. The firing (occurrence) of a transition takes place with a specified binding. Then, the tokens specified by the evaluation of the arc-expressions of the input arcs are removed from the corresponding input places, and the tokens specified by the evaluation of the arc-expressions of the output arcs are added to the corresponding output places.

An additional node called the *Declaration node*, is associated with the CPN, where all color set declarations, variables and functions are defined.

The hierarchy constructs are added to a CPN model, using the hierarchy mechanisms defined in [23], namely *substitution transitions (ST)*, page instances, and *place fusion (PF)*.

A *ST* represents an abstract view of an activity, and designates a page (i.e., a subnet), called a **subpage**, that contains the details of the activity. With respect to the subpage, the page containing the *ST* and the *ST*, are said to be a **superpage** and a **supernode** respectively. For the simulation, the subpage is substituted for the *ST* and executed. The Design/CPN package designates a ST by an annotation containing a **HS** inscription beside it. The hierarchy relationships are described in a box, next to it.

The places surrounding and connected to a *ST* are called *socket places*. Each socket place is assigned to a *port place* in the *subpage*. Each port place is identified by a **P** inscription in a box next to it.

It is possible to have many instances of an ST, designating the same subpage structurally. During simulation, for each instance of an ST, a subpage instance is created.

In addition in a hierarchical CPN, a special graph called a **page hierarchy** is used to give an overview of the whole net. Each node in the graph represents a node, and each arc represents a hierarchical relationship between two pages.

The concept of place fusion is realized by letting a set of places belong to the same fusion set, representing one conceptual place while represented on the net structure by many places at different locations. Three types of place fusion can be distinguished:

- Global fusion set: that includes places that share the same marking wherever they exist on the net. They are annotated by a **FG** mark.
- Page fusion set: that includes places that share the same marking on the same CPN page. They are annotated by a **FP** mark.
- Instance fusion set: that includes places that share the same marking on the same instance of a page. They are annotated by a **FI** mark.

Finally, the concept of time is enforced by letting each token carry a time stamp during simulation, which is determined by the delay associated to the transitions (or specified through arc expressions). For more detail about HTCPN models the reader is referred to [23], [24], [25].

4.2 Detailed description of the HTCPN model

Four different systems were considered, each one being modeled with a separate HTCPN model. They correspond to a symmetric system using either the SAS or the DAS, and an asymmetric one, also using either of the two schemes. All the HTCPN models have basically similar structures and differ only with some features related to the system differences mentioned above. We will describe the basic model structure for the symmetric system using the SAS, and describe the modifications that were made in the case of the other models, whenever it is appropriate.

Table 1. Nomenclature

Abreviation	Meaning
DLPDU	a Data Link Protocol Data Unit
CD	a Compell Data DLPDU
DT	a Data DLPDU
PT	a Pass Token DLPDU
LAS	the Link Active Scheduler
DLE	a Data Link Entity
DeT	Delegated Token
UP	Urgent Priority
NP	Normal Priority
TaP	Time-available Priority
MTHT	Maximum Token Holding Time
ATRT	Actual Token Rotation Time
ST	Substitution Transition

System configuration. A system of N stations is considered. It represents a typical automation system comprised of one controller (e.g., a Programmable Logic Controller (PLC)), and N-1 stations, representing sensors and actuators. The controller polls the sensors periodically, extracts data, process them through PID[4] algorithms, and sends commands to the actuators. The controller is supposed to control the link, and thus it will play the LAS role and host the LAS-DLE.

Thus, from the communication point of view, the LAS will execute, according to a predefined schedule, a periodic activity (based on a specified sampling period), of polling all the N-1 basic stations, by sending CDs to (j) stations (representing sensors), which reply by sending DTs, and then the LAS sends $(N$-1-$j)$ CDs to its LAS-DLE, that sends $(N$-1-$j)$ DTs to the remaining $(N$-1-$j)$ stations (representing actuators).

After the periodic phase is finished, the LAS allows the execution of the asynchronous phase as described in the protocol.

In the CPN model, 11 stations were used for the simulation that include the LAS station, 5 stations representing sensors, and 5 others representing actuators. Fig. 2, depicts a typical control system based on a Fieldbus network.

The protocol model was specified through three levels of abstraction. In the first level of the hierarchy, an abstract view of all the stations connected to the link is constructed (i.e., Fig. 4). The LAS, being an independent function (i.e., role), has been modeled separately from the LAS-DLE, to highlight the functionality of such role, and to emphasize the scheduling tasks behind it. The second level includes partially abstract views of the LAS role, of the LAS-DLE and of the basic DLEs, where only the synchronous activity related parts are detailed (i.e., Fig. 6, Fig. 9 and Fig. 10).

The activities related to the asynchronous phase are modeled as substitution transitions. These STs are detailed in the third level of the hierarchy (Fig. 7 (a), Fig. 7 (b), Fig. 11, and Fig. 13).

---

[4] Proportional Integral and Differential

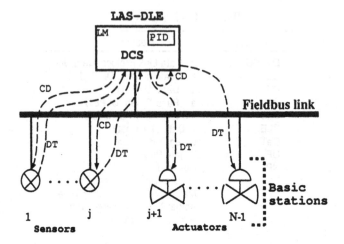

Fig. 2. A typical Fieldbus based control system architecture

The hierarchy page. The hierarchy page is depicted in Fig. 3. It shows the overall net structure of the HTCPN model, and the hierarchy levels. It has a node for each CPN page. An arc between two nodes indicates that the source node contains a *ST* (its name is inscribed on the arc), described in the destination node. In the first level we have the top page in the hierarchy described in Fig. 4.

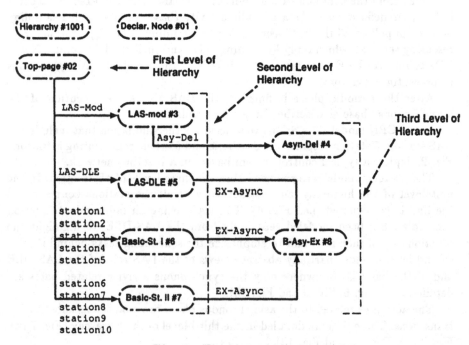

Fig. 3. The hierarchy page

In the second level, we have four nodes: one for the LAS model (i.e., the node *LAS-mod♯3*, in Fig. 6), one for the LAS-DLE (i.e., the node *LAS-DLE♯5*, in Fig. 9), one for the basic stations representing the set of sensors(station 1 through 5, (the node *Basic-St.I♯6*, in Fig. 10)), and one for the basic stations representing the set of actuators (station 6 through 10, (i.e., the node *Basic-St.II♯7* not shown)).

The third level contains two CPN pages corresponding to the asynchronous token delegation (i.e., the node *Asyn-Del♯4*, in Fig. 7(a) and Fig. 7(b)) and execution parts.

As all the stations behave in the same manner, for the asynchronous phase execution, they share the same node *B-asy-Ex ♯8* (i.e., Fig. 11 and Fig. 13), for which there are STs at the upper three super nodes (i.e., the nodes *LAS-DLE ♯5*, *Basic-St.I ♯6*, and *Basic-St.II ♯7*).

Top page. In Fig. 4, all the transitions are STs, and all the places are socket places. The places *CD1* and *DT1*, represent the sockets places through which the execution of synchronous services, is performed. The LAS sends a CD to a station, and that station replies by sending back Data i.e., a DT, and this for a set of stations, i.e., stations 1 to 5.

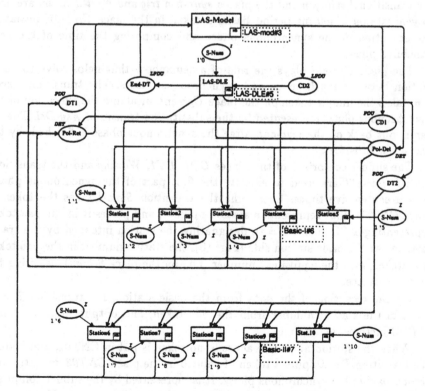

Fig. 4. Top-page in the hierarchy

The place *CD2* corresponds to the sending by the LAS of a CD to itself (i.e., to the LAS-DLE), which sends a DT to a specified station (the station address is indicated in the DT) through the place *DT2*, and then giving back the control to the LAS again through the place *End-DT*, to continue this operation for station 6 through 10. The place *End-DT* is used just for the sake of synchronization on the CPN model. In fact, the LAS detects the end of transmission and resumes activity thereafter, automatically. Thus, the second synchronous part of sending data to the actuators is finished.

The places *Pol-Del* and *Pol-Ret*, represents the sockets through which the asynchronous phases scheduled by the LAS take place. In this phase the LAS gives control to each station, in turn, to execute its asynchronous services by sending a PT (i.e., that creates a DeT), and then receiving the same token back.

The declaration node. An extract of the declaration node is depicted in Fig. 5. It displays most of the token colors and variables defined for the CPN model. The function definitions were omitted for the sake of brevity.

The LAS model. A detailed view of the page instantiated by the ST *LAS-Model* of Fig. 4, is depicted in Fig. 6. At the top left, the subnet constituted by the transition *Perio-gen* and the places *sync-tim trig* and *Next-Sch. act* are used to generate on a 400 ms period basis, a token in the place *CNTO1*, initiating the execution of the synchronous phase and computing the time of the next scheduled phase.

The place *Cyc-cnt* plays the role of a semaphore that helps solve the contention between synchronous and asynchronous phases. The transition *Perio-gen*, takes the unique token on the place *Cyc-cnt*, disabling the execution of the asynchronous phase represented by the substitution transition *Asy-Del*. The token is put back on the *Cyc-cnt*, after the synchronous phase is finished, by the transition *Init-Next*.

The subnet comprised of the places *CD1, DT1, Waiting* and the transitions *Sync-I, Rec-DT* are used to execute the first part of the synchronous phase. It is executed five times. Each time the transition *Sync-I* takes the token in place *CNTO1*, that represents a station number, and passes it in the *pdu* token representing a CD (i.e., $sn = i$), designating the station intended by the transmission. At the basic station side (Fig. 10), the station number in the *pdu* token is matched with the equivalent number (i.e., on the place *S-num*), selecting the proper instance.

On the reception of the data from the basic station, the transition *Rec-Dt* extracts the station number from the *pdu* token, and outputs it in the place *CNTO1* incremented with 1.

When the station number reaches 5, no token is put in CNTO1, and instead the transition *Rec-Dt* puts a token of value 6 in the place *CNTO2*, initiating the second part of the synchronous phase. It is represented by the subnet comprised of the places *CD2, Waiting II, End-DT*, and the transitions *Sync-II, Init-Next*.

It corresponds to the sending by the LAS of a CD to itself, in which the station number of the LAS-DLE, the data length (*dl*), and the number of the station that will receive data are specified. The station number is taken from the token on the place *CNT02*, placed therein at the end of the first part, described above.

```
(* Extract from the Declaration Node *)

color J = with j timed;   (* Color for the time triggers *)

color I = int;
var i, i1, i2, i3, i4, n1, n2, : I;

color D = real;
var d, d1, d2, d3, d4, d5 : D;

color B = bool with (T,F);
var b1, b2, b3 : B;

(* Color for the priority field *)
color P = with 0 | 1 | 2;

(* Color for the Delegated Token frame *)
color DET = record sn:I * dur:D * p:P declare mult timed;
var det, der : DET;

(* Color for PDU frames *)
color PDU = record sn:I * dl:D declare mult timed;
var pdu : PDU;

(* Color for PDU frames sent to the LAS-DLE *)
color LPDU = record sn:I * dl:D * n:I declare mult timed;
var lpdu:LPDU;

(* Color for the message *)
color MES = record dl:D * p:P * gt:D * n:I declare mult timed;
var mes : MES;

color Ar = product I * D;                (* for the MTHT array structure *)

color LM = int timed;                    (* token color for the basic stations *)
var S : LM;

color LAS = with l timed;                (* token color for the LAS station *)
var SM : LAS;

(* End Extract *)
```

Fig. 5. The declaration node

Upon reception of the CD, the LAS-DLE sends a DT to the designated station (in the field *n*, of the lpdu token), through the place CD2.

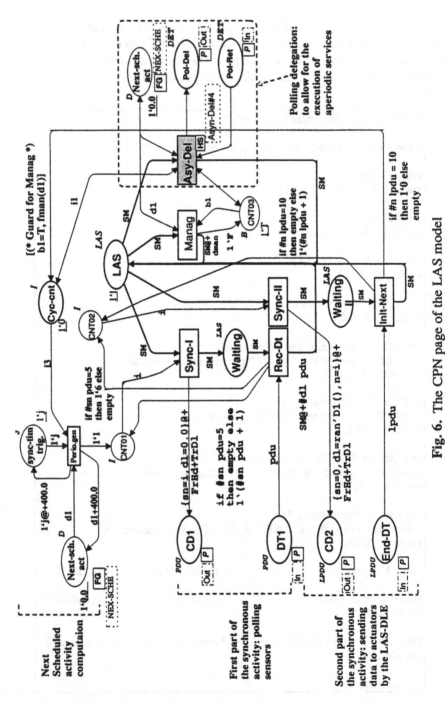

Fig. 6. The CPN page of the LAS model

This process is repeated 5 times until the station number reaches 10. Then the transition *Init-Next* will put a token in the place *Cyc-cnt*, enabling the asynchronous phase execution, and will put no token back in the control place *CNT02*, thus terminating the synchronous phase.

The places *CNT01*, *CNT02* are ensured to be 1-safe, since the synchronous phase is set to terminate before the current sampling period is finished.

Before starting the asynchronous phase, the management related activities, modeled by the timed transition *Manag* are executed following its firing, putting a boolean token F in *CNT03*, that ensures that at least once, the management activities are executed per token circulation cycle.

Furthermore, the function *fman()* in the guard of the transition *Manag*, ensures that no management is started when the next scheduled synchronous phase execution time gets closer.

The asynchronous phase is modeled by the ST transition **Asy-Del**, the port places *Pol-Del* and *Pol-Ret*, and the global fusion place *Next-Sch. act.* It is detailed in a lower level of the hierarchy (i.e., Fig. 7(a), Fig. 7(b)).

The asynchronous token delegation phase. Fig. 7(a), and Fig. 7(b) model the asynchronous token delegation process that represents the core of the scheduling mechanism. The CPN page was split into two parts for the sake of better understanding. They correspond to the node *Asyn-Del ♯4*, in Fig. 3.

In Fig. 7(a), the transition *compute duration* uses the places, *MTHA, Add Del-reqs, Delegated Dur, Next-sch. act, Station to-poll, Start time, Priority* and the functions *fdr ()* and *fprio ()*, to compute the next allocatable duration and the appropriate priority level. The transition's guard contains four functions (i.e., chnex1() to chnex4()) that are used to check whether the time left till the next scheduled activity, allows a full MTHT delegation, the requested additional delegation, or at least the minimum delegation time (i.e., mindel).

Also the ATRT is computed at this level using the value of the token on the place *Start time*, when the LAS-DLE station is selected (station number $i = 0$).

The computed values are used by the transition *Token Delegation* to initiate the token delegation to a specific station, setting up the level of priority. It updates, also, the corresponding values in the arrays of the tokens on the places *Add Del-reqs* and *Delegated Dur*.

In Fig. 7(b) the transition *Rec Req add del*, models the case when the station, that is returning the token, is requesting more delegation time. It uses the value of the array on the page fusion place *Delegated Dur*, while the transition *No add del Req*, models the other case.

The transition *Rec Req add del* records any eventual additional requests in the place *Add Del-reqs*, and checks the amount of delegated duration allocated so far, for a given station. If its MTHT is reached, its corresponding entry in the array of delegated duration is re-initialized to zero (using the function *initdel()*).

The same transition also checks whether the next station to poll is the current one or that the station number is increased based on whether or not it has received its full MTHT. For this sake the place *MTHA* is reproduced as a page fusion place in Fig. 7(b).

The token on the port place *CNT03* is set again to T (i.e., the boolean value True), after all the N stations have been served, and one cycle of token circulation is finished.

At that time also, an initialization of the array of delegated durations on the place *delegated Dur*, is started (using the place *Cnt10* and the transition *init*). The port places *Pol-Del* and *Pol-Ret* are mapped onto the socket places with the same name in the CPN page of Fig. 6, that are also mapped to similar socket places on the CPN page of Fig. 4.

To model the asymmetric case, first the color set associated with the place *MTHA*, is changed to the color *Ar* described in the declaration node (i.e., Fig. 5). Then, its marking is changed to an array structure where each entry corresponds to the station's number, followed by its MTHT.

Furthermore, to model the DAS the DET color set used for the DeT token-frame, is modified to include an additional field for unused duration (i.e., the *und* field). This duration is computed by the station receiving the DeT (in the subnet depicted in Fig. 13), and sent back within the DeT (i.e., in the returned token). Additionally, the place *Cnt20* and the transition *Init2* are added to ensure re-initialization. Fig. 8, depicts the modifications, described above, needed in Fig. 7(b), to model the DAS. The figure shows the transitions *Rec Req add del* and *No add del Req*, and the new arc connections. The remaining net structure is left unchanged and thus not shown. Depending on the station number included in the *der* token, the corresponding entry is selected from the array of MTHTs. This is enforced by the transitions' guards. The token color of the priority field is being added an additional value (i.e., 3), that informs the LAS that there was a left over bandwidth, stored in the *und* field. Depending on that value either of two transitions is enabled.

The *und* field is extracted from the *der* variable and checked, whether or not it is greater than a minimum value (that can be set to the slot time unit). In that case the *und* duration is added to the MTHT of the following station in sequence. This operations are performed along the output arc connecting the transition *No add del Req* to the place *MTHA*.

When a token circulation is finished (i.e., all the stations have received a full MTHT delegation), a re-initialization of the *MTHA* array is performed, using the place *Cnt20*, the transition *Init2* and the function *fmtht()*, defined in the global declaration node.

109

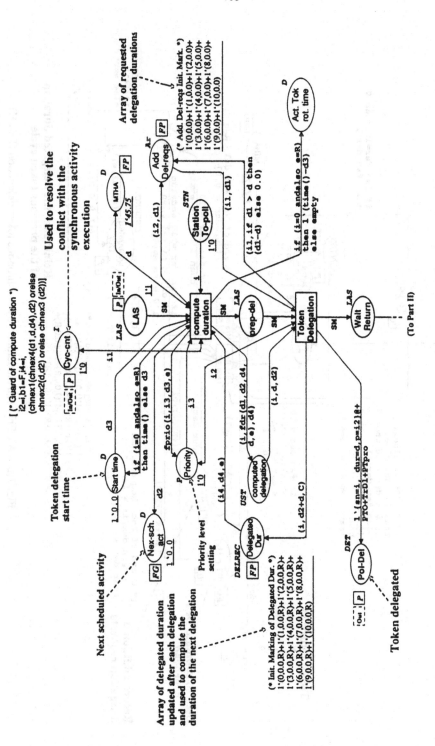

Fig. 7(a). A CPN page modeling the asynchronous token delegation (Part I)

110

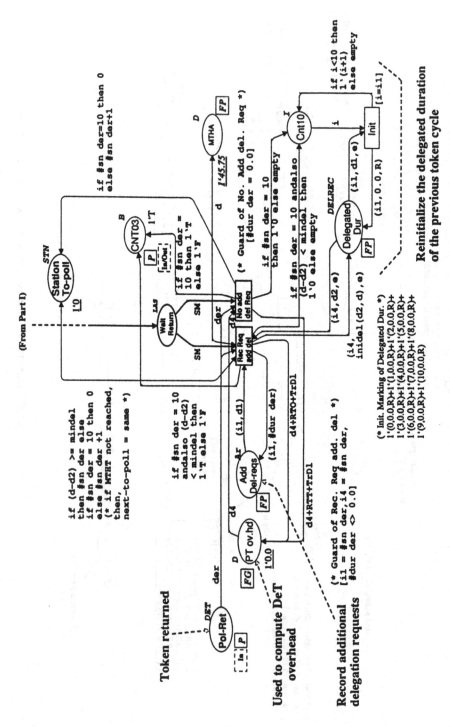

Fig. 7(b). A CPN page modeling the asynchronous token delegation (Part II)

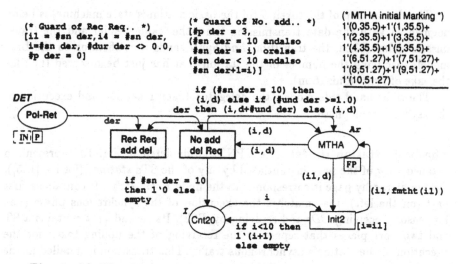

```
(* Guard of Rec Req.. *)
[i1 = #sn der,i4 = #sn der,
 i=#sn der, #dur der <> 0.0,
 #p der = 0]
```

```
(* Guard of No. add.. *)
[#p der = 3,
 (#sn der = 10 andalso
  #sn der = i) orelse
 (#sn der < 10 andalso
  #sn der+1=i)]
```

```
(* MTHA initial Marking *)
1'(0,35.5)+1'(1,35.5)+
1'(2,35.5)+1'(3,35.5)+
1'(4,35.5)+1'(5,35.5)+
1'(6,51.27)+1'(7,51.27)+
1'(8,51.27)+1'(9,51.27)+
1'(10,51.27)
```

Fig. 8. Modification needed to model the DAS (for async. tok. deleg.)

The LAS-DLE station model. The CPN page corresponding to the LAS-DLE model is depicted in Fig. 9. On the hierarchy page it is represented by the node *LAS-DLE*#5. It has two parts: the left part describes the reception of a *CD* from the LAS, and the sending of a *DT* to the station specified inside the *CD* frame (i.e., #n lpdu), that is saved in the place *St-num save*.

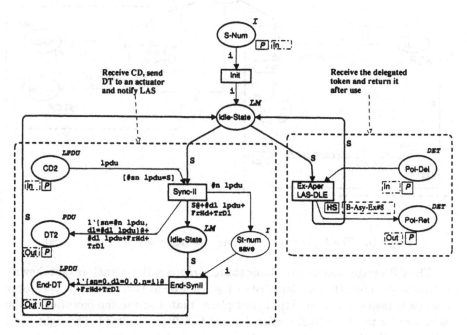

Fig. 9. The CPN page modeling the LAS-DLE station

The time stamp of the token S (of the basic station's state machine) is being augmented with the data transmission time, the frame header and the transmission delay. Then, the transition *End-SynII* sends a notification to the LAS model, specifying the number of the station that has just been served (i.e., for the sake of synchronization).

The right hand side models the delegated token reception and execution of the asynchronous phase.

The basic station model. The CPN page depicted in Fig. 10, represents a detailed view of the page instantiated by any of the STs *station*$_i$ (for i $\in$ [1..5]). On the hierarchy page it corresponds to the node *Basic-st.I#*6. It contains a first part (on the left), that matches the execution of the synchronous phase (i.e., for sensors receiving *CDs* and replying by *DTs*). Its second part contains a ST and two port places that represent the receiving of the polling token for the execution of the station's asynchronous traffic. This transition is detailed in the next paragraph.

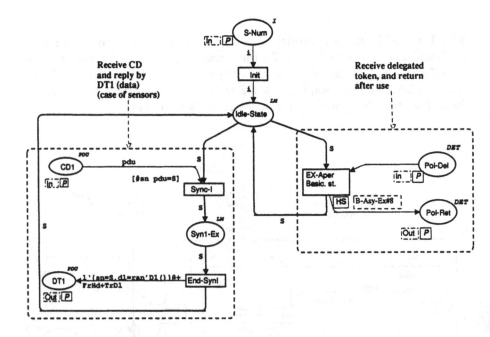

Fig. 10. The CPN page modeling the Basic station (for sensors)

The CPN page modeling the basic stations representing actuators (not shown), is basically similar the one depicted in Fig. 10, except that on its left part, only a single transition and one input port place, that describe the reception of the data sent by the LAS-DLE station (i.e., of Fig. 9), is depicted. The right hand side is exactly the same as in Fig. 10.

The message generation subnet. Fig. 11 depicts the CPN page modeling the message generation process, corresponding to a single station. It constitutes a part of the CPN page modeling the asynchronous phase execution (i.e., in the node *B-asy-Ex* ♯*8* of Fig. 3). During simulation, there is one page instance, for each stations.

Aperiodic messages (NP, TaP and UP messages) are generated in a sporadic fashion and thus their arrivals were modeled as Poisson processes with exponentially distributed inter-arrival times. Fixed length data are considered.

The urgent priority is assigned to alarms that occur very rarely[5], and as such will occupy a very small portion of the offered load (around 1 to 2 percent). We model the case when during the simulation period a set of stations among all the connected stations, generate few UP messages.

In the subnet modeling the UP message generation only the stations with odd addresses are being simulated to generate UP messages[6]. It is realized using the port place *S-Num* modeling the station number, and defined at the top page of Fig. 4, together with the arc expression from the transition *T-trig1* to the place *Tim-trig1*.

Next, we will describe only the part that models the TaP assigned messages. The NP class has a similar structure.

An exponentially distributed firing rate is enforced on the transition *TA-Ac-G* generating TaP messages. This is done by outputing a *j* colored token, while increasing its time stamp with a value generated using the function *fgt()*, after every message generation. This function is defined as:

$$fgt(a) = ln(ran'D0()) * -a$$

where *ran()*, is a uniformly distributed random number generator function. *D0()* is a real color set with a range of [0..1]. *a* is a mean exponential inter-arrival time, which is passed as an argument to the function. By changing it, the offered load is changed. The @ mark that follows the token *j*, represents the current time, maintained by a global clock of the system. The following message will not be generated until the added time stamp has expired.

The place *Trig3* and the transition *T-trig3* (that is fired only once at the beginning), are used for ensuring a proper initialization, so that the first message will not be generated at instant 0.0, but after some time. This lapse of time is generated randomly within a specific range (i.e., *DD3()*), chosen in accordance with the argument *r3*, passed to the function *fgt ()*.

In case of the UP class, the range *DD()* is chosen such as the generation of UP messages is started after sometime generated randomly within it, which ensures that, the system has reached minimum stabilization.

The place *Mes-idt* is used to generate an identifier for each message. This will be used to enforce the FIFO policy when executing the messages. The message generated in the place *TA-Mes*, is being added a data length (dl), a generation

---

[5] This is a realistic design option characterizing industrial applications
[6] We could also make random choices, however, the result is expected to be the same

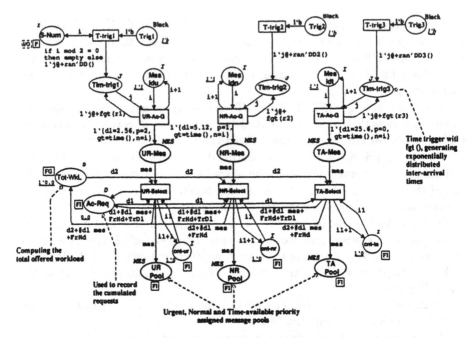

Fig. 11. A CPN page modeling the message generation process

time (gt) and an appropriate level of priority p (2 for urgent, 1 for normal and 0 for time-available).

The data length accounts only for the data part of the message received from the application layer, immediately above the DLL. The transition *TA-Select*, is used to update the counter place *cnt-ta*, which counts the number of messages generated so far. At this level also, the total workload and the accumulated requests' duration are updated (in the places *Tot-Wkl* and *Ac-Req*), whenever this transition fires.

Note that a message request duration accounts for the data portion (♯dl mes i.e., the *dl* field extracted from the variable *mes*), the Frame overhead **FrHd** and the expected transmission delay of the message **TrDl**. Finally, the generated messages are put in the pool output port place *TA Pool*. In the case of the UP and NP classes the places *cnt-ur* and *cnt-nr* are used as counters.

The asymmetric system is realized by modifying the CPN of Fig. 11. The modifications needed for the TaP message generation are depicted in Fig. 12. They concern the subnet comprised of the transitions *TA-Ac-G* and *T-trig3* and their surrounding places and arcs. Similar modifications are needed for the NP message class.

The modifications consist in adding the place *D-Trig3* and in extending the use of the port place *S-Num*, to divide the set of stations into two groups. The first having their address number less than 6, and the others.

In addition to the *j* token with the appropriate time stamp, the transition *T-trig3* generates a mean exponential inter-arrival time to each defined group of stations (i.e., *d1* or *d2* depending on the station number on the place *S-Num*), that is loaded in the place *D-Trig3*. It is used later on, as an argument (i.e., *r3*) for the *fgt ()* function generating the time stamp of the next message.

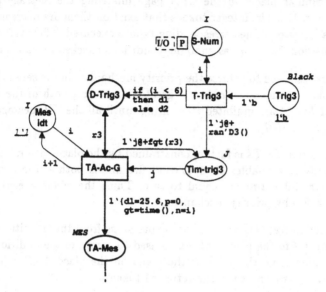

Fig. 12. Modification needed to model an asymmetrically loaded system

The CPN modeling the message execution subnet. The CPN page for the execution subnet is depicted in Fig. 13. This CPN constitute the second part of the node *B-Asy-Ext8* in Fig. 3. The places *Pol-Del* and *Pol-Ret* model the DeT reception and return. After reception of the DeT, the duration is extracted and loaded in the place *Del dur*, that is decreased whenever a message is executed, by the time needed to send the message, the frame overhead delay, and the transmission delay times. At the same time the duration is loaded in the place *Rec-del*.

Furthermore, the priority level is recorded in the place *Priority*, which is used as an additional constraint for the enabling of the message execution transitions. The place *End-acyc timer*, loaded also with the delegated duration value, represents a timer that detects the end of the delegated duration and triggers the firing of the transition *End AperII*.

This transition uses the content of the token on the place *Rec del*, and that of the token on the place *Ac-Req* to compute how much time is left from the previously allocated duration, and how much the station needs of additional delegation time to finish its accumulated requests.

The transition *End Aper* is enabled when either there are no more messages to service, or no messages with the appropriate priority level, or when the time left to execute a message in less than the time needed to do it. The guard of this transition on the top of the figure enforces this checking. The places *Fifo ur*, *Fifo nr*, *Fifo ta* are used to enforce the FIFO policy by executing the message that has an identifier equal to the token value on each of these places.

The places *cnt-ur*, *cnt-nr* and *cnt-ta* are in the same instance fusion sets containing similar places in the CPN page modeling the message generation subnet (in Fig. 11). The integer tokens that exist on them are decreased by one, whenever a message of the corresponding type is executed. Their values are used by the transition *End Aper* as enabling conditions checked in its guard.

They are also used to enforce the priority mechanism in the sense that, whenever the token on the place *cnt-ur* is greater than zero, both of the transitions *Ex-NR* and *Ex-TA* are inhibited from firing, until all the UP messages are executed (i.e., by *Ex-UR*).

The transition *Ex-TA* is further constrained by the place *cnt-nr*, which counts the NP messages available. Thus, it can't fire unless both of the tokens on the places *cnt-ur* and *cnt-nr* are equal to zero. Thus, the whole execution subnet enforces a FIFO by priority mechanism.

The function *fret ()* used in the arc expressions from the transition *End Aper* and *End AperII* to the place *Pol-Rel*, is used to compute the additional needed delegation time using the value of the token on the place *Ac-Req*. The needed delegation time is included in the returned token.

Besides, for modeling the DAS, we need to compute the left over duration, that were unused for lack of messages to execute at the station which has been delegated the token.

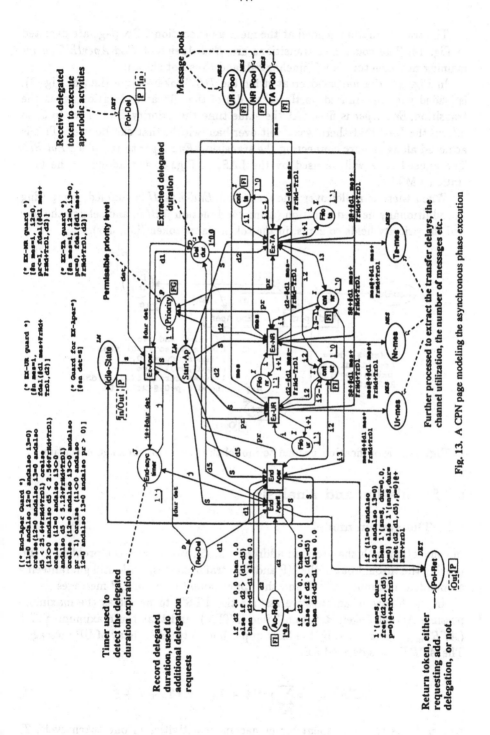

Fig. 13. A CPN page modeling the asynchronous phase execution

The transformation needed at the message execution CPN page are depicted in Fig. 14. The concerned transitions are *End-Aper* and *End-AperII*. The remaining net structure is left unchanged, and thus not shown.

In Fig. 14, the *und* field created for the *DET* token-frame (i.e., in Fig. 8), is loaded with the unused portion (i.e., *d5*) of the allocated duration, when the transition *End Aper* is fired. At the same time the priority field is set to 3, to inform the LAS that there was a left over bandwidth, that can be used. This is achieved along the arc connecting the transition *End Aper* to the place *Pol-Ret*. The unused time will be used by the LAS, in Fig. 8, and added to the next station's MTHT.

When there is no leftover, the transition *End-AperII* is enabled, computing the additionally needed duration (using the function *fret()*), and setting the *und* and the priority fields to zeros, in the token-frame token (i.e., *der*).

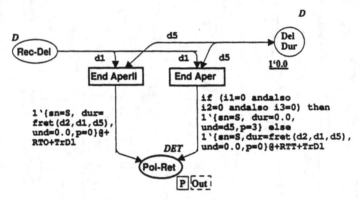

Fig. 14. Modification needed to model the DAS (for message execution)

5 Simulation and Analysis

5.1 The system model

In connection with the problem addressed in [5], we address the problem of improving the throughput of the NP and TaP traffic classes, while still guaranteeing a bounded maximum ATRT and the upper bound delay of UP messages.

In the following analysis we assume the TTRT to be set as the maximum possible Actual Token Rotation Time (ATRT), and that the maximum ATRT ($ATRT_{max}$) is bounded by the least upper bound transfer delay of UP messages. The $ATRT_{max}$ is derived as:

$$ATRT_{max} = \sum_{i=1}^{N} mtht_i + T_m + r * T_s + \Psi + \xi \tag{1}$$

where: T_m is the time spent for management activities in one token cycle; T_s represents the synchronous phase execution time. Ψ represents the total token overhead, and ξ accounts for the total idle time which is generated due to scheduling constraints (i.e., it is the collection of time gaps which are judged not sufficient to delegate the token for asynchronous phase execution without violating synchronous phase execution timings). Finally r represents the number of synchronous phases within $ATRT_{\max}$. It is bounded by:

$$0 < r \leq r_u = \left\lceil \frac{ATRT_{\max}}{P} \right\rceil \tag{2}$$

where P is the synchronous sampling period. Ψ is bounded by:

$$\Psi = (N + k) * \tau \qquad \text{where}, \quad 0 \leq k \leq r_u \tag{3}$$

where τ represents the token frame's sending, receiving and processing times. The minimum token overhead per $ATRT_{\max}$, is $N * \tau$. In addition, whenever a synchronous phase overlaps an asynchronous one, for a given station, the MTHT of that station is split into two parts, to allow the execution of the synchronous phase. Thus, that particular station, will receive its full MTHT, through two delegations. Each delegation will generate a token overhead (i.e., τ). Thus, at most we can have, $r_u * \tau$ additional token overheads per $ATRT_{\max}$. At this stage, for a symmetrically loaded system, applying an Equal Partition Allocation Scheme (EPAS) [6], the MTHT value common to all the stations can be derived as:

$$\forall i \in [1..N], mtht_i = \frac{ATRT_{\max} - (T_m + r * T_s + (N + k) * \tau + \xi))}{N} \tag{4}$$

In case an asymmetrically system is considered, the MTHT of each station is computed using an adapted form of the Normalized Proportional Allocation Scheme (NPAS) proposed in [5]. Using this scheme, the available asynchronous bandwidth is allocated to the stations according to a normalized form of an estimated load of asynchronous activities at each node, i.e.,

$$\forall i \in [1..N], mtht_i = \frac{R_i}{\sum_{i=1}^{N} R_i} * (ATRT_{\max} - (T_m + r * T_s + (N + k) * \tau + \xi)) \tag{5}$$

where R_i represents an estimated maximum delegation duration needed by station i, to serve its asynchronous traffic that may be generated within an $ATRT_{\max}$ time span.

When the NPAS cannot be fully applied, due to protocol's constraints, a procedure allowing its partial application, while guaranteeing compliance with the protocol's constraints, has been devised in [5], and was used during the simulation process.

As for ξ, in contrast to Ψ, it is accumulated prior to every synchronous phase execution, whenever the MTHT of a given station i, is not split into two parts. The idle time that may be generated prior to one synchronous phase is denoted as ζ. It can be bounded as:

$$0 \leq \zeta < mindel + \tau + \omega \tag{6}$$

Where *mindel*, represents the minimum delegation time, ω represents the token recovery time.

Thus, ξ can be bounded as:

$$0 \leq \xi \leq r_u * \zeta \qquad (7)$$

It can be noted that we have just given upper and lower bounds for the quantities Ψ and ξ, this because it is not easy to get an exact analytical derivation for them.

Thus, as stated earlier in [5], a simulation-based parameter calibration has been proposed to derive the appropriate asynchronous bandwidth that can be allocated without jeopardizing the delay bounds. Here, under similar constraints, we show that using the DAS, the overall performance of the protocol can be improved in terms of increased channel utilization, lower token overheads and lower delivery delays. Equations (4) and (5) were used to determine the MTHT share of each station.

The Dynamic Allocation Scheme (DAS). The DAS is based on the idea of using the leftover bandwidth from the MTHT delegation duration to a particular station, that is added to the share (i.e., the MTHT) of the following station in sequence. The leftover bandwidths are always summed up, except that at the beginning of each new token cycle, a re-initialization is performed, to keep it bounded. The DAS is defined as follows:

$$D_{mtht_i} = \begin{cases} mtht_i & \text{if } i = 1 \\ mtht_i + Res_{i-1} & \text{if } 1 < i \leq N \text{ and } Res_{i-1} \geq \theta \end{cases} \qquad (8)$$

where, D_{mtht_i} is the dynamically assigned delegation duration of station i, and θ represents the minimum usable leftover bandwidth.

On a pre-runtime basis, all the MTHTs are set using either the EPAS or the NPAS statically. During operation, these MTHTs are altered by the LAS, which keeps an array of them, as shown in the CPN page of Fig. 8.

5.2 Simulation

Only error free operation of the protocol is considered in order to assess its performance.

In this paper, we will report only the results obtained, considering a symmetrically loaded system. An asymmetrically loaded system, as described earlier, has been found to exhibit an almost similar behavior, with minor differences. Thus, it was omitted for the sake of brevity.

The UP message arrival rate is set to be the lowest among the three types, as the occurrences of alarms in the real world are rare. The NP message arrival rate is more frequent, and the TaP one is the most frequent. The load is increased by increasing the message's arrival rate of NP and TaP messages.

When the system reaches saturation, TaP message arrival rate is maintained at a fixed value, and the NP message is increased further. We have chosen this scheme, because as the TaP messages have the longest data length and the highest arrival rate, they will occupy most of the channel bandwidth, until its saturation. At that point more input from the TaP class will not cause any change to the network performance. Then, by further, increasing the input of the NP class, we can see the interaction between the two classes.

For UP messages, as they are matched with alarms, we simulate the case when during simulation a set of stations generate few UP messages at random time, and we record the maximum transfer delay (i.e., delivery delay) among all the generated UP messages. For the other two types of messages, we compute the mean transfer delay, and the mean channel utilization (i.e., the portion of bandwidth used to transmit only data, with the headers and transmission delays excluded).

The frame header overhead, the data length transmission time, and all transmission related overheads are recorded in time on the stamp of the token, modeling a message, whenever there is a need for that. That will advance the global clock accordingly. When the token reaches its destination, the difference between the arrival time and the generation time (recorded inside the token), is used to calculate the transfer delay incurred by that token.

Similarly, a special structure is reserved for the token-frame (i.e., DeT), and the overheads of token delegation transmissions are added to the time stamp associated with this token whenever necessary. These overheads are accumulated in a global fusion place (i.e., the place $PT\ ov.hd$ in Fig. 7(b)), to compute the mean token delegation overhead per second.

The simulation parameters are depicted in table 2.

Table 2. Simulation parameters

Number of Stations	N = 11 (1 LAS-DLE and 10 basic stations)
Transmission rate	31.25 kpbs (Slow H1 type Fieldbus)
Data length & Transmission times	UP mess. : 80bits, 2.56, ms NP mess. : 160 bits, 5.12 ms TaP mess: 800 bits, 25.6 ms
Minimum delegation Time (mindel)	12.096 ms (execution of 1 UP mess + all related delays)
$ATRT_{max}$	1980.0 ms, for b = 2000.0 ms (b represents the upper bound delay of UP mess.)
Frame header TransmissionTime	1.536 ms (for 48 bits)
Transmission delay	8.0 ms (including modem, station and propagation delays)
DeT sending delay	1.792 ms
DeT return	0.768 ms
Token processing and recovery safety margin	5.696 ms

$\tau = 22.656$ ms , $\omega = 1.6$ ms , $T_s = 209.0$ ms, $T_m = 25.0$ ms, $H_{max} = 36.136$ ms (Minimum MTHT permissible)
Observation time: 80,000 ms, P = 400 ms, r = k=5, j = 5.

For the idle time estimation, we first deduce the maximum one time, idle time delay possible. For that, ζ is set to its upper bound (minus a very small quantity ϵ) as: $\zeta = (mindel + \tau + \omega) - \epsilon$. Setting ϵ to 1 ms, we get $\zeta = 35.352ms$. Hence, ξ may vary in the following range: $0.0 \leq \xi \leq 5 * \zeta = 176.76ms$. Let $\xi_{max} = 176.76ms$, be the maximum possible total overhead per $ATRT_{max}$.

In the work presented in [5], the idle time calibration has been performed by taking different fractions of ξ_{max}. Here, we set ξ to $1/4 * \xi_{max}$, and deduce the whole allocable asynchronous bandwidth, and all the $mtht_i$. The MTHT setting is depicted in table 3.

Table 3. Asynchronous bandwidth and MTHT setting

Av. Asynch Bandwidth (ms)	MTHT (ms) (Sym. case)
503.314	45.75

The message arrival patterns are summarized in table 4.

Table 4. Message arrival pattern

NP. mess. arr. rate (mes./sec)	TaP. mess. arr. rate (mes./sec)
0.05 - 4	0.06 - 1.

The means displayed on the figures correspond to 10 independent simulation runs, generating 10 random variables. The means' standard deviations and the 95% confidence interval of the mean, were calculated using the upper $\alpha/2$ quantile of the t distribution with 9 degree of freedom. I.e., $\alpha = 0.05$, and $t_{0.025;9} = 2.26$.

The mean values and the confidence intervals for three representative values of load conditions ((a), (b) and (c)), corresponding to three performance levels on the curves (i.e., light loads, first level and second level of saturation), are shown below their respective graphs, in the figures 16, 18, 19, 20 and 21.

5.3 Results and analysis

In Fig. 15, the maximum transfer delay incurred by UP messages is plotted against the offered aperiodic traffic load, for the symmetric system. The main result of the plot is that the upper bound delay of urgent messages (i.e., $ATRT_{max}$), is not violated by any of the curves, either for those using the SAS or the DAS.

Furthermore, it can be seen that the maximum delay is greater when the DAS is used, than when the SAS is considered. This will be explained in Fig. 17.

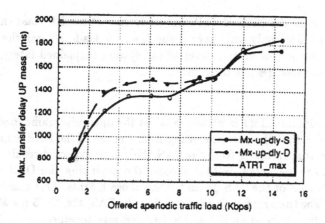

Fig. 15. Max. Transfer delay of UP mess. Vs. Aper. traffic load

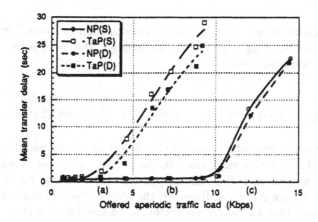

Curves	Curve (S)-NP			Curve (D)-NP			Curve (S)-TaP			Curve (D)-TaP		
Ref. Values	(a)	(b)	(c)	(a)	(b)	(c)	(a)	(b)	(c)	(a)	(b)	(c)
Mean	552.5	672.9	13293	563.7	697.1	12288	1814.8	20586	-	920.4	17554	-
Std. Dev.	9.32	7.53	157.9	4.4	6.68	146.9	106.8	260.9	-	34.25	355.9	-
95% Conf. Int of Mean	531.4 573.5	655.9 689.9	12937 13650	553.8 573.7	682.0 712.2	11956 12620	1573.5 2056.1	19996 21176	-	842.9 997.8	16749 18358	-

Fig. 16. Mean transfer delay of NP and TaP mess. Vs. Aper. traffic load

In Fig. 16, the mean transfer delay of NP and TaP classes is plotted versus the offered load of aperiodic traffic.

We can notice that the DAS is showing better performance in reducing the waiting time for the NP and TaP classes. In case of the NP class the DAS out-performs the SAS only for heavy loads (i.e., for loads greater then 10 Kbps) around the second saturation level of the system (explained later), when only the NP class messages get through.

The maximum ATRT is plotted against the aperiodic traffic load in Fig. 17. It can be seen that the curve resulting from the use of the DAS exhibits greater values than that resulting from the SAS. This is an expected result. The second thing that can be noted, is that around the beginning the DAS curve grows large, while that of the SAS doesn't reach those heights. This can be explained as follows: For the symmetric system using the SAS, the 45.75 ms MTHT value does not suffice to send an NP and a TaP message together.

Thus, whenever a NP message is executed (at a specific station), no other TaP can be sent, and at best another NP message can be sent, or two messages in all.

In that case there will be a loss of bandwidth (and of channel utilization). Moreover, as the NP class' input at the beginning of the curve is low, the probability of this bandwidth loss is high.

However, in case when the DAS is used the loss of bandwidth is recovered, for the subsequent station(s), and as such the ATRT growth happened.

Then, the curve generated using the DAS declines in the middle, and stabilizes for a while around a value close to that of the curve generated using the SAS. In this interval, the TaP throughput starts decreasing, being more and more blocked, by the NP class. However, the level of the NP class' input is not enough to ensure that the whole asynchronous bandwidth is used.

Toward the end of the interval, as the NP class' input (and throughput) increases, the TaP class is blocked further until its throughput is reduced to zero. At that level, the NP class fills the asynchronous bandwidth and generates big ATRT values which are near the upper bound (i.e., $ATRT_{max}$), for both curves.

Fig. 18 depicts the Mean ATRT variation with the traffic load, for the symmetric system using the DAS and the SAS. The plots show the higher values registered for the curves generated using the DAS, over those generated using the SAS. Moreover, they show two levels of saturation, as we explained before depending on the level of input of the NP and TaP classes respectively. The first level is reached when the TaP input is higher, and the second one is reached, when the NP class' input blocks completely the TaP class (as it has higher priority).

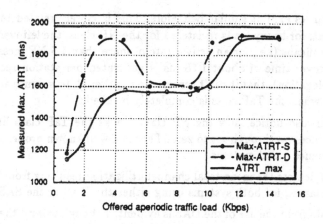

Fig. 17. Maximum ATRT registered Vs. Aperiodic traffic load

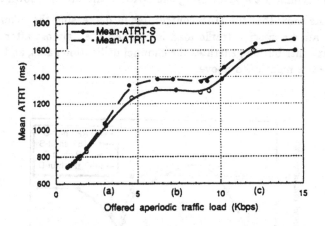

Curves	Curve (S)			Curve (D)		
Ref. Values	(a)	(b)	(c)	(a)	(b)	(c)
Mean	1022.0	1297.9	1595.7	1065.5	1372.0	1649.6
Std. Dev.	7.13	4.12	3.45	8.97	3.8	4.001
95% Conf. Int of Mean	1004.5 1039.6	1288.6 1307.3	1587.9 1603.5	1045.2 1085.8	1363.4 1380.6	1640.5 1658.6

Fig. 18. Mean ATRT Vs. Aperiodic traffic load

The mean channel utilization of the NP and TaP classes versus the aperiodic traffic load is plotted in Fig. 19. It can be noted that the DAS exhibits better behavior than the SAS, when the first level of saturation gets closer and afterward. The NP class seems not benefiting much from the usage of the DAS, as both curves exhibit similar behaviors. This is closely related to the MTHT setting that, for heavy loads registered for the NP class, the leftover bandwidth, is so insignificant (i.e., for an arrival rate of more than 1.5 NP message per second, a leftover time of about 1.78 ms is generated per station receiving a full MTHT delegation), that it can't serve, when gathered, to execute additional NP messages, when the TaP class is completely blocked.

We can also notice, how the NP class blocks the TaP class little by little until it reduces its throughput to zero. That is a well known property of multiple priority queues based systems.

Fig. 20 depicts the mean total channel utilization resulting from both the NP and TaP classes, for both systems using either the DAS or the SAS.

The gain from the use of the DAS is evident, although toward the end, it loses effectiveness. This can be explained with the same arguments as those used to explain the behavior displayed in Fig. 19, when only the NP class gets through.

Finally, the variation of the mean token delegation overhead with the traffic load for the symmetric system using the DAS or the SAS is shown in Fig. 21.

The plots show that the DAS contributes in decreasing the token delegation overheads, after a specific traffic load is reached (i.e., almost after the 3 Kbps level). That contribution generates the channel utilization gains, and the transfer delay decreases mentioned above.

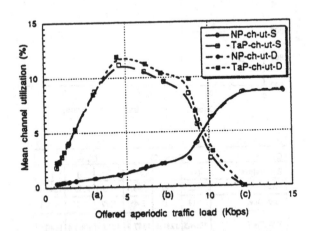

Curves	Curve (S)-NP			Curve (D)-NP			Curve (S)-TaP			Curve (D)-TaP		
Ref. Values	(a)	(b)	(c)	(a)	(b)	(c)	(a)	(b)	(c)	(a)	(b)	(c)
Mean	0.75	2.06	8.63	0.78	2.05	6.96	8.67	9.69	-	8.64	10.47	-
Std. Dev.	0.008	0.009	0.005	0.005	0.01	0.001	0.022	0.026	-	0.033	0.029	-
95% Conf. Int of Mean	0.69 0.82	2.0 2.09	8.58 8.67	0.74 0.82	1.98 2.12	8.63 8.76	8.51 8.84	9.5 9.88	-	8.4 8.88	10.26 10.68	-

Fig. 19. Mean channel utilization of NP and TaP classes Vs. Aper. traffic load

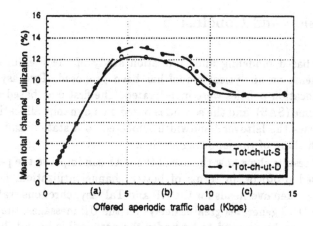

Fig. 20. Mean Total channel utilization Vs. Aper. traffic load

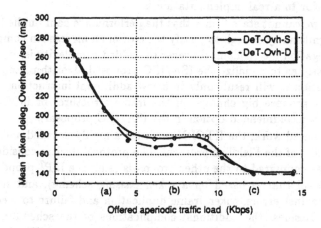

Fig. 21. Mean token delegation overhead per sec. Vs. Aper. traffic load

6 Discussion and Conclusion

A HTCPN based modeling and simulation study of the Fieldbus protocol, that complements an analytical model has been proposed. Two asynchronous bandwidth allocation policies were investigated. The first one, based on a static allocation scheme (SAS), and the second is based on a dynamic allocation scheme (DAS), that uses the leftover bandwidth allocated to a station and adds it to the bandwidth share of the following station.

The above results show how the use of the DAS can improve the performance of the proposed protocol, in terms of better channel utilization, less waiting delays and less token overheads for the NP and TaP asynchronous traffic classes. Although, the DAS generates greater delays for the UP messages, the maximum transfer delays can be ensured to be under the permissible bound, by a proper setting of the asynchronous allocatable bandwidth.

Furthermore, the results show that the DAS starts to be effective after a certain level of light load is reached. Below such a level, the DAS and SAS are almost similar.

Another feature that has to be stressed, is that the choice of MTHT value in connection with the time needed for the execution of messages, is critical for the DAS to be effective. As such, simulations like those presented in this study, constitute useful means for effective tuning of communication protocol's parameters, prior to a real implementation.

Although, we cannot state firmly that the performance gain implied by using the DAS is significantly enough to justify the cost of an increase in complexity of the token-delegation mechanism. However the added complexity is, in our opinion, not so costly as to penalize the ISA/IEC proposed algorithm, and in terms of implementation it will result only in a few additional instructions, knowing that we didn't propose big changes in the frame structures or in the MTHT array (that is a recommended option in the ISA/IEC proposal).

In the presented work, the HTCPN provided a highly detailed and accurate simulation model of the Fieldbus protocol. The protocol itself is considered very complicated, as compared to the other proposals (e.g., the ISP[7] proposal of the Fieldbus protocol that is based on a token passing scheme), and supposedly more robust against errors, token-frame duplication and failure to meet timing requirements. Besides, the additional complications of the scheduling mechanism implied by the use of the DAS made the protocol more difficult to model by a graphical tool. However, using the HTCPN model most of the protocols properties were modeled effectively.

The Design/CPN tool has rich simulation possibilities, allowing interactive, semi-automatic and automatic simulations. The interactive mode, helped in the debugging process, allowing correction of wrong behavior and verification of accuracy.

Before starting the automatic simulation mode, the interactive mode allowed

---

[7] Interoperable System Project

us to verify, on the fly, that all data are generated and processed properly [8]. In addition, it allowed us to verify that the asynchronous token delegation and circulation mechanism, and the message execution subnet are working correctly, in conformity with the specifications (e.g., correctness of the FIFO-by-priority mechanism, etc.).

The problem of *Transient (state) removal*, was solved by proper initialization of the time tags stamped to the initial generated data, and by a proper choice of the simulation run time.

Further research, can be directed toward the analysis of the sensitivity of the DAS to different data arrival patterns, and to data length variability.

Acknowledgment

The authors wish to thank Dr. Miyazawa of Kanagawa Industrial Technology Research Institute, for the useful discussions, we had with him on this paper.

References

1. P. Pleinevaux, J. D. Decotignie : Time Critical Communication Networks: Field Buses, IEEE Network Vol.2, No. 3, May 1988.
2. K. G. Shin and C. C. Chou : Design and Evaluation of Real-Time Communication for Fieldbus-Based Manufacturing Systems, IEEE Trans. on Robotics and Automation, Vol. 12, NO. 3, June 1996.
3. ISA and IEC commission : Field bus Data Link Layer Specification, Unapproved Committee Draft, 1995.
4. ANSI/IEEE std. 802.4-1995: Token-Passing Bus Access Method and Physical Layer Specifications, IEEE 1985.
5. A. Ben Mnaouer, T. Ito, H. Tanaka, W. K. Yoo, T. Sekiguchi : Asynchronous Bandwidth Allocation and Parameter Setting in the Fieldbus Protocol, in the IEE Japan, Transactions of Electronics, Information and Systems Society (C), pp. 962–970, July 1997.
6. G. Agrawal, B. Chen, W. Zhao and S. Davari: Guaranteeing Synchronous Message Deadlines with the Timed Token Medium Access Control Protocol, IEEE Trans. On Computers, Vol. 43, March 1994.
7. S. Cavalieri, A. Di Stefano, O. Mirabella: Pre-Run Time Scheduling to Reduce Schedule Length in the Fieldbus Environment, IEEE Trans. on Soft. Eng. Vol. 21, NO. 11, Nov. 1995.
8. P. Raja and N. Guevera: Static and Dynamic Polling mechanisms for Fieldbus Networks, ACM Operating Systems Review 27(1), July 1993.
9. P. Raja, K. Vijayananda and J. D. Decotignie : Polling Algorithms and their Properties for Fieldbus Networks, Proc. of the IECON'93, Vol. 1, pp. 530–534, Hawai, November 1993.
10. S. Cavalieri, A. D. Stefano, O. Mirabella: Optimization of acyclic bandwidth allocation exploiting the priority mechanism in the Fieldbus data link layer, IEEE Trans. on Indust. Elect. Vol. 40, NO. 3, June 1993.

---

[8] Every colored token's content can be visualized during interactive simulation

11. A. Di. Stefano and O. Mirabella : Evaluating the Fieldbus data link layer by a Petri Net-based simulation, IEEE Trans. on Industrial Electronics, Vol. **38**, No. 4, Aug. 1991.

12. Interoperable System Project Foundation : Draft For Review, Data Link Layer Specification, 92-29-01-Rev. 2.0.

13. G. Juanole and Y. Atamna : Modeling communications in the FIP (Factory Instrumentation Protocol) with the stochastic timed petri model, Proc. of ETFA'92, pp.336–341, Melbourne, Australia.

14. P. Huber and V. O. Pinci : A formal executable specification of the ISDN basic rate interface, in the proceedings of the 12th international conference on applications and tools of Petri Nets, Aarhus, Denmark, June 1991.

15. S. Christensen, L. O. Jepson : Modeling and simulation of a network management system using hierarchical colored Petri nets, Proc. of 1991 Europ. Simulation Multiconference, Copenhagen 1991, Society of Computer Simulation 1991, pp.47–52.

16. I. F. Akyildiz, G. Chiola, D. Kofman and H. Korezlioglu : Stochastic Petri Net Modeling of the FDDI Network Protocol, in Protocol Specification, Testing and Verification, XI, Elsevier Science Publishers B.V. (North Holland) 1991 IFIP.

17. S. Christodoulou, M. Zhou : A Petri Net Approach to Modeling and Performance Analysis of Fiber Data Distributed Interface (FDDI) Network, in proc. of the 1994 IEEE Symposium on Emerging Technologies and Factory Automation, pp. 373–380.

18. H. Clausen and P. R. Jensen : Validation and Performance Analysis of Network Algorithms by Coloured Petri Nets, In Petri Nets and Performance Models, Proc. of the 5th International Workshop, Toulouse, France 1993, IEEE Computer Society Press, 280–289.

19. H. Clausen and P. R. Jensen : Analysis of Usage Parameter Control Algorithms for ATM Networks, In S. Tohme and A. Casaca (eds.): Broadband Communications, II (C-24), Elsevier Science Publishers 1994.

20. D. J. Floreani, J. Billington, A. Dadej : Designing and verification a communication gateway using colored petri nets and design/CPN, 17th International Conf. on Application and Theory of Petri Nets, Lecture Notes on Computer Science, pp. 153–171, Osaka, June 1996.

21. J. B. Jorgensen, K. H. Mortensen: Modeling and analysis of distributed program execution in BETA using colored petri nets, 17th International Conf. on Application and Theory of Petri Nets, Lecture Notes on Computer Science, pp. 249–268, Osaka, June 1996.

22. G. Ciardo, L. Cherkasova, V. Kotov, T. Rokicki : Modeling a scalable high-speed interconnect with stochastic Petri nets, in Proc. of the Sixth International Workshop on Petri Nets and Performance Models, Durham, North Carolina, USA, Oct. 3-6, 1995.

23. K. Jensen : Colored Petri Nets, Basic Concepts, Analysis Methods and Practical Use, Vol.1 and Vol.2, Springer-Verlag 1992, 1994.

24. P. Huber, K. Jensen and R. M. Shapiro : Hierarchies in Coloured Petri Nets, Advances in Petri Nets 1990, LNCS, Vol. **483**, Springer, Berlin Heidelberg New York 1990, pp. 313–341.

25. K. Jensen, S. Christensen, P. Huber, M. Holla Design/CPN. A reference Manual, Computer Science Department, University of Aahrus, 1996.

26. J. Toksvig : Tool support for place flow analysis of Hierarchical CP-nets version 2.0, Technical report, Computer Science Department, University of Aahrus, 1993.

Parameter Region for the Proper Operation of the IEEE 802.2 LLC Type 3 Protocol: A Petri Net Approach

Hong-ju Moon*, Sang Yong Moon, and Wook Hyun Kwon

Control Information Systems Laboratory,
School of Electrical Engineering, Seoul National University
Shilim-dong, Kwanak-gu, Seoul 151-742, KOREA
Phone:+82-2-873-2279, Fax:+82-2-878-8933
hjmoon@kepri.re.kr
sangyong@cislrain.snu.ac.kr
whkwon@cisl.snu.ac.kr

Abstract. This paper derives the parameter region of the IEEE 802.2 LLC type 3 protocol to guarantee a proper operation. The protocol is modeled by a time Petri net and investigated using reachability analysis. Three necessary conditions on the parameters are derived, and then a reduced reachability graph is obtained by applying the necessary conditions to lessen the combinatorial state explosion. From the reduced reachability graph, a necessary and sufficient condition on the parameters is derived to guarantee the proper operation of the LLC type 3 protocol for a link. By using the condition, a procedure to set the parameters of the LLC type 3 protocol is provided at each station in a network.

Key Words: IEEE 802.2, LLC type 3 protocol, proper operation, parameter, Petri net

1 Introduction

The data link layer of any communication architecture must ensure orderly and correct delivery of packets between neighboring nodes in a network. Various protocols have been developed for this purpose. The LLC (logical link control)[1], which is based closely on the HDLC (high-level data link control), is used in the data link layer of MAP (manufacturing automation protocol)[2], TOP (technical and office protocol)[2], and other network protocols of OSI networks and many commercial network products as one of the international standards for the data link layer protocol.

The LLC sublayer can provide either connectionless (called type 1) or connection-oriented (type 2) service. It also provides acknowledged connectionless (type 3) service, which adds some simple functionalities to the basic datagram

---

* H. Moon is now with Korea Electric Power Research Institute, 103-16, Munji-dong, Yuseong-gu, Taejon, 305-380, Korea. hjmoon@kepri.re.kr, phone:+82-42-865-7633

scheme for granting reliable in-sequence data transfers. The type 3 service offers very simple, but also reliable, data transfer capabilities, and is suitable for the networks in which the complexity and the overheads introduced by the upper layers of the full OSI stack are not acceptable, as in the Mini-MAP network[2]. The LLC type 3 protocol is useful in factory automation environments where reliable data transfer is important, but fast data transfer is required and storage space is extremely limited.

The LLC type 3 protocol adopts the stop-and-wait scheme for the error control and the flow control as the alternating bit protocol[3]. The transmitter waits for an acknowledgement after a single transmission. If either a negative acknowledgement arrives or the acknowledgement timer expires, the frame is retransmitted till the maximum number of transmissions is reached. To couple transmissions and acknowledgements in a correct way, one bit sequence number is used. The proper operation of the LLC type 3 protocol must guarantee that a data frame or an acknowledgement be delivered in cases where there is no error during transmission, and not be duplicated.

The proper operation of the LLC type 3 protocol requires that the logical link parameters be set appropriately. When developing a system connected by a communication network, many communication problems may arise due to the inappropriate setting of the parameters in a network. Inappropriate parameters may cause loss of a frame normally transmitted without corruption, or duplication of frames for a single transmission. The values of the logical link parameters are determined on a system-by-system basis, and the parameter region for the proper operation must take into account the communication delay and the relations among the parameters for all possible links in a system.

While there have been some research on the LLC protocol[4–6], the setting of the parameters of the LLC type 3 protocol for proper operations has not yet been studied to the authors' knowledge. [4] and [5] investigated the usage and the performance of the LLC type 2 protocol. [6] discussed about a performance of communications with the LLC type 3 protocol.

Communication protocols are modeled and analyzed by Petri nets in many cases[3, 7, 8]. In this paper, the LLC type 3 protocol is modeled by Merlin's time Petri net[7] as in [8]. The operation of the protocol is investigated by the reachability analysis, which is very popular verification method for Petri nets[9, 10]. The symbolic verification approach[10, 11] is used for the efficient derivation of the proper parameter region. Symbolic reachability graphs are constructed to represent the possible operation of the LLC type 3 protocol depending on the parameter values. The state explosion of the reachability graph is a great problem especially in a time Petri net[10]. To lessen the combinatorial state explosion, this paper derives necessary conditions on the parameters to obtain a reduced reachability graph, and then the overall operation is investigated. By the investigation to the reduced reachability graph, the necessary and sufficient condition on the parameters is derived to guarantee the proper operation of the LLC type 3 protocol.

In Section 2, the LLC type 3 protocol is modeled. In Section 3, the allowable parameter region is derived to guarantee the proper operation. Section 4 concludes the paper.

2 Petri net model of the LLC type 3 protocol

In this section, the LLC type 3 protocol is modeled by Merlin's time Petri net (TPN)[7]. A formal definition of a TPN is given and the LLC type 3 protocol is modeled by a TPN for a link.

A TPN is a 4-tuple $M = (P, T, A, I)$ where: P is a finite set of *places*; T is a finite set of *transitions*; $A \subset (P \times T) \cup (T \times P)$ is a set of *arcs* connecting places and transitions; and $I : T \to \{[a, b] \subset [0, \infty)\}$ assigns a closed firing interval to each transition. If $I(t) = [0, 0]$ for a transition $t \in T$, the transition t is said to be an *immediate* transition. Other transitions are said to be *timed* transitions. The lower and upper bounds of the firing interval $I(t)$ for each transition $t \in T$, will be denoted by $I_l(t)$ and $I_u(t)$.

The *state* of a TPN M is specified by an ordered pair $x = (m, e)$, where $m : P \to N$ is a *marking* assigning a nonnegative integer to each place in M, and $e : P \to [0, \infty)$ is an *elapsed time*. $m(p) = n$ represents that there are n tokens in the place p. $e(p) = s$ represents that s has elapsed after the place p is marked. $e(p)$ becomes zero just after a token exits from the place p and $e(p)$ is not affected when a token enters the place p which has already been marked.

$^{(p)}t$ ($t^{(p)}$) denotes the set of input (output) places of a transition $t \in T$, i.e., $^{(p)}t := \{p \in P | (p, t) \in A\}$ ($t^{(p)} := \{p \in P | (t, p) \in A\}$). A transition $t \in T$ is *enabled* by a marking m when $m(p) \geq 1$ for all $p \in {}^{(p)}t$. A transition $t \in T$ can fire if $e(p) \geq I_l(t)$ for all $p \in {}^{(p)}t$, and t can be enabled without firing as long as $e(p) < I_u(t)$ for some $p \in {}^{(p)}t$. More than a single transition can fire simultaneously if each transition satisfies the firing condition and there is no conflict among those transitions for their input places. For a given state $x = (m, e)$, the firing of a transition set $T_e \subset T$ changes the current state x to a new state $x' = (m', e')$ defined by the following state transition equations:

$$m'(p) = m(p) - N_I(T_e, p) + N_O(T_e, p),$$

where $N_I(T_e, p)$ is the number of transitions $t \in T_e$ such that $p \in {}^{(p)}t$ and $N_O(T_e, p)$ is the number of transitions $t \in T_e$ such that $p \in t^{(p)}$, and

$$e'(p) = \begin{cases} 0 & \text{if } p \in {}^{(p)}t \text{ for some } t \in T_e; \\ e(p) & \text{otherwise.} \end{cases}$$

A state x in a TPN is said to be *reachable* from a state x_0 if there a sequence of transition sets exists, starting at x_0 such that the state eventually becomes x. The *reachability set* $R(M, x_0)$ of a TPN M denotes the set of all states that are reachable from the state x_0. The reachability set can be represented graphically by a reachability graph.

To simply denote a state $x = (m, e)$, the following notation will be used:

$$(k_1 p_1^{T_1}, k_2 p_2^{T_2}, ..., k_l p_l^{T_l}),$$

where $k_i = m(p_i)$, $k_i > 0$ and $\tau_i = e(p_i)$ $(1 \le i \le l)$. Here, l is the number of places which have a nonzero number of tokens at the state x. If $m(p) = 0$ for a place $p \in P$, the place p will not appear in the notation, and if a place $p \in P$ is not an input place for any timed transition, the elapsed time τ for the place p will not appear in the notation.

A TPN $M = (P, T, A, I)$ can be extended with *inhibitor arcs* to increase its modeling power. A TPN with inhibitor arcs is a 5-tuple $M_I = (P, T, A, A_I, I)$, where $A_I \subset P \times T$ is a set of inhibitor arcs. A transition $t \in T$ of a TPN M_I with inhibitor arcs is enabled by a marking m only when $m(p) \ge 1$ for all $p \in {}^{(p)}t$ and $m(p) = 0$ for all $p \in P$ s.t. $(p, t) \in A_I$.

The LLC type 3 protocol for a link is modeled by a TPN which consists of 3 subnets: the sender component, the receiver component, and the transmission line component. A link is uniquely identified by its unique network address built up by concatenating the user data-link service access point (LSAP) address of the sender/receiver node, with the MAC address of the station where the sender/receiver node resides. The state variables of the sender and the receiver for a link are dedicated to the link, and are not affected by the communications on other links.

The sender component waits for an acknowledgement after a single transmission. If either a negative acknowledgement arrives or the acknowledgement timer expires after the acknowledgement time value ($T1$), the frame is retransmitted as long as the maximum number of transmissions ($N4$) is not reached. For the retransmission mechanism, the sender component has the *retry count* to keep the number of transmissions for a specific data send. The receive sequence state variable ($V(RI)$) of the receiver component stores the sequence number of the frame to be received, and is used to detect duplicate frames. $V(RI)$ is destroyed after the expiration of the receive variable lifetime timer associated with the receive lifetime value ($T2$). The transmit sequence state variable ($V(SI)$) of the sender component stores the sequence number of the frame to be transmitted or the outstanding transmission. $V(SI)$ is used to relate a received acknowledgement to the outstanding transmission and allow the receiver to detect duplicate frames. $V(SI)$ is destroyed after the expiration of the transmit variable lifetime timer associated with the transmit lifetime value ($T3$). The values of those logical link parameters are determined on a system-by-system basis.

This paper considers only the acknowledged connectionless-mode service and the reply service is not considered. The reply service operates in the same way as the acknowledged data send service except that a reply is anticipated instead of an acknowledgement. Since the reply service is under the same timer mechanism with the acknowledged data send service, the proper operation of the reply service can be obtained automatically by guaranteeing the proper operation of the acknowledged data send service. This paper assumes transmission lines are unreliable, and that data frames are delivered in sequence or lost during transmission with the probability p_e. Any corrupted data frame is assumed to be thrown away by the lower-level transmission services, and regarded to be lost. The transmission delay is assumed to be bounded by $T_d(> 0)$. The proper operation of the LLC type 3 protocol requires that the logical link parameters $T1$, $T2$, and $T3$ be set appropriately, and this paper considers the setting of $T1$, $T2$, and $T3$ for the proper operation.

Figure 1 shows the TPN model M_S of the sender component. The graphical

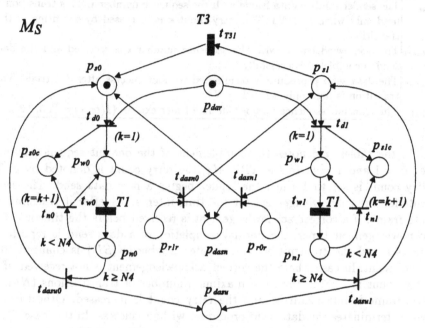

Fig. 1. TPN model M_S for the sender component.

representation of the TPN model follows conventional rules. A thin bar in the graph represents an immediate transition, and a thick bar in the graph represents a timed transition. The interpretations of the places and the transitions of the TPN M_S are explained in Table 1 and Table 2.

Table 1. Places of M_S.

p_{dar}	A data send request from a user has arrived.
p_{si}	The sender is ready to send a data frame with $V(SI) = i$.
p_{wi}	The sender is waiting for the acknowledgement with the sequence number i.
p_{ni}	The timeout of the acknowledgement timer has occurred for the data send with $V(SI) = i$.
p_{sic}	The data frame with the sequence number i is sent.
p_{rir}	The acknowledgement with the sequence number i has arrived.
p_{dasn}	The normal completion of a data send is reported to the user.
p_{dasu}	The unsuccessful completion of a data send is reported to the user.

Table 2. Transitions of M_S.

t_{di}	The sender sends a data frame with the sequence number i. The retry count k is set to 1 by the firing of this transition.
t_{wi}	The sender detects the timeout of the acknowledgement timer ($I(t_{wi}) = [T1, T1]$).
t_{ni}	The sender sends a data frame with the sequence number i. This transition is fired only when $k < N4$. The retry count k is increased by the firing of this transition.
t_{dasni}	The acknowledgement with the sequence number $\bar{i}$ is received and the data send is completed successfully.
t_{dasui}	The data send procedure is terminated without success after $N4$ trials. This transition is fired only when $k \geq N4$.
t_{T31}	The transmit sequence number lifetime timer expires ($I(t_{T31}) = [T3, T3]$).

In the tables, $\bar{i}$ denotes the complement of the one bit sequence number, i.e., $\bar{0} = 1$ and $\bar{1} = 0$. In the TPN M_S, the retry count is denoted by k. The retry count is set to 1 when the sender begins a new data send. The sender waits for the acknowledgement for the time interval $T1$ after it transmits the data frame. If a correct acknowledgement is received before the timeout of the acknowledgement timer, the normal completion of a data send is reported to the user and the transmit sequence state variable ($V(SI)$) is changed to its complement. In case where the correct acknowledgement is not received, if the retry count is smaller than the maximum number of transmissions ($N4$), the data frame is retransmitted and the retry count is increased. Otherwise, the sender terminates the data send procedure without success. In this case, $V(SI)$ is not changed. If $T3$ has elapsed after the last data send, $V(SI)$ is destroyed. Similarly, $V(SI)$ does not exist at the initial stage. These mean that $V(SI)$ is set to 0. p_{dar} is a source place, which a token can enter into at any instant. p_{dasn} and p_{dasu} are sink places, in which tokens exit immediately.

Figure 2 shows the TPN model M_R of the receiver component. In the graph, inhibitor arcs are used for convenience of the modeling. The interpretations of the places and the transitions of the TPN M_R are explained in Table 3 and Table 4.

Initially, the receive sequence state variable ($V(RI)$) does not exist and is destroyed if $T2$ has elapsed after the last data reception. If a data frame is received when $V(RI)$ does not exist, it is regarded as a non-duplicate data and the receiver acknowledges to the sender. In this case, the receipt of new data is reported to the user and $V(RI)$ is set to the complement of the sequence number of the received data frame. If a data frame is received when $V(RI)$ exists, the sequence number of the received data frame is compared with $V(RI)$. If the comparison shows equality, the received data frame is recognized as a non-duplicate one. In this case, the same operation occurs as the case where a data frame is received when $V(RI)$ does not exist. If the comparison shows inequality, the received data frame is recognized to be a duplication of the most

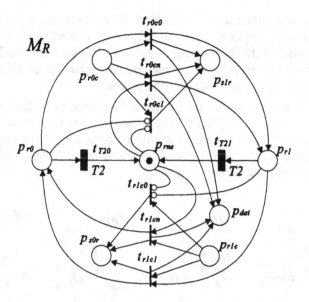

Fig. 2. TPN model M_R for the receiver component.

Table 3. Places of M_R.

p_{rne}	$V(RI)$ does not exist and the receiver is ready to receive a data frame with any sequence number.
p_{ri}	$V(RI)$ is i and the receiver is ready to receive a data frame with the sequence number i.
p_{ric}	A data frame with the sequence number i has arrived.
p_{sir}	An acknowledgement with the sequence number i is sent.
p_{dai}	The arrival of a new data is reported to the user.

Table 4. Transitions of M_R.

t_{rici}	The receiver receives a data frame with the sequence number i when $V(RI) = i$. In addition, the receiver sends the acknowledgement with the sequence number $\bar{i}$ and reports the arrival of new data to the user.
t_{ricn}	The receiver receives a data frame with the sequence number i when $V(RI)$ does not exist. In addition, the receiver sends the acknowledgement with the sequence number $\bar{i}$ and reports the arrival of new data to the user.
$t_{ric\bar{i}}$	The receiver receives a data frame with the sequence number i when $V(RI) = \bar{i}$. The receiver sends the acknowledgement with the sequence number $\bar{i}$, but does not report the arrival of the data to the user.
t_{T2i}	The receive sequence number lifetime timer expires when $V(RI) = i$ ($I(t_{T2i}) = [T2, T2]$).

recently received data frame. The arrival of the data frame is not reported to the user, but the reception of the data frame is acknowledged. In this case, the acknowledgement with the sequence number $V(RI)$ and the associated timer are not affected by the reception of a duplicate data frame. p_{r0c} and p_{r1c} are source places, and p_{s0r}, p_{s1r}, and p_{dai} are sink places.

The TPN model for the transmission line consists of four subnets: M_{0C}, M_{1C}, M_{0R}, and M_{1R}. They model the transmissions of the data frame with the sequence number 0 and 1, and the acknowledgement with the sequence number 0 and 1, respectively. Figure 3 shows the TPN models M_{0C}, M_{1C}, M_{0R}, and M_{1R} of the transmission line component.

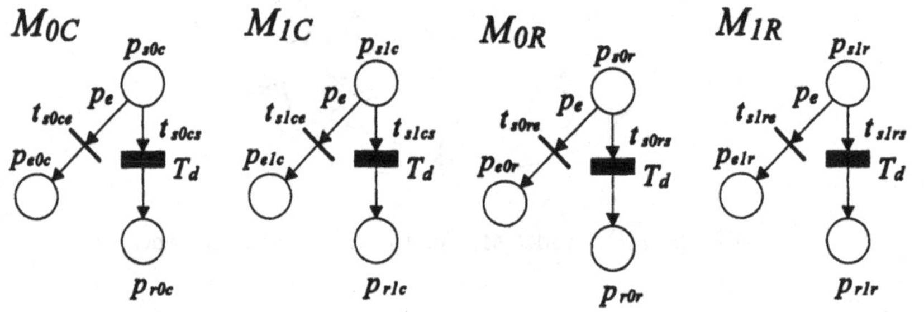

Fig. 3. TPN models M_{0C}, M_{1C}, M_{0R}, and M_{1R} of the transmission line component.

Table 5. Places of M_{0C}, M_{1C}, M_{0R}, and M_{1R}.

p_{sic}	The data frame with the sequence number i has been passed to the transmission line and is being sent to the receiver.
p_{ric}	The data frame with the sequence number i has been delivered to the receiver successfully.
p_{eic}	The acknowledgement with the sequence number i has been lost during transmission.
p_{sir}	The acknowledgement with the sequence number i has been passed to the transmission line and is being sent to the receiver.
p_{rir}	The acknowledgement with the sequence number i has been delivered to the receiver successfully.
p_{eir}	The acknowledgement with the sequence number i has been lost during the transmission.

The interpretations of the places and the transitions of the TPN M_{0C}, M_{1C}, M_{0R}, and M_{1R} are explained in Table 5 and Table 6. p_{s0c}, p_{s1c}, p_{s0r}, and p_{s1r} are source places, and p_{e0c}, p_{e1c}, p_{e0r}, and p_{e1r} are sink places.

The TPN model representing LLC type 3 protocol is obtained by merging TPNs M_S, M_R, M_{0C}, M_{1C}, M_{0R}, and M_{1R}. Figure 4 shows the external representation and their composition topology of the TPNs for the sender component, receiver component, and the transmission line component. The places

Table 6. Transitions of M_{0C}, M_{1C}, M_{0R}, and M_{1R}.

t_{sics}	The data frame with the sequence number i is transmitted correctly ($I(t_{sics}) = [0, T_d]$).
t_{sice}	The data frame with the sequence number i is lost. This transition fires with the probability p_e.
t_{sirs}	The acknowledgement with the sequence number i is transmitted correctly.
t_{sire}	The acknowledgement with the sequence number i is lost. This transition fires with the probability p_e.

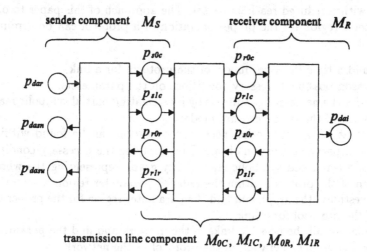

Fig. 4. The merged TPN model M_{LLC3} for the LLC type 3 protocol.

having identical names are merged into a single place. The merged TPN model for the LLC type 3 protocol is denoted by M_{LLC3}. Based on this TPN model, the parameter region for the proper operation of the LLC type 3 protocol is derived using reachability analysis.

3 Parameter region for the proper operation

The requirements for the proper operation of a protocol can be assumed in several ways[3, 12]. The following properties are considered as requirements for the proper operation of the LLC type 3 protocol:

1. An acknowledgement from the receiver must be noticed by the sender in the case where there is no error during transmission.
2. If a data frame is sent by a user, the corresponding user must receive it in the case where there is no error during transmission.
3. A data frame must not be duplicated by retransmission.
4. If a data frame is not delivered successfully, it must be noticed by the sender.

These requirements concern the safety of the operation of the protocol. Other properties such as liveness are presumed to be satisfied, and are not considered in this paper.

The above requirements can be achieved only when the parameters of the LLC type 3 protocol are set appropriately. The operations of the LLC type 3 protocol with various parameter values can be investigated by reachability analysis for the TPN model M_{LLC3}, presented in Section 2. However, the TPN model M_{LLC3} has many states, and its reachability graph is prone to combinatorial state explosion. Therefore, necessary conditions on the parameters for the proper operation are considered first, and then the overall operation is investigated with a reduced reachability set. The approach of this paper to obtain the parameter region for the proper operation of a protocol can be summarized as follows:

1. Build a time Petri net model of the protocol for a link.
2. Reason possible necessary conditions on the parameters.
3. Verify the necessary conditions by investigating partial symbolic reachability graphs of the time Petri net model.
4. Repeat (2)-(3) if another necessary condition can be found within the reduced parameter region obtained by applying the necessary conditions.
5. Build a reduced symbolic reachability graph representing the whole operation of the protocol within the reduced parameter region.
6. Investigate the graph and find the parameter region for the proper operation of the protocol for a link.
7. Consider all the possible links in the network and find the parameter region for the proper operation of the protocol in the network.

Now, a necessary condition on the timeout value of the acknowledgement timer is considered. As presented in the following lemma, the value must be restricted to prevent spurious timeout before arrival of a normal acknowledgement.

Lemma 1. *If $T1 \leq 2T_d$, the requirement (1) cannot be satisfied.*

Proof. The proof refers to the reachability graph shown in Figure 5. The figure shows only a part of the whole reachability graph of M_{LLC3}. The reachability graph represents the operation of M_{LLC3} which can occur when a data send request arrives at the state (p_{s0}, p_{rne}).

At the state $(p_{w0}^0, p_{rne}, p_{s0c}^0)$, the transition t_{w0} can fire if $T1 \leq T_d$. This is an inadequate early timeout, and the acknowledgement cannot be noticed properly. If the transmission delay τ_1 is smaller than $T1$, the state of M_{LLC3} is changed to the state $(p_{w0}^{\tau_1}, p_{rne}, p_{r0c})$, and thus, to the state $(p_{w0}^{\tau_1}, p_{r1}^0, p_{s1r}^0, p_{dai})$. Since the token in the place p_{dai} exits the place immediately, the state becomes $(p_{w0}^{\tau_1}, p_{r1}^0, p_{s1r}^0)$. The state $(p_{w0}^{\tau_1}, p_{r1}^0, p_{s1r}^0)$ is not drawn in the reachability graph for simplicity since it always follows the state $(p_{w0}^{\tau_1}, p_{r1}^0, p_{s1r}^0, p_{dai})$ in sequence and can be thought as this state.

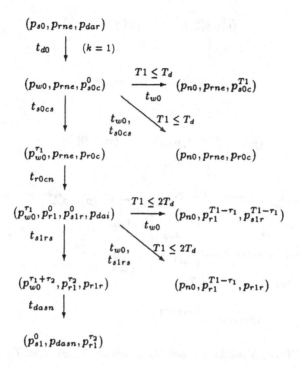

Fig. 5. A partial reachability graph of M_{LLC3} when $T1 \leq 2T_d$.

Let τ_2 denote the transmission delay of the acknowledgement at the state $(p_{w0}^{\tau_1}, p_{r1}^0, p_{s1r}^0, p_{dai})$. At this state, the transition t_{w0} can fire before the transitions t_{s1rs} and t_{dasn} occur if $T1 \leq \tau_1 + \tau_2$. This is also an inadequate early timeout, and the acknowledgement cannot be noticed properly. Since $\tau_1, \tau_2 \leq T_d$, the requirement (1) cannot be satisfied if $T1 \leq 2T_d$. $\qquad\square$

Next, it is shown that the timeout values regarding the destructions of the receive sequence state variable $V(RI)$ and the transmit sequence state variable $V(SI)$ must be restricted to maintain consistency of the sequence numbers in a link.

Lemma 2. If $T2 \geq T3$, the requirement (2) cannot be satisfied.

Proof. The proof refers to the reachability graph shown in Figure 6. The figure shows only a part of the whole reachability graph of M_{LLC3}. The reachability graph represents the operation of M_{LLC3} which can occur when a data send request arrives after some time has elapsed from the state $(p_{s1}^0, p_{r1}^{\tau_0})$. To simplify the reachability graph, the following notation is used:

$$p_{rxi}^\tau = \begin{cases} p_{rne} & \tau > T2; \\ p_{rne} \text{ or } p_{ri}^\tau & \tau = T2; \\ p_{ri}^\tau & \text{otherwise} \end{cases} \qquad i = 0, 1.$$

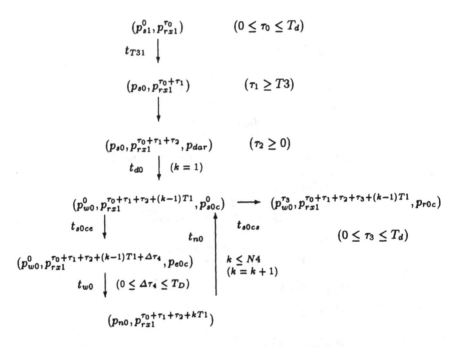

Fig. 6. A partial reachability graph of M_{LLC3} when $T2 \geq T3$.

The firing of the transitions t_{T20} and t_{T21} are not shown in the reachability graph.

$V(SI)$ is set to 1 after the acknowledgement is received from the receiver, and the transmission delay τ_0 $(0 \leq \tau_0 \leq T_d)$ has passed after $V(RI)$ is set to 1. If no data send request arrives during $\tau_1 (\geq T3)$ after $V(SI)$ was set to 1, $V(SI)$ is changed to 0, and the state of M_{LLC3} is changed from the state $(p_{s1}^0, p_{rx1}^{\tau_0})$ to the state $(p_{s0}, p_{rx1}^{\tau_0+\tau_1})$. If a data send request arrives after $\tau_2 (\geq 0)$ has elapsed, the state of M_{LLC3} is changed to $(p_{w0}^0, p_{rx1}^{\tau_0+\tau_1+\tau_2+(k-1)T1}, p_{s0c}^0)$ after the firing of the transition t_{d0}. $\tau_3 (0 \leq \tau_3 \leq T_d)$ denotes the transmission delay for the delivery of the data frame to the receiver.

At the state $(p_{w0}^{\tau_3}, p_{rx1}^{\tau_0+\tau_1+\tau_2+\tau_3+(k-1)T1}, p_{r0c})$, $p_{rx1}^{\tau_0+\tau_1+\tau_2+\tau_3+(k-1)T1}$ is either p_{rne} if $\tau_0 + \tau_1 + \tau_2 + \tau_3 + (k-1)T1 > T2$ or $p_{r1}^{\tau_0+\tau_1+\tau_2+\tau_3+(k-1)T1}$ otherwise. Therefore, the data frame cannot be received properly if $\tau_0 + \tau_1 + \tau_2 + \tau_3 + (k-1)T1 \leq T2$. Since $\tau_0 \geq 0$, $\tau_1 \geq T3$, $\tau_2 \geq 0$, $\tau_3 \geq 0$, and $k \geq 1$, the minimum of $\tau_0 + \tau_1 + \tau_2 + \tau_3 + (k-1)T1$ is $0 + T3 + 0 + 0 \cdot T1 = T3$. Therefore, the requirement (2) cannot be satisfied if $T2 \geq T3$. □

Thirdly, the timeout value for the receive sequence state variable $V(RI)$ must be restricted to prevent duplicated reception for a single data send.

Lemma 3. *If $T2 \leq (N4-1)T1 + T_d$, the requirement (3) cannot be satisfied.*

Proof. The proof refers to the reachability graph shown in Figure 7. The figure shows only a part of the whole reachability graph of M_{LLC3}. The reachability graph represents the operation of M_{LLC3} which can occur when a data send request arrives at the state (p_{s0}, p_{rne}). The data frame is delivered to the remote node successfully at the first attempt, but the acknowledgement is lost during transmission, and consequently the sender retries to send the data frame. In addition, all the data frames are lost during transmission till the $(N4-1)$-th transmission. The reachability graph represents the operation that the data frame is delivered to the remote node at the $N4$-th transmission. Let τ_1 denote the transmission delay at the first transmission of the data frame and τ_2 denote the transmission delay at the $N4$-th transmission of the data frame. At the $N4$-th transmission, the state of M_{LLC3} becomes $(p_{w0}^{\tau_2}, p_{r1}^{(N4-1)T1-\tau_1+\tau_2}, p_{r0c})$ when $(N4-1)T1 - \tau_1 + \tau_2 < T2$ or otherwise $(p_{w0}^{\tau_2}, p_{rne}, p_{r0c})$. If the state is $(p_{w0}^{\tau_2}, p_{rne}, p_{r0c})$, the transition t_{r0cn} fires. This means that the receiver recognized the data frame as a duplicate one.

Since $-T_d \leq -\tau_1 + \tau_2 \leq T_d$, it may be $(N4-1)T1 - \tau_1 + \tau_2 \geq T2$ if $T2 \leq (N4-1)T1 + T_d$. Therefore, the requirement (3) cannot be satisfied if $T2 \leq (N4-1)T1 + T_d$. $\qquad\square$

By applying Lemma 1, Lemma 2, and Lemma 3, the parameter region is restricted and a reduced reachability set is obtained. The allowable parameter

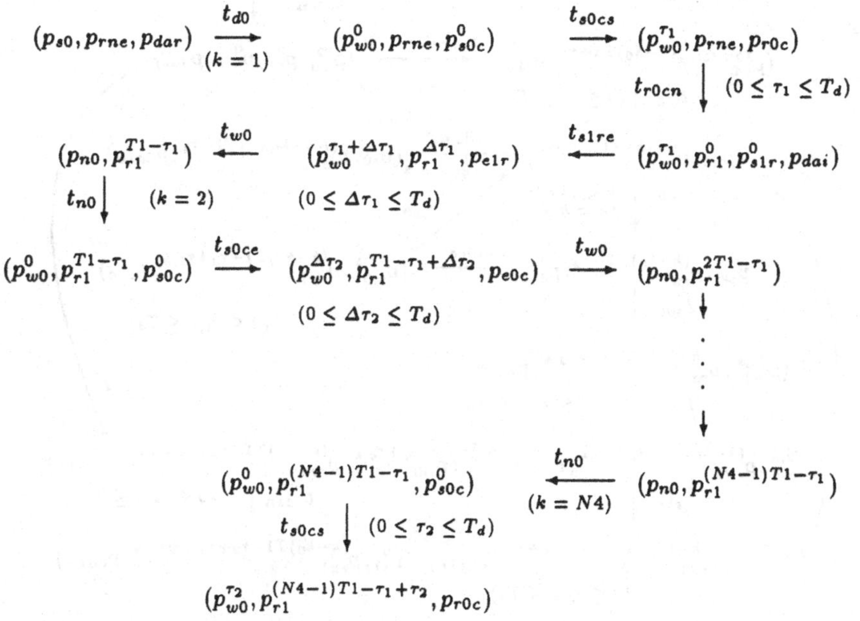

Fig. 7. A partial reachability graph of M_{LLC3} when $T2 \leq (N4-1)T1 + T_d$.

region to guarantee the proper operation is derived for a link by investigating the reduced reachability set as follows:

Theorem 1. *The requirements (1), (2), (3), and (4) can be achieved if and only if the following conditions on the parameters are satisfied:*

1. $T1 > 2T_d$.
2. $T2 < T3$.
3. $T2 > (N4 - 1)T1 + T_d$.

Proof. The *only if* relation of the theorem is proved by using Lemma 1, Lemma 2, and Lemma 3. Since the requirements (1), (2), or (3) are not all achieved if one of the conditions in the theorem is not satisfied, the *only if* relation of the theorem holds.

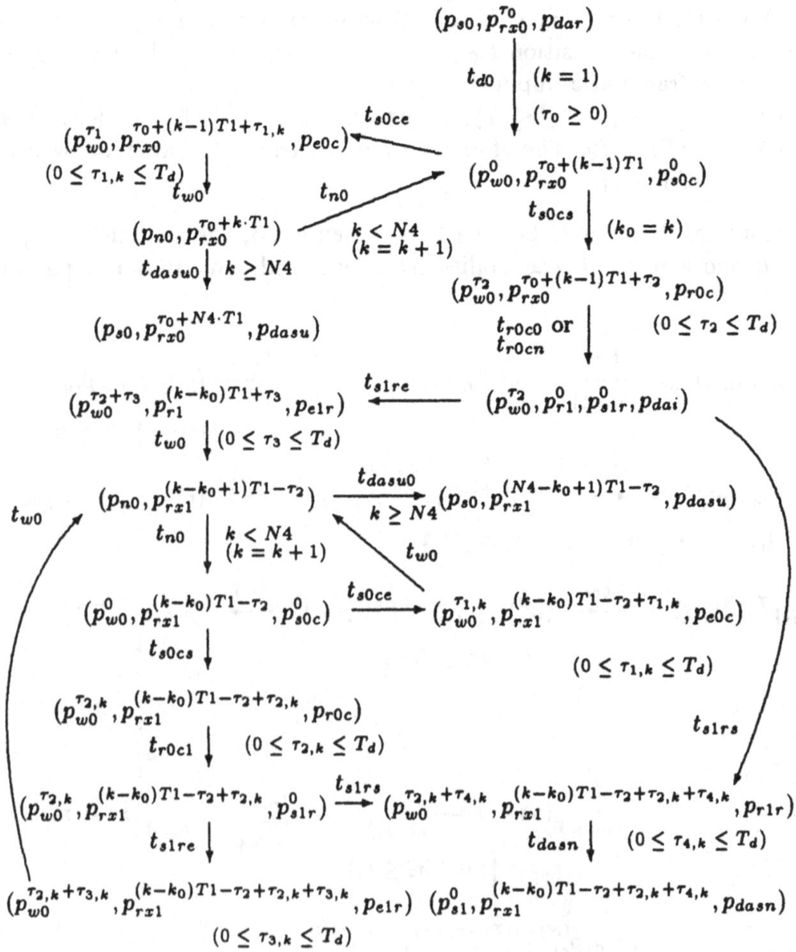

Fig. 8. The reachability graph of M_{LLC3} with appropriate parameters when the initial $V(SI)$ equals 0.

The *if* relation of the theorem is proved with reference to the reachability graphs shown in Figure 8 and Figure 9. The reachability graphs shown in Figure 8 and Figure 9 represent the operation of the LLC type 3 protocol with the parameters adjusted as above, when $V(SI) = 0$ and $V(SI) = 1$, respectively. Normally, $V(SI)$ and $V(RI)$ coincide, and the state of M_{LLC3} is (p_{s0}, p_{rx0}^{τ}) or $(p_{s1}^0, p_{rx1}^{\tau})$ ($\tau \geq 0$) when there is no data send.

The reachability graph shown in Figure 8 represents the situation that the data send request arrives when τ_0 ($\tau_0 \geq 0$) has elapsed after $V(RI)$ has been set to 0. In the reachability graph, k_0 is used to denote that the data frame is delivered to the remote node successfully for the first time at the k_0-th transmission.

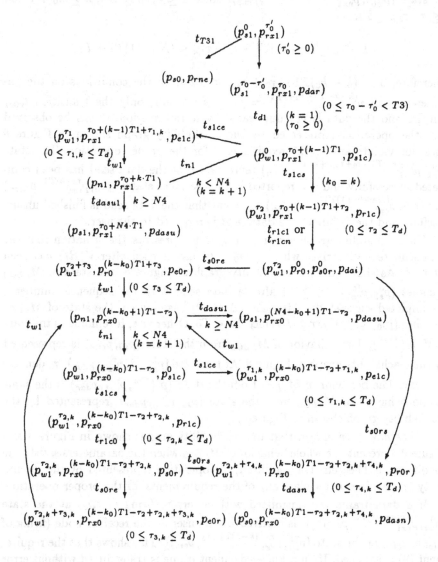

Fig. 9. The reachability graph of M_{LLC3} with appropriate parameters when the initial $V(SI)$ equals 1.

k is used to denote the transmission delays at the k-th trail for a data send. $\tau_{1,k}$ and $\tau_{3,k}$ $(0 \leq \tau_{1,k}, \tau_{3,k} \leq T_d)$ denote the transmission delays when a data frame or an acknowledgement is lost during transmission, respectively. $\tau_{2,k}$ and $\tau_{4,k}$ $(0 \leq \tau_{2,k}, \tau_{4,k} \leq T_d)$ denote the transmission delays when a data frame or an acknowledgement is delivered successfully, respectively. τ_2 denotes τ_{2,k_0} and τ_3 denotes τ_{3,k_0}.

At the state $(p_{w0}^{\tau_2}, p_{rx0}^{\tau_0+(k-1)T1+\tau_2}, p_{r0c})$, $p_{rx0}^{\tau_0+(k-1)T1+\tau_2}$ may be p_{rne} if $\tau_0 + (k-1)T1 + \tau_2 \geq T2$ and $p_{rx0}^{\tau_0+(k-1)T1+\tau_2}$ may be p_{r0} if $\tau_0 + (k-1)T1 + \tau_2 \leq T2$. In each case, the transitions t_{r0c0} and t_{r0cn} are fired respectively. Now consider the state $(p_{w0}^{\tau_{2,k}}, p_{rx1}^{(k-k_0)T1-\tau_2+\tau_{2,k}}, p_{r0c})$. Since $1 \leq k, k_0 \leq N4$, $k \geq k_0 + 1$ and $0 \leq \tau_2, \tau_{2,k} \geq T_d$,

$$T1 - T_d \leq (k - k_0)T1 - \tau_2 + \tau_{2,k} \leq (N4 - 1)T1 + T_d.$$

Therefore, $0 < (k - k_0)T1 - \tau_2 + \tau_{2,k} < T2$ from the conditions on the parameters, and $p_{rx1}^{(k-k_0)T1-\tau_2+\tau_{2,k}} = p_{r1}$. As a result, only the transition t_{r0c1} can fire and the data is not duplicated by retransmission. It can be observed that the operation represented by the reachability graph shown in Figure 8 does not violate any of the requirements for the proper operation. The state $(p_{s1}^0, p_{rx1}^{(k-k_0)T1-\tau_2+\tau_{2,k}+\tau_{4,k}}, p_{dasn})$ represents that the data send has been completed successfully and it is reported to the user. The states $(p_{s0}, p_{rx0}^{\tau_0+k \cdot T1}, p_{dasu})$ and $(p_{s0}, p_{rx1}^{(N4-k_0+1)T1-\tau_2}, p_{dasu})$ represent that the data send is finished unsuccessfully and the failure of the data send is reported to the user.

The reachability graph shown in Figure 9 represents the situation that the data send request arrives when τ_0 $(\tau_0 \geq 0)$ has elapsed after $V(RI)$ has been set to 1. As shown in the reachability graph of Figure 8, the state of M_{LLC3} becomes $(p_{s1}^0, p_{rx1}^{\tau_0'})$ $(\tau_0' \geq 0)$ after a data send with the sequence number 0 is completed successfully. After $\Delta\tau$ $(\Delta\tau \geq 0)$ has elapsed, the state of M_{LLC3} becomes (p_{s0}, p_{rne}) if $\Delta\tau \geq T3$ since $T2 < T3$. Otherwise, it remains as the state $(p_{s1}^{\Delta\tau}, p_{rx1}^{\tau_0'+\Delta\tau})$. The behavior of M_{LLC3} from the state (p_{s0}, p_{rne}) is represented by the reachability graph of Figure 8. At the state $(p_{s1}^{\tau_0-\tau_0'}, p_{rx1}^{\tau}, p_{dar})$, τ_0 denotes $\tau_0' + \Delta\tau$. The behavior of M_{LLC3} from the state $(p_{s1}^{\tau_0-\tau_0'}, p_{rx1}^{\tau}, p_{dar})$ is the same as the behavior of M_{LLC3} from the state $(p_{s0}, p_{rx0}^{\tau_0}, p_{dar})$, represented by the reachability graph shown in Figure 8.

It has now been shown that the reachability graphs shown in Figure 8 and Figure 9 represent the whole behavior of M_{LLC3} when the parameters satisfy the conditions in Theorem 1. In addition, the operation represented by the reachability graphs does not violate any of the requirements for the proper operation.

If a data frame is transmitted without error (firing of t_{sics} at the state $(p_{wi}^0, p_{rxi}^{\tau_0+(k-1)T1}, p_{sic}^0)$), it is received by the user at the receiver node (firing of t_{rici} or t_{ricn} at the state $(p_{wi}^{\tau_2}, p_{rxi}^{\tau_0+(k-1)T1+\tau_2}, p_{ric})$). This shows that the requirement (2) is satisfied. If an acknowledgement frame is transmitted without error (firing of t_{sirs} at the state $(p_{wi}^{\tau_2}, p_{ri}^0, p_{sir}^0, p_{dai})$ or $(p_{wi}^{\tau_{2,k}}, p_{rxi}^{(k-k_0)T1-\tau_2+\tau_{2,k}}, p_{sir}^0)$), the acknowledgement from the receiver is noticed by the sender (firing of t_{dasn} at the state $(p_{w\bar{i}}^{\tau_{2,k}+\tau_{4,k}}, p_{rxi}^{(k-k_0)T1-\tau_2+\tau_{2,k}+\tau_{4,k}}, p_{rir}))$. This shows that the requirement (1) is satisfied. Once a data frame is received by the user at the receiver node (firing of t_{rici} or t_{ricn}), the receiver does not receive the retransmitted data frame (no firing of t_{rici} or t_{ricn}). This shows that the requirement (3) is satisfied. Whenever a data send procedure is terminated without firing of t_{dasn}, the transition t_{dasui} fires, and the failure of the data send is noticed by the sender. This shows that the requirement (4) is satisfied.

It has now been shown that the requirements (1), (2), (3), and (4) are achieved when the parameters satisfy the conditions in Theorem 1. This proves the *if* relation of the theorem. □

The state $(p_{s0}, p_{rx1}^{(N4-k_0+1)T1-\tau_2}, p_{dasu})$ and the state $(p_{s1}, p_{rx0}^{(N4-k_0+1)T1-\tau_2}, p_{dasu})$ represent the situation that the data frame is delivered to the remote node, but its acknowledgement is not delivered to the sender successfully. In these cases, $V(SI)$ and $V(RI)$ do not coincide. Therefore, the user at the sender node must reset the link and resynchronize $V(SI)$ and $V(RI)$ for the link when the failure of a data send is reported.

From this condition, it can be observed that the proper operation cannot be guaranteed when the LLC type 3 protocol is applied with a medium access method which does not provide a bounded transmission delay such as CSMA/CD (carrier sense multiple access with collision detection).

Till now, only a single link has been considered. Since all the links established in a station are affected by a single parameter set of the LLC type 3 protocol, the parameters in a station should be set with consideration for the parameters of the other stations that may establish links with it. Let $S_n = \{i|1 \leq i \leq n\}$ denote the set of stations in a network and $L = \{(j,k)|(j,k) \in S_n \times S_n$ and $j \neq k\}$ denote the set of ordered pairs of stations that may establish links. (j,k) denotes that a station j may establish a link with a station k to send data to the station k. $T1(i)$, $T2(i)$, $T3(i)$, and $N4(i)$ denote the parameters of the LLC type 3 protocol at a station $i(\in S_n)$, and $T_d(i,j)$ denote the transmission delay between stations i, j $(\in S_n, i \neq j)$ $(T_d(i,j) = T_d(j,i)$ for any $i,j \in S_n)$. The parameters of the LLC type 3 protocol at each station in the network can be chosen as follows using Theorem 1.

Step 1 Choose $N4(i)$s large enough to overcome the loss of a frame during transmission for each $i \in S_n$.

Step 2 Choose $T1(i)$s to satisfy the following condition for each $i \in S_n$:

$$T1(i) > 2T_d(i,j) \text{ for all } j \in S_n, \text{ s.t. } (i,j) \in L.$$

Step 3 Choose $T2(i)$s to satisfy the following condition for each $i \in S_n$:

$$T2(i) > (N4(j) - 1)T1(j) + T_d(i,j) \text{ for all } j \in S_n, \text{ s.t. } (j,i) \in L.$$

Step 4 Choose $T3(i)$s to satisfy the following condition for each $i \in S_n$:

$$T3(i) > T2(j) \text{ for all } j \in S_n, \text{ s.t. } (i,j) \in L.$$

In practice, there should be safety margins in the above conditions on the parameters. The safety margins shall take various factors into account, including the timer resolution and the variations in the processing time.

The approach used in the paper works well for finding the parameter region to guarantee the proper operation of the LLC type 3 protocol. This approach can be applied to other protocols by joint use of reasoning and enumerative techniques. Especially with authoritative knowledge and experience for the protocol, the approach can be used efficiently to find proper parameter values.

4 Conclusion

This paper derived the allowable parameter region of the IEEE 802.2 LLC type 3 protocol to guarantee proper operation using a Petri net approach.

The protocol was modeled by a time Petri net and investigated using the reachability analysis. To lessen the combinatorial state explosion of the reachability graph, three necessary conditions on the parameters were derived and a reduced reachability graph was obtained. The overall operation was investigated with the reduced reachability graph, and a necessary and sufficient condition on the parameters was derived to achieve the proper operation. Using the condition, a procedure was presented for setting the parameters at each station in a network to guarantee the proper operation.

The result provides a way to set the parameters of the LLC type 3 protocol appropriately to guarantee proper operation when developing a system connected by a communication network based on the LLC type 3 protocol. The result shows that proper operation cannot be guaranteed with the LLC type 3 protocol when it is applied with a medium access method which does not provide a bounded transmission delay such as CSMA/CD. This study also shows that to obtain the proper operation of the LLC type 3 protocol, the user at the sender node must reset the link and resynchronize the sequence state variables when the failure of a data send is reported.

The approach used in the study can be applied to other protocols efficiently especially with authoritative knowledge and experience for the protocols. It will be also useful to investigate the parameter region of the LLC type 2 protocol using the approach used in this study for the LLC type 3 protocol.

References

1. *ISO/IEC 8802-2: Logical Link Control*, IEEE, Inc., 1994.
2. A. Valenzano, C. Demartini, and L. Ciminiera, *MAP and TOP Communications: Standards and Applications*, Addison-Wesley, 1992.
3. Ichiro Suzuki, "Formal Analysis of the Alternating Bit Protocol by Temporal Petri Nets," *IEEE Trans. on Software Engineering*, Vol. 16, No. 11, Nov. 1990, pp. 1273-1281.
4. Ernst W. Biersack, "Performance of the IEEE 802.2 Type-2 Logical Link Protocol with Selective Retransmission," *IEEE Trans. on Comm.* Vol. 41, No. 2, Feb. 1993, pp. 291-294.

5. H. K. Pung, "Effects of window flow control on the 802.2 Type-II logical link performance in ArbNet," *Computer Communications*, Vol. 16, No. 7, July 1993, pp. 403-412.

6. Abd E. Elnakhal and Helmut Rzehak, "Design and Performance Evaluation of Real Time Communication Architectures," *IEEE Trans. on Industrial Electronics*, Vol. 40, No. 4, Aug. 1993, pp. 404-411.

7. P. Merlin and D. J. Faber, "Recoverability of communication protocols," *IEEE Trans. Commun.*, Vol. COM-24, No. 9, Sept. 1976, pp. 1036-1043.

8. B. Berthomieu and M. Diaz, "Modeling and Verification of Time Dependent Systems Using Time Petri Nets," *IEEE Transactions on Software Engineering*, Vol. 17, No. 3, Mar, 1991, pp. 259-273.

9. Nancy G. Leveson and Janice L. Stolzy, "Safety Analysis Using Petri Nets," *IEEE Trans. on Software Engineering*, Vol. SE-13, No. 3, March 1987, pp. 386-397.

10. Giacomo Bucci and Enrico Vicario, "Compositional Validation of Time-Critical Systems Using Communicating Time Petri Nets," *IEEE Trans. on Software Engineering*, Vol. 21, No. 12, December 1995, pp. 969-992.

11. Rajeev Alur, Thomas A. Henzinger, and Pei-Hsin Ho, "Automatic Symbolic Verification of Embedded Systems," *IEEE Trans. on Software Engineering*, Vol. 22, No. 3, March 1996, pp. 181-201.

12. R. L. Schwartz and P. M. Melliar-Smith, "From State Machines to Temporal Logic: Specification Methods for Protocol Standards," *IEEE Trans. on Commun.*, Vol. COM-30, No. 12, Dec. 1982, pp. 2486-2496.

Timed Petri Net Models of ATM LANs

M. Reid and W.M. Zuberek

Department of Computer Science, Memorial University of Nfld
St.John's, Canada A1B 3X5

Abstract. The Asynchronous Transfer Mode (ATM) is a fast packet–switching communication method using small fixed-length cells. A model of an ATM LAN is presented which provides a realistic representation of data transmission by modeling both the ATM network and the applications running over it. Colored Petri nets are used to create a compact model that is capable of representing a variety of different protocols at a high level of detail. The model is designed to allow easy reconfiguration or addition of details at different levels of the system. Simulation is used to evaluate the performance of the modeled system, and some results are compared to actual data gathered from the campus network at Memorial University.

1 Introduction

The basic premise of ATM (Asynchronous Transfer Mode) [8] is that information of all types (i.e., video, audio, data) is divided into small fixed-length data units (cells), which can then be sent across switching networks to be recombined at the receiving end. The challenge for an ATM communication system designer is to ensure that this cell–based communication system can provide correct behavior for a video stream as well as a phone call.

Since many of the performance promises of ATM depend on its ability to transmit cells at high rates with little or no loss, much of the research has focused on switch design [4], with the input data represented by fairly simple models. However, ATM systems are now starting to appear as data network backbones. One particular method of using ATM to transmit data is LAN emulation, which attempts to simulate an Ethernet in such a way that the user is unaware that the ATM network exists. There has not been a great deal of study on how such an ATM backbone will behave under a real network load, or how ATM will affect the applications using it.

This paper presents a model of an ATM LAN that provides a realistic model of data transmission over ATM. It does so by modeling the applications that are running over the ATM LAN, as well as the ATM network itself. Most current research uses simple stochastic state–based models or queueing models to describe an ATM network; such approaches, however, ignore the synchronization aspects of network protocols. This synchronization can easily be represented in Petri net models [21, 14]. Furthermore, Petri net models can also represents the applications directly in terms of different protocols and numbers of active users, and this permits the designer to estimate the impact of these variables on an ATM backbone.

An analysis of network behavior on the campus of Memorial University has shown that while network protocols can exhibit quite complex behavior under specific situations, much of this behavior does not appear on a LAN in relatively non–congested periods of operation. If the net behavior under congested conditions is less important than detecting when congestion might occur, then many elements of protocol behavior can be ignored. The remaining behavior is remarkably similar across a set of protocols carried at Memorial's network.

To capture this similarity of behavior while still permitting the individual protocols to act independently, the model is based on timed Petri nets with deterministic and exponentially distributed firing times [25]. For modeling systems that contain similar components, colored Petri nets are quite convenient because similar components can essentially be superimposed on one another, and distinguished by token attributes (called colors). The colored tokens can operate independently from one another or interact, as required by the modeled system. The structure of a colored net model represents the basic behavior common to all components (in this case, protocols), while the different colors are used to model the differences from the common model, such as temporal characteristics or packet sizes of individual protocols.

The model presented in this paper is modular, with well defined boundaries between modules, and with modules corresponding to easily identifiable entities of the modeled LAN. The modules can be modified independently from each other, and re–configured into different network models to reflect different structures of physical entities and protocol stacks in the modeled networks. The modular structure of the model can also be used to obtain approximate, preliminary results from a simplified model, in which some of the modules are replaced by simple elements with 'typical' or 'average' properties.

The approach to building the resulting composite model, the attention paid to modularity, and the focus on capturing interactions between applications and protocols become visible and important only when the entire network, not just its component entities, is considered. This composite model provides a very useful template for simple modeling of communication networks. Following this template will require only minor modeling effort, while the returns in terms of results can be significant if the model is validated properly and the results interpreted thoughtfully.

Although analytical solutions are possible for many classes of Petri nets, the model uses a number of extensions which prevent such analysis. Therefore, simulation is used to evaluate the performance of the model [26]. Once the model is validated (by comparing its performance to the real data collected by monitoring the campus network), it can be used for different performance studies, including:

- investigation of the effect of an increased load of a particular protocol on the network load by other protocols,
- investigation of the effects of additional network elements (bridges, switches) on the performance of the network,

- identification of network bottlenecks,
- optimization of network structure to maximize its performance, and so on.

Some simple results are included as an illustration of such studies.

The description of the proposed net model is rather informal as the emphasis of this paper is on a structured modeling approach rather than on the derivation of a specific model. Consequently, some more detailed aspects of the model are not addressed here; more details and other simulation results can be found in [20].

The paper is organized in five sections. Section 2 describes the modeled environment which corresponds to the campus network at Memorial University. Section 3 recalls basic concepts of timed Petri nets and then discusses the Petri net model of the network. Some performance results are presented in Section 4, while Section 5 contains a short discussion of the proposed approach and a number of concluding remarks.

2 Modeled Environment

The protocols and data used in this model correspond to the communication network implemented on the Memorial University campus. The protocols can be represented by the protocol stack shown in Fig.2.1.

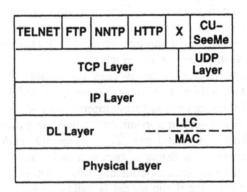

Fig.2.1. Protocol stack.

The protocols (from top) include:

- Application protocols, which usually act as the direct point of contact between the user and the network; these protocols include:
  - TELNET, providing a remote terminal connection from one host to another; TELNET is normally used for text–based interactive computing; RLOGIN, which performs a similar function as TELNET, is combined with TELNET in this study;

- FTP, probably the most commonly used application for transferring files from one host to another;
- NNTP, used for transferring USENET articles from host to host;
- HTTP, used to transmit most of the information on WWW, although some other protocols are used as well;
- X WINDOWS: X is a client–server based system for the management of remote graphics displays;
- CU–SEEMEE, a video conferencing application designed (at Cornell University) to work over the Internet.

– TCP/IP, a suite of networking protocols that have gained wide acceptance as the basis for the global Internet. The TCP protocol is built on a connectionless datagram service (the Internet Protocol, IP) with primarily two higher level services – UDP, which offers a connectionless service, and TCP, which provides a reliable connection–oriented service. Most higher–layer applications use one of these two services to transmit data between hosts.

– Ethernet, a Data Link (DL) layer Media Access Control (MAC) system designed for Local Area Networks (LANs), based on a bus topology with control of the medium distributed among the stations attached to the bus. Access to the transmission medium is by a method commonly known as CSMA/CD (Carrier Sensing Multiple Access / Collision Detection); each station can detect if another station is transmitting, and if so, it refrains from attempting to send the packet for a random period of time and then tries again. The Logical Link Control (LLC) sublayer is the other part of the divided Data Link Layer [8].

ATM is a method of transferring information (data, voice, video) using small fixed–length cells. ATM is connection–oriented; a data transfer between two entities over an ATM network will follow a path determined (by a signaling protocol) before the transfer begins. Each data connection represents a different virtual path. Although cells are always in sequence on any given virtual channel, the channels are multiplexed together through switching devices and underlying media. The ATM protocol stack is shown in Fig.2.2.

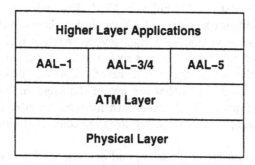

Fig.2.2. ATM protocol stack.

A number of higher–level protocols, called the ATM Adaptation Layer (AAL), provide different classes of service to upper level applications [8]. For example, AAL–1 provides a constant bit rate time division multiplexor service suitable for voice transmission, while AAL–5 provides variable rate service for data blocks of varying size for LAN traffic. In all cases, the information is eventually broken into 53–byte cells (5–byte header and 48–byte payload) at the ATM layer. The use of small fixed–length cells permits the design of very fast ATM switching devices.

The ATM Adaptation Layer is divided into two sublayers, the Segmentation And Reassemly (SAR) sublayer and the Convergence sublayer (CS). The SAR sublayer brakes down the original frame or other data unit into cells to be sent by the ATM Layer, and also reconstructs the original data unit from a sequence of cell at the other side of the connection. The CS provides the mechanism for mixing the different requirements of voice, video and data by defining a number of classes of service, each with appropriate parameters for the service. These are used to provide the proper quality of service (QoS) parameters on the connection.

Most existing applications do not interface directly with the ATM. Since they are designed to use more traditional protocols such as IP or IPX, there has been a considerable amount of work in designing interfaces that allow these protocols to operate over ATM. The IP–over–ATM standards of the IETF and MPOA [2] are examples of work in this area.

Another attempt to interface traditional protocols and communication systems with ATM is LAN emulation [8], which simulates a MAC layer (either Ethernet or Token Ring) over an ATM network. Any application or protocol that would normally operate over an Ethernet or a Token Ring network can work without modification on a LAN emulation network; the presence of ATM is hidden from the upper level applications.

Each host that is a part of an emulated LAN performs certain services for the emulated LAN. Most of these services are activated when a client (which can be a single computer or a bridge between an ATM and another type of network) first joins an emulated LAN or sends a broadcast packet. A data transfer between two hosts is usually carried over a single ATM VC (virtual circuit) without the involvement of the LAN emulation services.

At Memorial University campus, the new backbone replaces the Ethernet and routers with an emulated LAN built on ATM. The end–point LANs attach to ATM/Ethernet bridges which act as LAN emulation clients. If the two end–point LANs are in the same virtual LAN (i.e., several separate LANs that 'look' like one big LAN), then the data path is as shown in Fig.2.3; the data flows between the two ATM/Ethernet bridges via a direct ATM VC. However, if traffic is between two end–point LANs that are not in the same virtual LAN, then the data path is as shown in Fig.2.4; the path includes a router as an intervening device. An increasing number of (new) hosts are connected directly to the ATM medium.

A considerable variety of modeling and analysis methods have been applied to ATM, the main focus being the representation of input traffic to a switch or

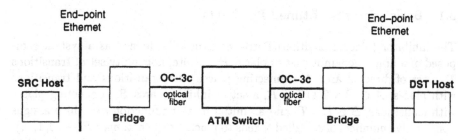

Fig.2.3. Typical data path – same virtual LAN.

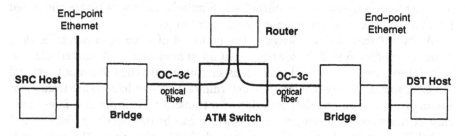

Fig.2.4. Typical data path – different virtual LANs.

network of switches. A summary of such methods can for instance be found in [23].

A popular input model used in ATM analysis is the MMPP (Markov Modulated Poisson Process) [17]. This and related models are used in [24] and [7] to estimate cell loss probabilities in ATM networks. Queueing models are used to study transmission delays in ATM networks [15, 16], and also buffer allocation within an ATM switch [12].

Discrete–event simulation has also been used to analyze ATM performance. A simulation comparison of ATM, Frame Relay, and DQDB is given in [19]. A simulator, specifically developed for ATM–based systems, is described in [1].

The issues involved in operating traditional protocols over ATM have generated a number of studies of IP performance over ATM. The issue of protocol overhead has been examined in [3, 17], while [18] analyzes the effect of TCP/IP and system design on IP–ATM performance. [22] evaluates two strategies for effective discard of packets in an ATM environment, while [13] examines in detail a deadlock situation that can occur with TCP over ATM.

The approach presented in this paper uses a simple behavioral model of ATM. A good conformance of simulation results and real measurements indicates that even this simple model is quite satisfactory for many performance studies.

3 Petri Net Model

This section first briefly recalls basic concepts of timed Petri nets, and then describes the timed net model of an ATM network.

3.1 Basic concepts of timed Petri nets

The inhibitor (place/transition) Petri net is usually defined as a system composed of a finite, nonempty set of places P, a finite, nonempty set of transitions T, a set of directed arcs A, connecting places with transitions and transitions with places, $A \subset P \times T \cup T \times P$, a set of inhibitor arcs B, connecting places with transitions, $B \subset P \times T$, and an initial marking function m_0 which assigns nonnegative numbers of so called tokens to places of the net, $m_0 : P \rightarrow \{0, 1, ...\}$. Usually the set of places connected by (directed) arcs to a transition is called the input set of a transition, and the set of placed connected by (directed) arcs outgoing from a transition, its output set. Similarly, the set of places connected by inhibitor arcs to a transition is called its inhibitor set.

A place is shared if it belongs to the input set of more than one transition. A net is conflict–free if it does not contain shared places. A shared place is (generalized) free–choice if all transitions sharing it have the same input sets and inhibitor sets. Each free–choice place determines a class of free–choice transitions sharing it. It is assumed that selection of a transition for firing in a free–choice class of transitions is a random process which can be described by (free–choice) probabilities assigned to transitions in each free–choice class. Moreover, it is usually assumed that the random choices in different free–choice classes are independent one from another.

A shared place is guarded if for any two transitions sharing it there exists another place which belongs to the input set of one of these two transitions, and the inhibitor set of the other transition. If a place is guarded, at most one of the transitions sharing it can be enabled by any marking function.

A shared place which is not free–choice and is not guarded, is a conflict place. The class of enabled transitions sharing a conflict place depends upon the marking function, so the probabilities of firing conflicting transitions must be determined in a dynamic (i.e., marking–dependent) way. A simple but usually satisfactory approach is to use relative frequencies of transition firings assigned to conflicting transitions [9]; the probability of firing an enabled transition is then determined by the ratio of transition's (relative) frequency to the sum of (relative) frequencies of all enabled transitions in a conflict class. Another generalization is to make such relative frequencies (and probabilities of firings) dynamic, depending upon the marking function, for example, by using the number of tokens in a place rather than a fixed, constant number as the relative frequency.

In ordinary nets the tokens are indistinguishable, so their distribution can be described by a simple marking function $m : P \rightarrow \{0, 1, ...\}$. In colored Petri nets [11], tokens have attributes called colors. Token colors can be quite complex, for example, they can describe the values of (simple or structured) variables or the contents of message packets. Token colors can be modified by (firing) transitions and also a transition can have several different occurrences (or variants) of its firings, for different combinations of colored tokens.

The basic idea of colored nets is to 'fold' an ordinary Petri net. The original set of places is partitioned into a set of disjoint classes, and each class is replaced by a single place with token colors indicating which of the original places the tokens belong to. Similarly, the original set of transitions is partitioned into a set of disjoint classes, and each class is replaced by a single transition with occurrences indicating which of the original transitions the firing corresponds to.

In order to study performance aspects of Petri net models, the duration of activities must also be taken into account and included into model specifications. Several types of Petri nets 'with time' have been proposed by assigning 'firing times' to the transitions or places of a net. In timed nets, firing times are associated with transitions (or occurrences), and transition firings are 'real-time' events, i.e., tokens are removed from input places at the beginning of the firing period, and they are deposited to the output places at the end of this period (sometimes this is called a 'three-phase' firing mechanism as opposed to a 'one-phase', instantaneous firings of nets without time or stochastic nets).

In timed nets, all firings of enabled transitions are initiated in the same instants of time in which the transitions become enabled (although some enabled transition cannot initiate their firings). If, during the firing period of a transition, the transition becomes enabled again, a new, independent firing can be initiated, which will 'overlap' with the other firing(s). There is no limit on the number of simultaneous firings of the same transition (sometimes this is called 'infinite firing semantics'). Similarly, if a transition is enabled 'several times' (i.e., it remains enabled after initiating a firing), it may start several independent firings in the same time instant.

The firing times of transitions can be constant (or deterministic), or can be random variables described by a probability distribution function. Exponentially distributed firing times (sometimes also called stochastic or Markovian firing times) are particularly popular because of the memoryless property of models with such firing times.

The firing times of some transitions may be equal to zero, which means that the firings are instantaneous; all such transitions are called immediate (while the other are called timed). Since the immediate transitions have no tangible effect on the (timed) behavior of the model, it is convenient to first fire the (enabled) immediate transitions, and then (still in the same time instant), when no more immediate transitions are enabled, to start the firings of (enabled) timed transitions. It should be noted that such a convention introduces the priority of immediate transitions over the timed ones, so the conflicts of immediate and timed transitions should be avoided. Similarly, the free-choice classes of transitions must be 'uniform', i.e., all transitions in each free-choice class must be either immediate or timed.

There are three basic approaches to analysis of timed Petri net models. For some classes of nets, structural methods, and in particular invariant analysis, can provide performance characteristics. In other cases, when the model is bounded and not excessively complex, the (exhaustive) reachability analysis can be used to find the stationary probabilities of states, and to derive performance characteristics from these stationary probabilities. Finally, the most general (but also the least flexible) approach to analysis of net models is to use discrete-event simulation. The last approach is used to obtain some performance characteristics of net models developed in this paper.

3.2 Complete Model

The system modeled in this paper is described in the reference model shown in Fig.3.1; this model can be thought of as a "cross–section" through the backbone configuration (Fig.2.3).

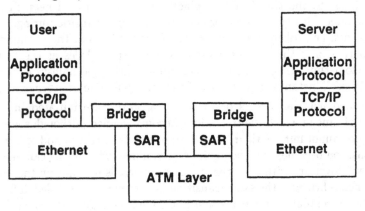

Fig.3.1. Reference model.

The model follows the layered structure of the reference model. Each layer is represented by a module of the net, which allows to change a particular layer independently of the other layers. Moreover, it was decided to originate all transfers from one side of the network, which is a typical pattern for much of Memorial's traffic (personal computers and workstations requesting various services from central servers).

The complete model is shown in Fig.3.2, with sections corresponding to the layers of the reference model. The user and application protocols from Fig.3.1 are represented by single transitions with corresponding delays.

The complete colored model can be considered as a stack of identical nets superimposed one upon another, with different nets representing different application protocols. The superimposed nets are independent at the SRC (source) and DST (destination) processes at each end, but dependencies exist between the layers at the intervening network sections, representing resources such as an Ethernet which can only transmit one frame at a time. In each layer, each transition should be regarded of as a collection of occurrences corresponding to different application protocols. Again, each transmission is independent of the others at the User/Application and TCP/IP layers, but the transmissions must be serialized at the network level.

3.3 User/Application Level

The basic model at this level is a user running an upper level application (such as TELNET or FTP). The user is assumed to spend some time thinking, after

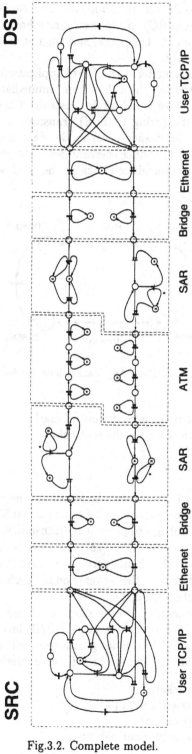

Fig.3.2. Complete model.

which a request is sent to the remote computer. This causes a data transmission from the originating host (SRC) to the destination host (DST). The SRC host then waits for a reply from the DST host, and when it is received, the SRC host returns to the thinking state.

Fig.3.3 sketches the net model at the User/Application level (as usual, timed transitions are represented by 'thick' bars, and immediate transitions by bars; moreover, inhibitor arcs have small circles instead of arrowheads). The think time is represented by the (firing time of) transition S_TNK, while the "Network" transitions (with the dashed arcs) model data transmissions. The DST process (transition D_TNK) introduces certain delay (to process the request) before replying with a data transmission back to the SRC process.

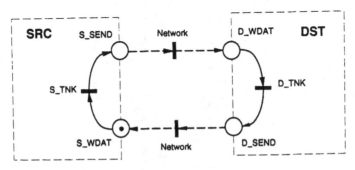

Fig.3.3. User/Application level model.

The token in S_WDAT represents the initial marking of the model, with each protocol represented by a token of different color.

3.4 TCP/IP Level

Fig.3.4 shows a net model of the User/Application level with a TCP/IP model. The model assumes that packets cannot be lost, and that SRC and DST processes will always respond fast enough to prevent re–transmissions. It is quite straightforward to add timeout mechanisms [25] which would retransmit the lost or distorted packets. On the other hand, very low probability of such events results in quite insignificant influence of these events on the performance of the network, so they are not represented in Fig.3.4.

While the User/Application level is only concerned with the data flowing between the SRC and DST processes, the TCP/IP level deals with data and acknowledgements, since TCP provides guaranteed delivery. The packets transmitted between SRC and DST can be broadly characterized into four types:

- SRC to DST: data packets;
- SRC to DST: acknowledgement packets;
- DST to SRC: data packets;
- DST to SRC: acknowledgement packets.

Most network applications send a group of data packets, wait for a group of packets in reply, send another group, and so on. Tab.3.1 shows typical data group sizes for various protocols.

Tab.3.1. Mean size of data groups (in packets).

direction	TELNET	FTP	NNTP	X
SRC to DST	1.06	11.6	1.02	1.70
DST to SRC	1.46	9.8	5.75	1.72

For real network processes, the end of a group can be detected through the data contained in the packets. A TELNET session, for example, echoes keystrokes until it recognizes the end of the line, at which point it processes the command. The model does not have any information about the content of the packets, so a different mechanism is provided to signal the end of a data group; this is done by creating special packet types, LST, to represent the last packet in a data group (one in each direction). The 'regular' data packets are denoted as MID packets.

So, each process sends a certain number (possibly zero) of MID packets, followed by a single LST packet (it is assumed that the receiving process is sending back an acknowledgement packet, ACK, for each data packet received). When the receiving process detects an LST packet, it knows that the sender has finished the data group and is now ready to receive data packets in reply.

In the model, one color is used for the User/Application level, and further six colors are used for the six packet types (i.e., MID (type '1'), LST (type '2') and ACK (type '3') packets in the direction from SRC to DST, and another three types, MID (type '4'), LST (type '5') and ACK (type '6'), for packets in the opposite direction). This distinguishing of packet types also allows the model to represent behavior based on packet type. For example, the transmission delay of a packet through the network, which is often dependent on the size of the packet, can be based on the packet size for each type rather than the overall average packet size.

In Fig.3.4, the control tokens cycling through places S_IDLE, S_SEND and S_WDAT represent the User/Application level sketched in Fig.3.3. When the control token is in place S_SEND, both transitions S_SMD and S_SLT are enabled. This (generalized) free–choice structure is described by choice probabilities assigned to the two transitions; for example, by assigning choice probability 0.8 to S_SMD (which represents sending MID(1) packets) and 0.2 to S_SLT (which represents sending LST(2) packets), each firing of S_SLT will correspond (on average) to 4 firings of S_SMD. Data in Tab.3.1 are used to determine the values of these choice probabilities for different protocols.

When S_SMD fires, a token (representing a MID(1) packet) enters N_SEND_S, which models the host that transmits a data packet through the next layer of the reference model. A control token is also removed from place S_WIND. If no tokens are present in S_WIND, the number of unacknowledged packets in the sliding window is at maximum, and no more data packets can be transmitted until some acknowledgements arrive. Tokens are also deposited in places S_WACK

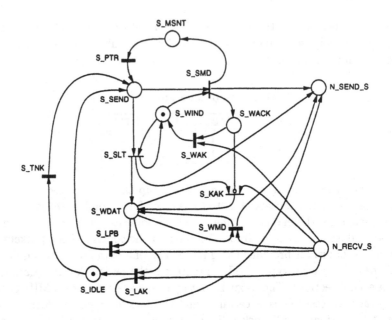

Fig.3.4. TCP/IP source (SRC) model.

and S_MSNT. The token in S_WACK waits until an acknowledgement packet (ACK) arrives from the DST process, and then transition S_WAK fires and deposits a control token back in place S_WIND (i.e., the sliding window moves forward by one packet). The firing of transition S_PTR returs a token to place S_SEND (after a delay representing the time needed for assembling a new packet of data). This cycle continues sending MID(1) packets to the DST process until S_SLT fires and sends an LST(2) packet (which indicates the end of data group).

The size of the data group is determined by the probability assigned to S_SLT (this probability can be different for each protocol); if p is the choice probability of S_SLT, then the size of the data group is a (discrete) random variable X with geometric probability density function, so:

$$\text{Prob}\{X = k\} = (1 - p)^{k-1}p, \quad k = 1, 2, ...$$

When S_SLT fires (sending an LST(2) packet), a token is deposited into S_WDAT. This represents the User/Application level in "wait" mode; a command or request has been sent to the DST process and a reply should arrive (after some delay) in place N_RECV_S as a sequence of MID(4) packets followed by an LST(5) packet.

ACK(6) packets are accepted by transition S_KAK. The inhibitor arc from S_WACK to S_KAK enforces the priority for ACK(6) tokens.

A MID(4) packet arriving in place N_RECV_S is accepted by transition S_WMD. Its firing deposits a token (an ACK(3) packet) in place N_SEND_S to acknowledge the arrival of a MID(4) packet. An arriving LST(5) packet indicates the end of the data group. At this point, the SRC process can either return an acknowl-edgement or start the transmission of the next data group, which implies that the acknowledgement is piggybacked onto the first data packet. This is modeled by a free–choice structure of transitions S_LPB and S_LAK. Piggybacking is de-scribed by the choice probability of S_LPB; this probability can be different for each protocol.

Once an arriving MID(4), LST(5) or ACK(6) packet is handled by the SRC process, SRC returns to the start state, either through the transition S_TNK (which indicates that the User/Application level requires some 'thinking' time), or directly to place S_SEND, starting transmitting the next data group.

The DST process is a mirror image of the SRC process. However, it typi-cally has different timing properties (the responses of the DST process represent software replies to a command or request).

3.5 UDP model

UDP provides connectionless service between two hosts. Much of the function-ality required in a TCP model is not needed for UDP; UDP leaves reliability of service to the higher layers of the protocol stack. As a result, a UDP model does not need the windowing, acknowledgements or piggybacking described in the previous section.

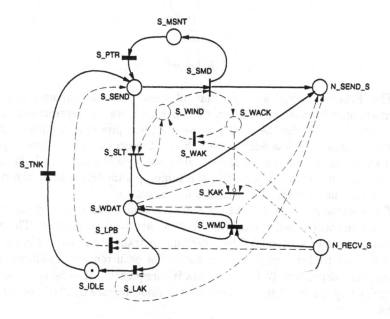

Fig.3.5. UDP source (SRC) process.

A UDP process is modeled as a subnet of the TCP process. Fig.3.5 shows a UDP model, with the unused TCP portions showed by dashed lines. The UDP model basically exchanges data groups, with the SRC process sending a series of MID(1) packets followed by a single LST(2) packet (UDP does not use acknowledgements, so the timing between packets depends on the application). As for the TCP protocol, the size of the data group is modeled by a geometrically distributed random variable.

For UDP, once the DST process receives an LST(2) packet, it begins transmitting its own data groups, finishing with an LST(5) packet. The SRC process then begins again.

3.6 LAN level

The Ethernet level of the reference model is described by a simple net (Fig.3.6) that performs two basic functions: (i) it adds a transmission delay to each packet, and (ii) it introduces 'serialization' (and blocking) as only one packet of one protocol can be transmitted at the same time.

It should be noticed that with the exception of the windowing mechanism at the TCP/IP level, each session of each protocol has, until now, been able to operate independently of the other sessions.

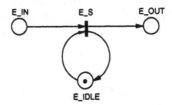

Fig.3.6. Ethernet model.

This Ethernet model assumes that all sessions are on unique hosts. That is, interactions between sessions on the same host are not represented. This is reasonably accurate for the SRC processes if these processes originate on PCs or workstations. It is less accurate for the DST processes, since these typically represent a smaller number of servers. However, delays caused by buffering on the hosts are included by default in the timing parameters used in the model. The collision mechanism of Ethernet is not modeled.

The controlling place E_IDLE in Fig.3.6 contains one token which assures that only one occurrence of transition E_S can fire at the same time. The actual transmission times are modeled by deterministic transitions, with different firing times for each packet type of each protocol. The occurrence probabilities of E_S are marking–dependent [26], so the relative frequencies of chosing a color are determined by the numbers of tokens of that color in place E_IN of the Ethernet model.

3.7 ATM level

The ATM level of reference model has two subsections: the AAL layer which provides the segmentation and reassembly (SAR) functionality (i.e., dividing packets into cells and combining cells into packets), and the cell switching functionality of the ATM layer itself.

The AAL layer is modeled in two sections. The first (Fig.3.7) shows the segmentation part of the layer – it takes a token (in the input place) representing the arrival of a packet, and generates a series of tokens representing a number of cells for transmission over an ATM switching network. The second section (Fig.3.8) represents the reverse process; it inputs a number of cells and re–creates the original packet for transmission to the next layer.

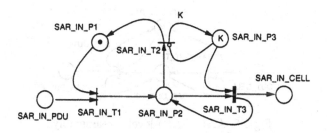

Fig.3.7. AAL (SAR) level – segmentation.

These SAR layer models take advantage of the fact that ATM is connection-oriented – cells for a particular packet must arrive in order, and there cannot be any interleaving of cells. Because of very low probability of cell losses and misinsertions, and insignificant influence of such events on the performance of the network, these very infrequent events are not represented here.

In the segmentation section, a token deposited in place SAR_IN_PDU represents the arrival of a packet at the AAL level. Place SAR_IN_P1, containing one control token, ensures that only one packet is segmented at a time. Place SAR_IN_P3 controls the number of cells that are generated for each individual packet type. For example, if packet MID(1) of the TELNET protocol is, on average, divided into 5 cells, then 5 tokens of the color representing the TELNET MID(1) packets are placed in SAR_IN_P3. When a token is deposited in SAR_IN_P2 (place SAR_IN_P2 is guarded), transition SAR_IN_T3 will fire a number of times corresponding to the color of the token in place SAR_IN_P2, generating a specific number of cells for each packet type. When the last cell has been generated, all tokens of this color have been removed from SAR_IN_P3, so transition SAR_IN_T2 can fire, which removes the token from SAR_IN_P2, replaces the required number of tokens in SAR_IN_P3 (to allow segmentation of another packet of this type), and returns the control token to place SAR_IN_P1, to start the segmentation of the next packet (of any type). It should be noticed that the values of K are associated with colors, so the numbers of cells can be different for different application protocols.

The reassembly section (Fig.3.8) re–creates a packet form a stream of cells. Tokens, which represents the arrival of cells at the SAR level, are received in place SAR_OUT_CELL. Place SAR_OUT_P1 contains a number of colored tokens (for each packet type) equal to the number of cells the packet is divided into.

As cells arrive, they are accepted by transition SAR_OUT_T1. Place SAR_OUT_P2 (also a guarded place) contains one control token to ensure that only one cell is processed at a time. When all cells of a packet have arrived (which is indicated by removal of all tokens of the corresponding color from place SAR_OUT_P1), transition SAR_OUT_T2 fires, depositing a single token of the reassembled packet's color in place SAR_OUT_PDU. SAR_OUT_T2 also deposits the required number of tokens into place SAR_OUT_P1 in preparation for the next stream of cells of that color.

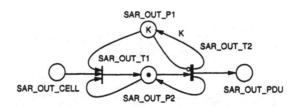

Fig.3.8. AAL (SAR) level – reassembly.

The color of the token arriving to SAR_OUT_PDU indicates the type of the application protocol the cell belongs to, and it selects the matching color of tokens in SAR_OUT_P1, so, cells of different application protocols can be reassembled in different ways (consistent with the segmentation process in Fig.3.7).

3.8 ATM switch

The ATM switching fabric is modeled by a series of delays as shown in Fig.3.9. The delays represent the latencies of the two OC-3c SONET links and the ATM switch (Fig.2.3). The use of these three delays rather than one longer delay represents the concurrency of the stream of incoming cells (transition LINK1_T1), outgoing cells (transition LINK2_T1) and the switching process (transition SW_T1).

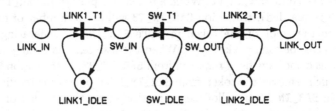

Fig.3.9. ATM switch fabric.

3.9 Other considerations

An important aspect of the model is that it should be able to represent multiple simultaneous interactions of a set of protocols. The model does not represent the contents of the packets being transmitted, so it is not obvious that the multi–session versions, obtained by simply using multiple control tokens in the corresponding modules, provide accurate representations of the real behavior. With multiple control tokens, it may be difficult to match a particular control token with the other tokens it generates. For example, when S_SLT fires (Fig.3.4) sending an LST(2) packet to the DST process and placing a control token in S_WDAT, there is no difference between an arriving LST(5) packet caused by this particular control token, and caused by an earlier or later firing of S_SLT.

In order to verify that such a multi–session model is valid, a different model was developed, in which unintensional interference of different sessions was eliminated by assigning a different set of colors to each session. The single protocol with n concurrent sessions (multi–session model) was compared to a multi–protocol model, i.e., a model with n unique single–session protocols, each with parameters identical to the multi-session model. The multi–protocol model more closely resembles the reality, since the unique protocols completely separate the n sessions in the same way that connection ID's and sequence numbers separate packets on a real network (the multi–protocol model uses many colors and is not feasible for large number of session). It was found that the multi–session model gave slightly higher results for the numbers of packets and bytes per second, but the results were not statistically significant when compared using standard hypothesis testing [10].

Another consideration is the priority of acknowledgements. In Fig.3.4, an ACK(6) packet deposited in place N_RECV_S is acknowledging either a previously transmitted MID(1) packet, or an LST(2) packet. TCP normally uses sequence numbers to differentiate between these two cases. However, since the model does not reflect that level of detail, a simpler mechanism is needed. In a single-session case, where there is only one user per protocol, so the location of the control packet determines which data packet is being acknowledged. That is, if the control token is in place S_WACK, representing a transmitted MID(1) packet, then the acknowledgement is for that packet. Otherwise, if the control token is in place S_WDAT, then the model accepts the acknowledgement for a transmitted LST(2) packet.

In the multi–session version of the model, however, there is no simple way to match a particular ACK(6) packet with the MID(1) or LST(2) packet that generated it. Two solutions were developed to address this. The first was to introduce an inhibitor arc from S_WACK to S_KAK, which effectively gives MID(1) packets a priority over any arriving ACK(6) packet. The other option was to add an additional packet type, ACK(7), and use an ACK(6) as acknowledgements for MID(1) packets, and ACK(7) as acknowledgements for LST(2) packets. The second solution, while somewhat more realistic, also adds two colors per protocol and two occurrences per transition for the intervening layers in the reference model.

The simulation results for the first option showed slightly higher values for throughput and average burst rate. When compared using hypothesis testing

[10], the difference was not found to be significant. Therefore it was decided to use the first option.

4 Model Performance

Many parameters of the model (firing times of transitions, probabilities of firings in free-choice classes, relative frequencies of firings for conflict classes of transitions) can be established on the basis of real, physical data characterizing the Memorial's network. The Ethernet backbone of the campus network was monitored for a week (the data were collected using the TCPDUMP public domain TCP/IP monitoring program), and packet and byte counts per protocol (as determined by TCP/IP port numbers) were recorded. Tab.4.1 shows the relative frequencies for the most common protocols.

Tab.4.1. Packet and byte counts for the most common protocols.

Protocol	packets	bytes
SHELL	13.8 %	33.7 %
TELNET	40.2 %	11.5 %
NNTP	5.99 %	12.1 %
NFS	7.75 %	9.72 %
SMTP	3.80 %	9.37 %
X	11.1 %	7.21 %
LOGIN	8.25 %	3.97 %
FTP (data)	1.46 %	3.76 %
HTTP	1.43 %	1.59 %
OTHER	6.24 %	4.14 %

Since many of the model parameters are estimated from timing data taken at the network level, it was necessary to take two trace sets for each protocol; one for the source processes and one for the destination processes (the "Monitor" boxes in Fig.4.1 show the data collection points in the analyzed system).

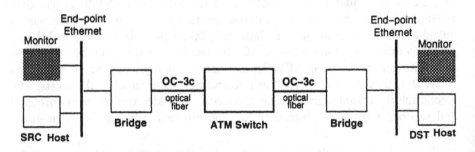

Fig.4.1. Data collection configuration.

The parameters for the SRC and DST processes were estimated by analyzing the pairs of adjacent packets in the recorded traces. An example of a matrix describing the "next packet" probability density for the TELNET protocol is shown in Tab.4.2 (for example, entry [2,4] is the probability that a packet of type "2" is followed by a packet of type "4", in the same user session).

Tab.4.2. "Next packet probability" matrix for TELNET protocol.

TELNET	1	2	3	4	5	6
1	0.105	0.035	0.009	0.000	0.000	0.851
2	0.000	0.000	0.004	0.106	0.707	0.183
3	0.016	0.585	0.012	0.200	0.186	0.001
4	0.000	0.000	0.860	0.066	0.074	0.000
5	0.006	0.367	0.627	0.000	0.000	0.000
6	0.102	0.096	0.002	0.506	0.294	0.000

A similar matrix was generated for timing information. Other parameters, also extracted from the recorded traces for the TELNET protocol, are shown in Tab.4.3.

Tab.4.3. Values of model parameters for TELNET protocol.

parameter	average value	units
SRC group size	1.051	packets
DST group size	1.456	packets
SRC window size	1.173	packets
DST window size	1.117	packets
packet size (1)	60.74	bytes
packet size (2)	60.04	bytes
packet size (3)	60.00	bytes
packet size (4)	401.8	bytes
packet size (5)	167.9	bytes
packet size (6)	60.00	bytes
thinking time	2.456	sec
reply delay	0.047	sec
packet interarrival time	0.8188	sec

The results obtained by simulation of the developed model [26] correspond quite well to the recorded data. For example, Fig.4.2 compares the latency values reported in [5] (the solid line) with the simulated results (the dashed line); both plots exhibit linear relationship between the frame sizes and the average latencies, and both plots are quite close one to another.

Fig.4.3 compares the distribution of packet sizes for FTP protocol; the solid line shows the distribution of the data collected from the network, with two 'peaks', one for the very short frames, and another one for long frames.

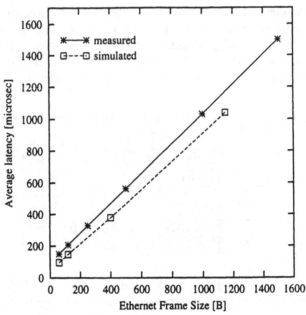

Fig.4.2. Simulated vs real ATM latency.

The simulated results also have a characteristic peak for very short frames, but there is a double peak for long frames; this double peak should be converted into a single peak, similar to the measurement data, by tuning model parameters.

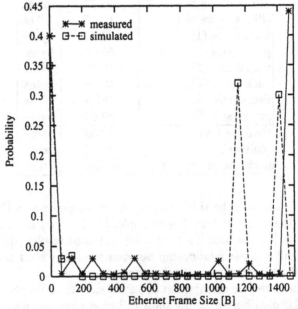

Fig.4.3. Packet size distribution for FTP protocol.

Fig.4.4 shows an example of information that can be obtained from the model [20]; it compares the maximum ATM burst rate for two protocols, using a 10 msec window. The figure indicates that even a low impact protocol like TELNET (the solid line) can create short bursts of cells at high speed, far above the average behavior. It also shows that the load of protocols can be limited by activities of other protocols – in this case, increasing (with the number of users) load generated by TELNET restricts the effective load of FTP (represented by the dashed line).

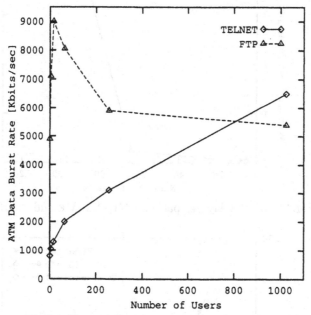

Fig.4.4. Network load by protocol – max ATM data rate.

Fig.4.5 shows the effect of TELNET and FTP protocols on CU–SEEME; as the number of users in the system increases (it is assumed that all users have the same network traffic characteristics), the average delay of CU–SEEME packets initially increases very slowly, but after certain number of users, this average delay grows rather quickly. The 'critical' number of users is equal to 64 (the "knee" point of the delay curve), so the limit on the number of users in order to avoid excessive delays should be 64 (for the assumed traffic characteristics).

The model can be extended in a number of ways by very simple modifications. For example, multiple Ethernets (at one or both sides of the ATM link) can be represented by increasing the number of (initial) tokens in E_IDLE (Fig.3.6); this allows multiple simultaneous firings of transition E_S, up to the number of tokens in E_IDLE. Fig.4.6 shows that as the number of Ethernets increases, the average load per network decreases. The ATM load initially increases but then stabilizes, which indicates a bottleneck in the system (a bridge could be this bottleneck).

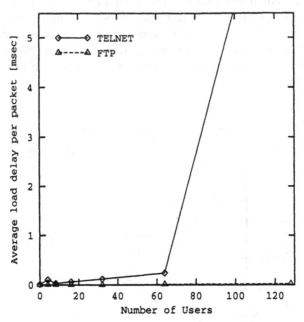

Fig.4.5. Load delay per packet – CU–SEEME and TELNET/FTP.

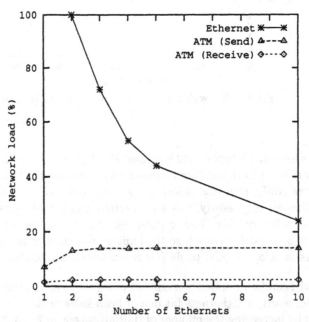

Fig.4.6. Effect of multiple Ethernets on ATM load.

5 Concluding Remarks

A timed Petri net model of an ATM LAN provides a reasonably accurate representation of the behavior of the original system, so the model can be used for detailed investigations of various parts of the system. The model can easily be extended by introducing additional elements, for example, bridges or routers between bridges. The ability to represent multiple distinct protocols can be extended past the basic concept to study the interactions between protocols, timing delays and where those delays occur.

The proposed model provides good conformance to protocol behavior under normal load, it directly represents network traffic in terms of users and protocols, its structure can easily be modified, and it allows to study properties of different network configurations (without actually implementing them).

Although only a very simple model of the workload has been used for obtaining the results presented in this paper, the simulation results conform to the real, measured quantities quite well. Moreover, more elaborate models of the workload can easily be introduced by adding a few additional net elements to the model [6].

The model has some limitations. It becomes less accurate when it passes the point of congestion because it does not represent many effects which affect the behavior of the network if congestion is taken into account (a more complex Ethernet model would be needed and retransmission of lost packets would be required).

In extending the proposed model, some additional constructs would be of benefit. The ATM layer is complicated by the requirement that cells remain in strict order, yet can be buffered at various points in the system. As well, current work on hierarchical net structures would be beneficial to this model, simplifying the introduction of mechanisms that cannot be layered easily (e.g., lost packet mechanism).

The complete model is quite complicated but it has a very modular structure. It would be interesting to check if structural methods could be successfully applied to analysis of this type of net models.

Acknowledgments

A number of interesting comments and constructive remarks of two anonymous referees are gratefully acknowledged.

The Natural Sciences and Engineering Research Council of Canada partially supported this work through Research Grant A8222.

References

1. Ajmone Marsan, M., Cigno, R.L., Munafo, M., Tonietti, A., "Simulation of ATM computer networks with CLASS"; in: "Computer Performance Evaluation: Modelling Techniques and Tools", pp.159–179, Springer Verlag 1994.

2. Alles, A., "ATM internetworking"; Technical Report, CISCO Systems Inc. 1995.

3. Armitage, G.J., Adams, K.M., "How inefficient is IP over ATM anyway?"; IEEE Network, vol.9, no.1, pp.18–26, 1995.

4. Awdeh, R.Y., Mouftah, H.T., "Survey of ATM switch architectures"; Computer Networks and ISDN Systems, vol.27, no.12, pp.1567–1613, 1995.

5. Bradner, S., " Bradner Reports – Catalyst 5000 switch"; Technical Report, Cisco Systems Inc, Sept. 1995.

6. Chen, P-Z., Bruell, S.C., Balbo, G., "Alternative methods for incorporating non-exponential distributions into stochastic timed Petri nets"; Proc. 3-rd Int. Workshop on Petri Nets and Performance Models (PNPM'89), Kyoto, Japan, pp.187–196, 1989.

7. Descloux, A., "Stochastic models for ATM switching networks"; IEEE Journal on Selected Areas in Communications, vol.9, no.3, pp.450–457, 1991.

8. Goralski, W.J., "Introduction to ATM networking"; McGraw Hill 1995.

9. Holliday, M.A., Vernon, M.K., "Exact performance estimates for multiprocessor memory and bus interference"; IEEE Trans. on Computers, vol.36, no.1, pp.76-85, 1987.

10. Huntsberger, D.V., Billingsley, P., "Elements of statistical inference" (5-th ed.), Allyn and Bacon 1981.

11. Jensen, K., "Coloured Petri nets"; in: "Advanced Course on Petri Nets 1986" (Lecture Notes in Computer Science 254), Rozenberg, G. (ed.), pp.248-299, Springer Verlag 1987.

12. Lin, A.Y.M., Silvester, J.A., "Queueing analysis of an ATM switch with multi-channel transmission"; Performance Evaluation Review, vol.18, no.1, pp.96–105, 1990.

13. Moldeklev, K., Gunningberg, P., "How a large ATM MTU causes deadlocks in TCP data transfers"; IEEE-ACM Trans. on Networking, vol.3, no.4, pp.409–422, 1995.

14. Murata, T., "Petri nets: properties, analysis and applications"; Proceedings of IEEE, vol.77, no.4, pp.541–580, 1989.

15. Ohba, Y., Murata, M., Miyihara, H., "Analysis of interdeparture processes for bursty traffic in ATM networks"; IEEE Journal on Selected Areas in Communications, vol.9, no.3, pp.468–476, 1991.

16. Onvural, R.O., "On performance characteristics of ATM networks"; Proc. Super-Comm/ICC '92, pp.1004–1008, 1992.

17. Onvural, R.O., "Asynchronous Transfer Mode Networks: Performance Issues"; Artech House 1994.

18. Perloff, M., Reiss, K., "Improvements to TCP performance in high–speed ATM networks"; Communications of the ACM, vol.38, no.2, pp.91–109, 1995.

19. Petr, D.W., Frost, V.S., Neir, L.A., Demirtjis, S., Braun, C., "Simulation comparison of broadband networking technologies"; SIMULATION 64, pp.42–50, 1995.

20. Reid, M., "Modeling and performance analysis of ATM LANs"; M.Sc. Thesis, Department of Computer Science, Memorial University of Newfoundland, St. John's, Canada A1B 3X5, 1997.

21. Reisig, W., "Petri nets – an introduction"; Springer Verlag 1985.

22. Romanow, A., Floyd, S., "Dynamics of the TCP traffic over ATM networks"; IEEE Journal on Selected Areas in Communications, vol.15, no.4, pp.633–641, 1995.

23. Stamoulis, G.D., Anagnostou, M.E., Georgantas, A.D., "Traffic source models for ATM networks: a survey"; Computer Communications, vol.17, no.6, pp.428–438, 1994.

24. Yamada, H., Sumita, S., "A traffic measurement method and its application for cell loss probability on ATM networks"; IEEE Journal on Selected Areas in Communications, vol.9, no.3, pp.305–314, 1991.

25. Zuberek, W.M., "Timed Petri nets – definitions, properties and applications"; Microelectronics and Reliability (Special Issue on Petri Nets and Related Graph Models), vol.31, no.4, pp.627–644, 1991.

26. Zuberek, W.M., "Modeling using timed Petri nets – event–driven simulation"; Technical Report #9602, Department of Computer Science, Memorial University of Newfoundland, St. John's, Canada A1B 3X5, 1996.

Performance Evaluation of Polling-Based Communication Systems Using SPNs

Boudewijn R. Haverkort

Laboratory for Distributed Systems, RWTH-Aachen, D–52056 Aachen, Germany
http://www-lvs.informatik.rwth-aachen.de/

Abstract. In this paper we present stochastic Petri nets (SPNs) that can be used for the evaluation of polling mechanisms. Polling mechanisms (or systems) appear in many forms in computer-communication systems: the well-known token-ring and token-bus network access schemes such as present in IEEE P802.4/5 and in FDDI, and the scheduling mechanisms in switching fabrics, e.g., for ATM systems, operate along the lines of polling systems. Polling systems have been studied for many years now, and many analytical techniques have been developed to study them. It seems, however, that a number of system aspects can not be covered adequately by such analytical approaches. Most notably are time-dependent polling variants where the amount of service a station receives per visit is time-limited, load-dependent polling strategies where the ordering of station visits is dependent on the loading of the stations, as well as non-Poisson arrival processes. Therefore, we present SPN-based models that allow us to cope with these system aspects. The two major problems that appear when taking the SPN approach are the size of the underlying CTMC and the use of non-exponential timing. The latter problem is not really addressed in this paper; rather it is circumvented by employing the well-known method of stages, thus even worsening the state-space size problem. The first problem is coped with, by presenting two decomposition approaches and by presenting a subclass of SPNs that allows for an efficient matrix-geometric solution, thus avoiding the explicit generation of the overall state space.

1 Introduction

In the design of communication systems, performance evaluation plays an important role. With the increasing complexity of communication systems and the way in which they are used, it has become more and more difficult to construct models that are analytically tractable. This has lead numerous researchers to turn to simulation as the method to evaluate the performance of communication systems. However, simulation does suffer from some drawbacks, most notably its relatively high cost, especially in obtaining accurate estimates for small quantities such as blocking probabilities, and the error-prone process of coding simulation programs.

In between the closed-form analytical and the simulation approach, i.e., in between with respect to both the modelling capabilities and the evaluation costs, lies the numerical approach based on stochastic Petri nets (SPNs) [1, 8, 9, 32]. SPNs allow for a very flexible construction and numerical solution of, possibly large, continuous-time Markov models of communication systems. Advantage of such an numerical approach over simulation is that rare-events are less of a problem. Also, by its formally well-established semantics, SPN models are easily constructed and less error-prone than simulation programs. As an example of this, once an SPN has been specified, properties such as liveness and deadlock freeness can be verified automatically. Of course, this increases the confidence one can have in the model, especially in comparison to hand-coded simulation programs. In some cases, particular structural properties of the SPN can even be exploited in the solution process; see e.g., our work on mean-value analysis of so-called product-form SPNs [10], and our work on matrix-geometric SPNs (see also Section 6). On the other hand, by the fact that a numerical solution is employed, more generality in the models can be achieved than with the closed-form analytical approaches.

In this paper we will present an overview of SPN models, some new and some being presented (by others) before, that are useful in evaluating the performance of many communication systems. In particular, we will focus on the class of SPN-based polling models. As will be illustrated below, such models find their application when studying token-based networks of various kinds, as well as when studying the behaviour of various ATM-based multiplexers.

Polling models have been the subject of study for some years now. Over the 15 years especially, the amount of published articles on polling models has increased tremendously; Takagi [38, 39, 40] reports about 250 publications in the period 1986–1990 only! Overviews can be found in the theses by Groenendijk [14] and Weststrate [43], the survey by Sidi and Levy [30]. It must be stated here, however, that almost all of the above studies refer to analytical approaches for solving polling models only. Although these approaches are very worthwhile, it seems that for some applications, the assumptions made to allow for the analytical solutions are not always realistic. The aim of this paper is therefore to show how SPN-based polling models can be used to address various performance questions that arise during the design of computer-communication systems. In doing so, the emphasis is on SPN-based polling models that incorporate system aspects that can not easily be coped with by the analytical approached mentioned above, such as time-based or load-based polling mechanisms.

We are not the first to work in this direction. Seminal papers in this field have been published by Trivedi, Ibe and Choi [23, 6, 24] and by Ajmone Marsan et al. [2, 3, 4]. It should be note that the latter group of others have focussed on *multi*-server polling models; we do restrict ourselves her on single server polling models.

Throughout this paper, especially in Sections 3 through 5, we will use the terminology and notation as proposed by Ciardo et al. [7] in their definition of a class of SPNs: the so-called stochastic reward nets. The presented models have been evaluated with the tool SPNP.

This paper is further organized as follows. In Section 2 we provide the necessary background in the field of polling models. In Section 3 we then address count-based polling models. We then proceed to time-based polling models, that find their application in modelling for instance the IBM token ring or the FDDI ring, in Section 4. We then address, in Section 5, SPNs for non-cyclic and load-dependent polling models, as they find their application in evaluating ATM-based multiplexers. We finally report about a recently developed class of SPNs that allows for a matrix geometric solution so that the underlying CTMC need not be generated explicitly, in Section 6. Such SPNs can then be used nicely to describe and evaluate ATM multiplexers with intricate scheduling mechanisms under complex workloads, thereby avoiding the explicit generation of the overall state space. The models presented in Sections 4 through 6 have been developed by us. Section 7 concludes the paper.

2 Background on polling models

We introduce some basic terminology in Section 2.1. We then discuss the important issues of scheduling and visit ordering in Section 2.2 and 2.3 respectively.

2.1 Basic terminology

In polling models, there is a single server which visits (polls) a number of queues in some predefined order. Customers arrive at the queues following some arrival process. Upon visiting a particular queue, queued customers are being served according to some scheduling strategy. After that, the server leaves the queue and visits the next queue. Going from one queue to another takes some time which is generally called the switch-over time.

Throughout this paper we address polling models with N stations, modelled by queues Q_1 through Q_N. We use queue indices $i, j \in \{1, \cdots, N\}$. At queue i customers arrive according to some arrival process with rate λ_i. In most analytical models, the arrival processes are Poisson processes; in SPN-based models this need not be the case. The service requirements of customers arriving at queue i are characterized by the first and second moment, $E[S_i]$ and $E[S_i^2]$ respectively. The total offered load is given by $\rho = \sum_{i=1}^{N} \rho_i$, with $\rho_i = \lambda_i E[S_i]$. The mean and variance of the time needed by the server to switch from queue i to queue j are denoted $\delta_{i,j}$ and $\delta_{i,j}^{(2)}$ respectively.

When the queues are assumed to be unbounded, under stability conditions, the throughput of each queue equals the arrival rate of customers at each queue. The main performance measure of interest is then the customer waiting time for queue i, i.e., W_i. Most analytic models only provide insight in the average waiting time $E[W_i]$. When the queues are bounded, however, the throughput and blocking probability at the stations are also of interest. For these cases, however, the analytical approaches fall short, which is also the case when the arrivals do note take place according to a Poisson process.

2.2 The visit order

We distinguish four different visit orders: a cyclic ordering, a Markovian ordering, an ordering via a so-called polling table, and a load-based polling order.

Cyclic polling. In a cyclic visiting scheme, after having served queue i, the server continues to poll station $i \oplus 1$ where $\oplus$ is the modulo-N addition operator such that $N \oplus 1 = 1$. As a consequence of this deterministic visit ordering only "neighbouring" switch-over times and variances are possibly non-zero, i.e., $\delta_{i,j} = \delta_{i,j}^{(2)} = 0$ whenever $j \neq i \oplus 1$. For ease in notation we set $\delta_i = \delta_{i,i\oplus1}$ and $\delta_i^{(2)} = \delta_{i,i\oplus1}^{(2)}$. In Figure 1 we show the basic polling model with cyclic visit ordering.

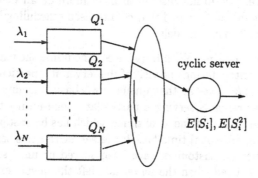

Fig. 1. Basic polling model with cyclic visit ordering

Markovian polling. In a Markovian polling scheme (also called random polling), after having polled queue i, the server continues to poll queue j with probability $p_{i,j}$.

Tabular polling. Finally, an ordering via a polling table $T = (T_1, T_2, \cdots, T_M)$ establishes a cyclic visit ordering of the server along the queues, however, these cycles may contain multiple visits to the same queue. The server starts with visiting queue Q_{T_1}, then goes to Q_{T_2}, etc. After having visited Q_{T_M} the server visits Q_{T_1} again and a new cycle starts.

Load-based polling. In some systems the choice of the next queue to be visited depends on the actual queue lengths. In such systems, the server might for instance try to minimize the variance in queue lengths, or might try to keep one queue length below a certain threshold at all costs, and at the same time use its spare time to serve other queues.

2.3 The scheduling strategy

The scheduling strategy defines how long or how many customers are served by the server once it visits a particular queue. Two main streams in scheduling strategies can be distinguished: count-based and time-based scheduling.

Count-based scheduling. With count-based scheduling the maximum amount of service that is granted during one visit of the server at a particular queue is based on the *number of customers* served in the polling period. Among the most well-known scheduling disciplines are the following: exhaustive (E): the server continuously serves a queue until it is empty; gated (G): the server only serves those customers that were already in the queue at the time the service started; k-limited (K): each queue is served until it is emptied, or until k customers have been served, whichever occurs first; decrementing (D): when the server finds at least one customer at the queue it starts serving the queue until there is one customer less in the queue than at the polling instant; or thresholding (T): the server continues to serve a queue until its remaining length is below a certain threshold. There exist various extensions and additions to the service strategies mentioned above. We do not discuss them in more detail here but refer to the survey in [43]. It should be noted that count-based scheduling has been the topic of most analytical polling models.

Time-based scheduling. With time-based scheduling the maximum amount of service that is granted during one visit of the server at a particular queue is based on the *time already spent* at that queue. Two basic variants exist: local time-based: the server remains serving a particular queue until either all customers have been served or until some local timer, which has been started at the polling instant, expires; and global time-based: the server remains serving a particular queue until either all customers have been served or until some global timer, which has been started when the server last left the queue, expires. In fact, the first mechanism can be found in the IBM token ring (IEEE P802.5) and the IEEE P802.4 token bus [5, 41], whereas the second one can be found in FDDI [5, 25, 41]. Time-based scheduling is less easy to capture in an analytical model, however, in an SPN-based modelling approach, it does give us less problems.

3 Count-based cyclic polling models.

In this section we present basic count-based cyclic polling models in Section 3.1. We then continue with two approximate approaches: a folding strategy is presented in Section 3.2 and a decomposition approach is presented in Section 3.3. We finally address other than cyclic visit orderings in Section 3.4.

3.1 Basic polling models

Using SPNs we can construct polling models of a wide variety. Since the models are solved numerically, there is no intrinsic difference between asymmetric and symmetric models. As a consequence of this, fairly intricate systems can be modelled. However, the choice for a numerical solution of a finite Markov chain implies that only finite-buffer (or finite-customer) systems can be modelled, and that all timing distributions are exponential or of phase-type. Both

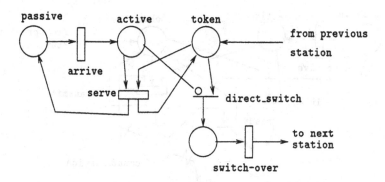

Fig. 2. SPN model of a station with exhaustive scheduling strategy

these restrictions do not imply fundamental problems, however, from a practical point of view, using phase-type distributions or large finite buffers, results in large Markovian models which might be too costly to generate or solve. The recent developments in the use of so-called DSPNs [13, 27, 31, 32] allow us to use deterministically timed transition as well, albeit in a restricted fashion. This might avoid some of the above problems.

Ibe and Trivedi discuss a number of SPN-based cyclic polling models [23]. A few of them will be discussed here. First consider the exhaustive service model of which we depict a single station in Figure 2. The overall model consists of a cyclic composition of a number of station submodels. Tokens in place passive indicate potential jobs; after an exponentially distributed time, they become active, in the sense that they move to place active where they wait until they are served. When the server arrives at the station, indicated by a token in place token, two situations can arise. If there are jobs waiting to be served, the service process starts, thereby using the server and transferring jobs form place active to passive via transition serve. After each service completion, the server returns to place token. Transition serve models the jobs service time. If there are no jobs waiting (anymore) to be served (place active is empty), transition serve is disabled and the server is transferred via the immediate transition direct-switch to place switch. After the switch-over (transition switch-over), the server arrives at the place token of the next station.

Using this model, an underlying Markov chain can be constructed and solved automatically. Suppose, for station i, the initial number of customers in its place passive equals n_i and the arrival rate of customers λ_i. Then, from the SPN analysis we obtain $E[N_{q,i}]$, the expected number of customers in place active, and α_i, the probability that place passive is empty. The effective arrival rate of jobs to place active is $(1 - \alpha_i)\lambda_i$ since only when passive is non-empty, the arrival transition is enabled. Using Little's law, the expected waiting time then equals

$$E[W_i] = \frac{E[N_{q,i}]}{(1 - \alpha_i)\lambda_i}. \tag{1}$$

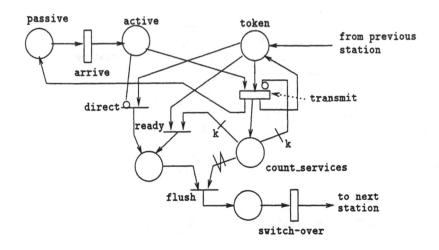

Fig. 3. SPN model of a station with k-limited scheduling strategy

In Figure 3 we depict a similar model for the case the scheduling strategy is k-limited. Here we have to distinguish three main cases: no jobs buffered (transition direct fires); k services have taken place (transition flush fires; see below); and l ($1 \leq l < k$) services have taken place and there are no jobs left anymore (transition ready fires). The "zig-zag" (z) symbol used in the arc from count_services to flush is an arc with variable multiplicity 9another special feature of the class of SPNs supported by SPNP). Transition flush can fire irrespective of the number of tokens in place count_services. Whatever that number is, flush takes them all. Notice that in case $k = 1$, a simpler model can be used. Also notice that the inhibitor arc from count_services to transmit may be omitted: once k tokens are present in count_services transition transmit will not fire since the immediate transition ready is enabled. Also transition direct might be enabled at the same time; either of these two may fire, since both lead to the same next tangible marking. For this model, similar expressions for the waiting times as above are easily derived.

An advantage of this approach is that asymmetric models are as easy to solve as symmetric models and that different scheduling strategies can be easily mixed. Also other scheduling strategies can be implemented. This advantage is, however, directly related to the main disadvantage: the state-space size increases rapidly with the number of stations and the number of buffer places accounted for, simply because possible model symmetries are not exploited. Ibe and Trivedi report state spaces of up to 200000 states for a model with only 3 stations, 20 buffer places per station and exhaustive scheduling strategy (see also [23, Table XIII—IX].

For the exhaustive scheduling strategy, it is possible to estimate the number of states in the underlying CTMC. When there are at most M customers per station ("the buffer size is M") this results in $O(M)$ different states for a single station. Given N stations, the number of states increases to $O(M^N)$. Furthermore, the single server can be actively present at any of these N stations, or it can reside in N different switch-over states, thus enlarging the number of states to $O(2NM^N) = O(NM^N)$.

To cope with the problem of very large Markovian models, one can go at least two ways. One can try *to exploit symmetries* in the model in order to reduce the state space. This can sometimes be done in an exact way, in other circumstances only approximately. Another way to go is *to decompose* the model (divide and conquer) and to analyze the submodels in isolation. Again, this can sometimes be done exactly, in other cases it is only approximate. In the next two subsections, we report on two approximate approaches.

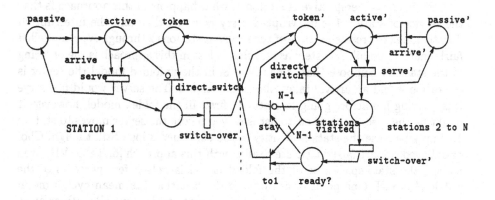

Fig. 4. Folding an N-station model to an approximate 2-station model

3.2 Exploiting symmetries: folding

When the model is highly symmetric it is possible to exploit this fact by folding states together which are statistically equivalent or near equivalent. Preferably though, we perform such a folding operation already at the SPN-level, so that we do not have to bother about the underlying Markov chain with its very many states.

As an example of this, consider a cyclic polling model with N stations (with 1-limited scheduling, although this is not a prerequisite). Furthermore assume that stations $2, \cdots, N$ behave statistically the same. About station 1 we do not make any assumptions. Instead of modelling stations 2 through N separately, we can also fold them together, i.e., model them as one "super" station with an increased arrival rate and which is visited $N - 1$ times after another, each time including a switch-over, before a visit to station 1 occurs.

This approach is illustrated in Figure 4 where basically two stations are depicted, however, the station on the right "models" stations 2 through N. Station 1 operates as usual: once it is visited by the server (a token in place token), either a single message is transmitted (transition serve fires), or the server moves to the next station (transition direct_switch fires). Transition arrive' models all arrivals at these stations (its rate equals the sum of the original arrival rates at stations 2 through N). Place active' now models a shared job buffer of stations 2 through N. If there is a job buffered, it can be served (transition serve', after which an artificial switch-over takes place (transition switch-over'). Then, after the switch-over delay, the server arrives in place ready?. There it is checked whether there have already been $N - 1$ services or polls in the superstation. If so, the servers moves to station 1 (transition to1 fires), otherwise transition stay fires and the server arrives in place token'. This continues until either $N - 1$ services or pollings (via transition direct-switch') have taken place.

Seen from the perspective of station 1, what happens in this approach is that the service of station 1 is interrupted every now and then and the interruption duration, i.e., the residence time of the server at stations 2 through N, is modelled fairly much in detail. A system aspect that is lost in this approach is the ordering of the stations. Suppose that, when $N = 8$, in the unfolded model the server is at station 4 and an arrival takes place at station 2. The server would not serve this arriving job before going to station 1 first. In the folded model, however, it might be the case that this customer is served before the server moves to station 1, simply because the station identity of the customer is lost. Still though, Choi and Trivedi report fairly accurate results with this approach [6, Table VII] even though the state-space size of the folded model is only a few percents of the unfolded model. One point of criticism is the fact that the mean cycle-time in the polling model, i.e., the time it takes the server to cycle along all stations once, is used as an accuracy criterion. It is known that for infinite-buffer systems this value is independent of the scheduling strategy and the service ordering (for 1-limited or exhaustive scheduling polling models, it equals $\Delta/(1 - \rho)$ where Δ is the total switch-over time in a cycle). Therefore, one might question the use of this accuracy criterion.

Ibe *et al.* [24] apply the above strategy also in the analysis of client-server architectures with token ring and Ethernet communication infrastructures. Here they report error levels of at most 1% for the mean response times perceived at the stations (a better criterion), when state space sizes are reduced to less than 10% of the original [24, Tables IV—V].

Again for the exhaustive scheduling strategy, it is possible to estimate the number of states in the underlying CTMC. When there are at most M customers per station this results in $O(M)$ different states for a single station. Since we have summarized all but one station into one special station, the effective number of stations is now 2. The number of customers in the special station will be taken equal to the sum of that number in the original situation (in the worst case), i.e., $O(NM)$. Furthermore, when the single server is present in the special station, the number of polls is "remembered", thus yielding another multiplicative factor

N. Thus, the overall number of states is $O(M \cdot NM \cdot N) = O(M^2 N^2)$.

When we would have dealt with Markovian polling with $p_{i,j} = 1/N$, the folding strategy will yield exact results. In such a case, the folding technique corresponds to the mathematically exact technique of state lumping in CTMCs. The software tool UltraSAN [9] supports this kind of lumping automatically.

3.3 Model decomposition: fixed-point iteration

For asymmetric models the folding procedure normally tends to inaccurate results. Instead, one can employ a procedure in which all the stations are analyzed individually, thereby taking into account the "server unavailability" due to service granted at other stations. Since these latter quantities are unknown in advance, one can initially guess them. Using these guesses, a more exact approximation can be derived which can again be used to obtain a better approximation, etc.

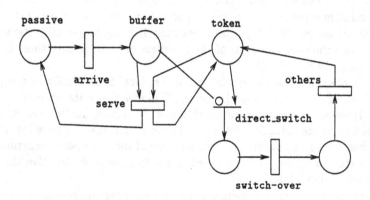

Fig. 5. Approximate model M_i for a single station in a polling model

Using such a fixed-point iteration technique, Choi and Trivedi derive results for large asymmetric polling models with acceptable accuracy [6]. We briefly describe this approach below.

One starts with an N-station polling with exhaustive services; we do not assume anything regarding numerical parameters or symmetries. For every station i, a model M_i, similar to the one presented in Figure 5, is solved. In M_i, transition others models the delay of all stations j other than i. What happens during this delay is not important for station i; what matters for station i is that it takes a particular time before the server returns. Initially, a reasonable guess is done for the rate of others, e.g., $\sum_{j \neq i} \delta_j$. From M_i one can calculate the probability $p_i(k)$ of having $k \in \{0, \cdots, k_i\}$ customers queued in place buffer

(k_i is the initial number of tokens in place `passive`). The expected delay the token perceives when passing through station i then equals

$$d_i = \sum_{k=0}^{k_i} d_i(k)p_i(k),$$

where $d_i(k) = \delta_i + k/\mu_i$. If the individual stations use more complex scheduling mechanisms, it might be difficult to obtain an explicit expression for $d_i(k)$. Therefore, Choi and Trivedi propose to derive $d_i(k)$ numerically as the time to absorption in an acyclic Markov chain. This should, however, only be done when a direct expression can not be easily found.

When all the d_i have been calculated, the server unavailability perceived by station i equals $D_i = \sum_{j \neq i} d_j$. The reciprocal value $1/D_i$ can then be used as the new guess for the rate at which transition `others` completes in model M_i. The procedure stops when succeeding estimates for D_i are within some accuracy bound.

From the analysis of M_i with the converged value $1/D_i$ for the rate of `others`, various performance measures can easily be derived as before. Choi and Trivedi report relative errors on the mean response time per node of less then 1% for low utilizations (up to 50%), up to less than 10% for larger utilizations, when compared to the exact analysis of the complete models (in case these complete models can still be solved).

The state spaces of the smaller models reduce dramatically in comparison with those of an overall model; per small model the state space size is now $O(M)$. However, it should be noted that now N of such smaller models have to be solved in an iterative fashion. The number of iterations required is normally small (less than 10 usually). The solution time of the fixed-point iteration model was reported to be only a few percent of the time required to solve the overall models ([6, Table VIII].

There are two intrinsic assumptions in the above approach: (i) the server unavailability time perceived by a station is exponentially distributed, and (ii) the arriving server sees the station it arrives at in equilibrium, which implies that the server arrives as being part of a Poisson process. Despite these two assumptions, which might not be true in practice, the results obtained so far with this approach are not bad at all. In fact, we experimented recently with non-exponential server unavailability times, based on the first two moments of the residence time in the stations, and we did not obtain better results [33] than reported by Choi and Trivedi.

3.4 Polling models with Markovian and tabular visit order

Instead of having a single switch-over transition to shift the server from station i to $i \oplus 1$, one can imagine that starting from station i there are more switch-over possibilities. This can very easily be incorporated in the SPN models by introducing immediate transitions that express the choices to go from one station to the next. The state space does not increase in size, however, the number of non-zero entries in the generator of the underlying CTMC does increase. Also, folding techniques cannot be applied when the routing probabilities are station dependent. In case the routing possibilities are independent of the originating station (the rows in the matrix P are the same), folding techniques can be applied, as illustrated in [3]. The fixed-point iteration technique can, in principle, still be applied. However, depending on the structure of P the variance in the "server unavailability" can vary enormously which might worsen the convergence of the iteration, i.e., the assumption on exponentially distributed server unavailability times might be less appropriate.

Tabular visit orderings can also be described using SPNs. The model parts for the actual stations remain unchanged. An extra place has to be added in which the number of tokens represents the index in the table of the current station visited. Upon service completion at a station, this index is increased, modulo the number of table entries, simply by adding a token in this place (the modulo operation can be performed via an immediate transition that removes all tokens as soon as their number equals the number of table entries). Again using this marking, a decision can be made which station to visit next using as many immediate transitions as that there are table entries. Using the enabling functions as provided by SPNP [7] (or the predicates and gate functions as provided in UltraSAN [9]) exactly one of these transitions can be made enabled for a given number of tokens; thus the tabular behaviour can be included in the model.

4 Time-based cyclic polling models

Many communication systems operate with some form of time-based scheduling. Therefore, in this section, we present SPN models that closely reflect such systems. Local timing is addressed in Section 4.1, whereas global timing is discussed in Section 4.2.

4.1 Local timing

Although count-based scheduling mechanism are interesting to study, there are some communication systems that do behave slightly differently. In particular, the IBM token ring (IEEE P802.5) and the standardized token bus (IEEE P802.4), use polling schemes where the time the server stays at a particular station is bounded by some time [5, 41]. In these systems, once the server arrives at a station, a count-down timer is started at an initial, station-dependent value THT (token holding time). Then the transmission of packets starts. The transmission of packets then can continue until either all packets are served, or until the timer has expired. In the latter case, the current packet being transmitted may be finished after which the server has to move to the next station.

Time-based polling models are difficult to grasp analytically although some approximations do exist (see e.g., [42]). Most studies try to approximate the

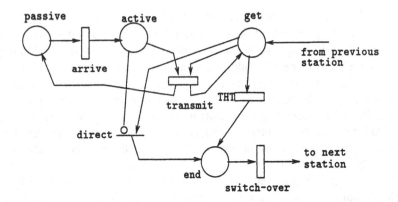

Fig. 6. SPN-based station model with local, exponentially distributed THT

time based scheduling by some form of k-limited scheduling [14]. Recently, de Souza e Silva *et al.* proposed a numerical approach based on embedded CTMC as well [37]; their approach seems to be closely related to the numerical approach employed when using DSPNs (see below).

We can easily model such local time-based polling models using SPNs, as already proposed in [16]. Consider the SPN as depicted in Figure 6. It represents a single station of a polling model. Once the token arrives at the station, i.e., a token is deposited in place get, two possibilities exist:

1. There are no customers buffered: the token is immediately released and passed to the next station in line, via the immediate transition direct;
2. There are customers buffered: these customers are transmitted and simultaneously, the token holding timer is started. In two ways the transmission process can end:
   - The token holding timer expires by the firing of transition THT with rate $1/tht$, in which case the token is passed to the next downstream station, thereby leaving behind one or more queued customers;
   - All customers are served via transition transmit before the token holding timer expires: the token is simply forwarded to the next downstream station.

Instead of using a single exponential transition to model the token holding timer, one can also use a more deterministic Erlang-J distributed token holding timer as depicted in Figure 7. There the number of exponential phases making up the overall Erlang distribution is present in the model via the multiplicity of the arc from count to expire and the rate of transition THT which equals J/tht. It should be noted that every time transmit fires, transition THT, modelling 1 of the J phases in the clock, is resampled. Due to the memoryless property of the negative exponential distribution, this is no problem. Of course, the number of

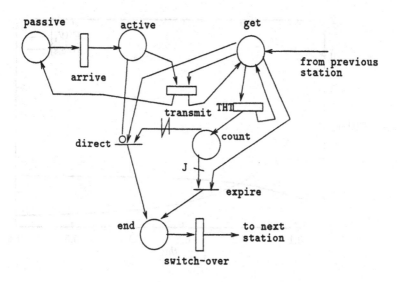

Fig. 7. SPN-based station model with local, Erlang-J distributed THT

clock "ticks", i.e., the number of tokens in count, since the arrival of the server at the station is not altered by a firing of transmit.

Using Erlangian clocks does increase the state space size (see the example below). To avoid this, one can follow the DSPN approach with deterministically timed THTs. Since at most one timer is active at any time, at most one deterministic transition is enabled at any time, so that an embedded Markov chain can be constructed to compute the steady-state probability distribution [31, 32] (to do so, the model needs to be adapted in order to avoid the "clock resampling" mentioned above). This has been done in [27] but also here the state-space size increased enormously with the number of stations and the number of customers per station ; for practically sized models, simulations were employed. The approach proposed by De Souza e Silva *et al.* is based on similar principles [37]. As before, one can also employ the earlier discussed folding and/or fixed-point techniques.

The influence of the THT in a symmetric model. Consider a 3-station cyclic polling model as depicted in Figure 7 with $J = 2$. The system is fully symmetric but for the THT values per station: we have tht_1 varying whereas $tht_{2,3} = 0.2$. The other system parameters are $\lambda = 3$, $E[S] = 0.1$ and $\delta = 0.05$ (all delays are assumed to be exponentially distributed). Setting a limit of at most 12 customers per station, the number of tangible states is 18759.

In Figure 8 we depict the average waiting times perceived at station 1 and stations 2 and 3 (the latter two are almost the same) when we vary tht_1 from 0.05 through 0.6 seconds. As can be observed, with increasing tht_1, the performance of station 1 is increased at the cost of stations 2 and 3.

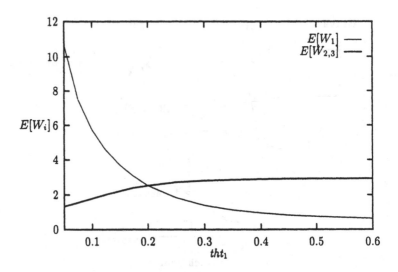

Fig. 8. The influence of tht_1 on the average waiting times in a symmetric system

Erlang-J distributed THT. Consider a symmetric cyclic polling model constituting of $N = 3$ stations of the form as depicted in Figure 7. As parameters we have $\lambda = 2$, $E[S] = 0.1$ and $\delta = 0.05$, and $tht = 0.2$ for all stations (all delays, except the THT, are assumed to be exponentially distributed).

In Table 1 we show, for increasing J, the state space size K and the number of non-zero entries η in the Markov chain generator matrix **Q** which is a good measure for the required amount of computation (the computation times varied from 100 to 350 seconds on a Sun Sparc 1+ using SPNP [7]), the expected waiting time, and the expected queue length. As can be observed, when the THT become more deterministic, i.e., when J increases, the performance becomes better. This is due to the fact that variability is taken out of the model. The simplest model is obtained for $J = 1$; its result can be used as an upper bound.

Table 1. The influence of the variability of the THT

J	K	η	$E[W]$	$E[N_q]$	J	K	η	$E[W]$	$E[N_q]$
1	4488	20136	0.778	1.731	5	12648	58056	0.648	1.482
2	6528	29616	0.699	1.580	6	14688	67538	0.642	1.471
3	8568	39096	0.671	1.527	7	16728	77016	0.638	1.463
4	10608	48576	0.657	1.499	8	18768	86496	0.635	1.457

The influence of the THT in an asymmetric model. Consider a 2-station polling model as depicted in Figure 7 with $J = 4$, i.e., we model the THT fairly deterministic. Furthermore, we have $E[S_i] = 0.5$ and $\delta_i = 0.1$. The asymmetry exists in the arrival rates: $\lambda_1 = 0.8$ and $\lambda_2 = 0.2$. The system is moderately loaded: $\rho = 0.5$.

In Figure 9 we show $E[W_1]$ and $E[W_2]$ as a function of *tht*, which is the same in both stations. For small values of the THT, the system behaves approximately as a 1-limited system and in the higher loaded station (1) a higher average waiting time is perceived. When the THT becomes very large, the system behaves as an exhaustive service system in which station 1 dominates and station 2 suffers. Indeed, for *tht* = 100 (not in the figure), we find the (approximate limiting) values $E[W_1] = 0.616$ and $E[W_2] = 0.947$ (which can also be computed differently). The performance perceived at station 2 is worse than at station 1, despite its lower load. When increasing the THT, $E[W_1]$ monotonously decreases: the larger the THT the more station 1 profits. For station 2 this is not the case. When the THT increases, station 2 first profits from the increase in efficiency that is gained. However, when the THT grows beyond 1.5, station 2 starts to suffer from the dominance of station 1.

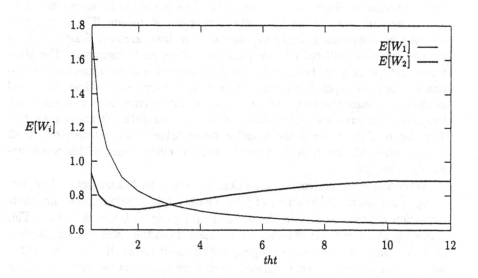

Fig. 9. The influence of the THT on the average waiting times in an asymmetric system

4.2 Global timing

In the fiber-optic FDDI ring, the timers that are employed are not local but global [5, 25, 41]. This means that a station starts a timer when it releases the token. Once it gets the token back, it is then allowed to transmit packets until its previously-set timer expires. In this way, the time it takes for the token to circle along all the stations can be bounded, thus providing facilities for guaranteed synchronous bandwidth allocation. A number of basic timing properties of FDDI access mechanism have been studied by Jain [25], Johnson [26] and Sevcik and Johnson [36].

The global time-based mechanism is very difficult to incorporate in analytical polling models. Apart from approximate approaches, e.g., those by Tangemann [42], we are not aware of any analytical results. Still though, the behaviour of such systems can easily be described using SPNs. It should be noted that the DSPN approach for modelling the timers is not applicable in this case anymore since in each of the stations that we model one timer is active. That is, there are multiple timers running concurrently, i.e., there is more than one deterministic transition enabled in most of the markings so that the "normal DSPN embedding" does not work anymore. Whenever a fixed-point iteration scheme would be used, every model would again comprise a single station only, so that again only one deterministic transition is enabled at any time so that then the DSPN approach is applicable again [31, 32].

As an example, we present an SPN model of a single station in a globally-timed system, as illustrated in Figure 10. The place world serves as an finite source and sink of jobs; initially it contains K tokens. The arrival process of jobs comprises a simple exponential transition arrive. Tokens in place buffer represent buffered jobs (or packets) waiting for transmission. The place timer_not_yet_expired contains a token whenever the global timer for this station has not yet expired. As can be seen from the figure, this timer is started at the time instance when the server leaves the station (transition start_timer), just before the server switches to the next downstream station (transition switch). The length of the timer is the so-called target token rotation time (ttrt). If all stations obey this timer, the (network) token will cycle around within a precomputable time [26, 36].

When the server arrives in place token and the timer has not yet expired, two things can happen. If the queue of jobs is empty (place buffer) the immediate transition empty fires so that the server starts going to the next station. This transition also takes care of removing all tokens from the timer-related places so that the timer can be started properly for the next round. If the place buffer does contain tokens, the thus represented jobs are served via the timed transition transmit in an exhaustive manner, i.e., the servicing of jobs continues until either all jobs have been served, or until the timer expires. The timer expires via the timed transition tick_tack. Notice that we have drawn a single transition here; we could also use Erlang-J distributed timers. Once the timer has expired, transition expired becomes enabled which starts moving the server to the next station. Also in this case, the new timer is started.

When a complete FDDI ring is modelled, multiple station submodels have to be connected to one another via the switch transitions and the places token. It should be noted that in the numerical experiment given below, in which we connected three stations, we use one place to model the environment of the system, i.e., one place from which all arrivals come and to which all served customers depart. The number of tokens in this place initially equals K.

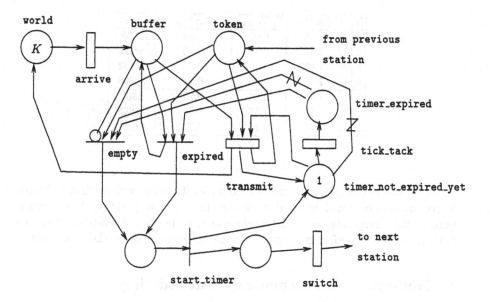

Fig. 10. An SPN-based polling model of a globally-timed cyclic server

As an example of a numerical evaluation, we studied a three-station symmetric system where $ttrt = 0.55$, where arrivals take place according to a Poisson process with rate 3 jobs/sec (as long as the finite source is non-empty) and where services take an exponential amount of time, with mean value 0.1 sec. Switch-overs are also exponentially distributed with mean 0.05 sec. The timer is modelled as an Erlang-J distribution, with $J = 1, \cdots, 4$. The number of tokens in place world, is taken to be 6 or 8. In Table 2 we list the obtained expected waiting time ($E[W_{\text{buffer}}]$) and the expected buffer occupancy ($E[\text{buffer}]$) for these scenarios. We also list the number of tangible states of the underlying CTMC and the probability that place world is non-empty. As can be observed, the state number increases rapidly. It is again observed that with increasing clock-determinism the performance measures of interest, i.e., the average buffer filling and the average waiting time, slightly decrease. Most gain is already reached when modelling the clock as an Erlang-2 distribution, instead of an exponential distribution. We furthermore observe that the probability that world is non-empty is considerable, and remains to be so when we increase K; moreover, it is almost not affected by larger values of K. We would need much more tokens in place world to increase this probability to (almost) 1 so that we model a truly infinite-source system; this would, however, increase the state space even more.

Table 2. Performance results for a globally-timed, three station symmetric polling model

K	J	$E[\texttt{buffer}]$	$E[W_{\texttt{buffer}}]$	$\Pr\{\texttt{NonEmpty}\}$	# states
6	1	1.46	0.62	0.68	2688
6	2	1.42	0.58	0.70	9828
6	3	1.41	0.56	0.70	24192
6	4	1.40	0.56	0.71	48300
8	1	1.96	0.80	0.73	5400
8	2	1.92	0.77	0.74	19845
8	3	1.91	0.75	0.74	48960

From the above simple numerical example, it becomes clear that although SPNs provide in principle good modelling facilities, in practice things might become unmanageable. We will therefore need to perform fixed-point approximations to derive performance measures for realistically-sized FDDI systems.

5 Non-cyclic, load-dependent visit ordering

In Section 5.1 we discuss a practical performance issue that illustrates the need for load-dependent polling models. We discuss two load-dependent polling schemes in Sections 5.2 and 5.3 together with the corresponding SPN models. Some numerical results are presented in Section 5.4. A more elaborate treatment of these models has been presented in [17].

5.1 Problem sketch: multiplexing in ATM systems

In ATM systems traffic streams from many sources have to be routed to many different destinations; this takes place in ATM switches. An important part of these switches is formed by simple multiplexers, i.e., by devices that can merge multiple cell (mini-packet) streams on a single outgoing line. During this multiplexing, some cells might need to be buffered for some time. In ATM systems it is customary to distinguish between cells belonging to a high priority class (real-time cells) and to cells belonging to a low priority class (see also [22]). The high-priority class than comprises cells that are delay sensitive, e.g., while they form part of some digitized audio or video stream, whereas the low-priority cells comprise delay-insensitive data, e.g., parts of files.

The above priority distinction can be taken into account in the multiplexing process. Within ATM multiplexers, often two (logical) buffers are used, one for each priority class. One can then grant absolute priority to one class, at the expensive of the other. Alternatively, one can also try to grant just that amount of priority service to the high-priority class, such that its delay requirements are just met. The remaining capacity can then be used to satisfy the non-delay sensitive traffic the best. Such a scheme can be interpreted as a polling scheme where the polling order depends on the load of the two involved queues (see Figure 11). It is important to note here that the modelled switch-overs between the two classes (or queues) required in the polling process do not take time. Two of these dynamic priority schemes will be presented in the sections below, together with the corresponding SPN models.

Since we are using Markovian SPNs, i.e., with exponentially distributed transition delays, we cannot directly incorporate deterministic timing delays, such as arise when modelling the transmission of ATM cells. Two ways can be followed to overcome this disadvantage: one can either use phase-type distributions, in particular Erlang distributions to approximate deterministic behaviour, or one can employ SPN-classes that allow for deterministic timing as well (DSPNs). Disadvantage of the former approach is the increase in state space size whereas in the the latter approach the modelling freedom is slightly restricted and the generation and evaluation of the underlying Markov chain becomes more costly. Throughout this paper we adhere to the former approach. However, the models presented here can, with suitable tools for DSPNs, also be used in the latter approach.

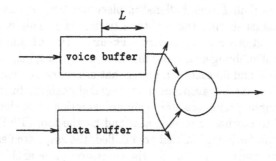

Fig. 11. An ATM multiplexer as a 2-station polling model

5.2 Threshold priority policy

In the threshold priority policy (TPP) [29], two buffers are used for the two traffic classes. A predetermined threshold L is associated with the real-time buffer.

When the queue length in the real-time buffer is less than or equal to L, the server cyclically polls and serves the two buffers (as long as a queue is not empty). On the other hand, when the queue length in the real-time buffer exceeds L, the server continues transmission only from the real-time buffer until its queue length is reduced to L. The value of the threshold L gives the degree of preferential treatment of the real-time traffic. When $L = 0$, real-time traffic is given an absolute priority. In case $L = \infty$, both traffic classes are served alternatingly when not empty, yielding a normal 1-limited scheduling. By selecting L between these two extremes, one may provide an adequate quality of service to both real-time and nonreal-time traffic.

In Figure 12 we depict the corresponding SPN-based polling model. On the left side, we see the arrival streams coming into the buffers for the two traffic classes. The arriving tokens, via the transitions arrive_rt and arrive_nrt, originate in separately-specified arrival process SPN models. The departing cells, indicated with realtime- and nonreal-time departures, flow back to these workload models. In this figure we do not show the arrival processes any further. In the evaluations that follow, we have used Poisson arrival processes for the low-priority data traffic, and Markov modulated Poisson processes (MMPPs) for the high-priority real-time traffic. In particular, we use an implementation of Saito's model for digitized video [35].

The server is represented by the single token that alternates between places try-rt and try-nrt. After a cell of one class is served (via either transition serve-rt or serve-nrt) the server polls the other class. When nothing is buffered for a particular traffic class, the server also polls the other class, via the transitions empty-rt and empty-nrt. However, depending on whether there are more or less than L cells buffered in place buff-rt, it can be decided that the server remains serving the real-time traffic class. This is enforced by the immediate transitions rt-rt, rt-nrt, nrt-nrt, nrt-rt1 and nrt-rt2. Apart from the normal enabling conditions for these transitions, i.e., at least a token in every input place and no tokens in places that are connected via an inhibitor arc to the transition, these transitions have so-called enabling functions associated with them (enabling functions are again an extension of of classical SPNs that are supported in stochastic reward nets and by the tool SPNP). Whenever all normal condition for being enabled are fulfilled, the associated enabling function is validated and must yield true for the transition to be enabled. The enabling functions for the five immediate transitions are given in Table 3; they are taken such that the TPP is exactly enforced. They also assure that infinite firings of immediate transitions cannot occur.

It should be noted that the polling of the two buffers does not cost time. All the involved transitions are of immediate type. We are also *not* claiming that the polling procedure in a real multiplexer is *implemented* in the same way as *modelled* here; we are only claiming that the above sketched model has the same functional behaviour as the real multiplexer with respect to the service ordering.

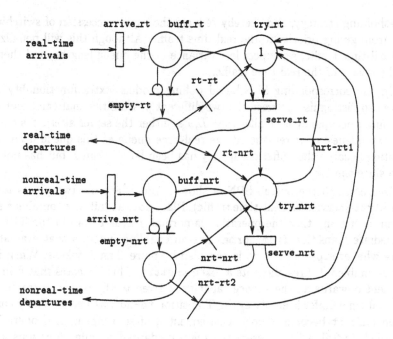

Fig. 12. The TPP as an SPN-based polling model

Table 3. Enabling functions for the immediate transitions in the TPP model

transition	enabling condition
rt-rt	$(\#\text{buff\_rt} > L)$ or $((\#\text{buff\_rt} > 0)$ and $(\#\text{buff\_nrt} = 0))$
rt-nrt	$(\#\text{buff\_nrt} > 0)$ and $(\#\text{buff\_rt} \leq L)$
nrt-nrt	$(\#\text{buff\_rt} = 0)$ and $(\#\text{buff\_nrt} > 0)$
nrt-rt1	$(\#\text{buff\_rt} > L)$
nrt-rt2	$(\#\text{buff\_rt} > 0)$

5.3 Exhaustive threshold priority policy

When the TPP is used in combination with Poisson arrival processes, it can
lead to acceptable performance for both traffic classes, as shown in [29, 17].
However, when used in combination with MMPP real-time traffic, it is highly
likely that by the bursty nature of the sources, once the real-time buffer exceeds
L, the buffer will fill rapidly. Instead of again polling the nonreal-time buffer as
soon as this buffer occupancy is smaller than L, one could give priority to the
real-time traffic for the duration of the burst. This can be done by serving the
real-time buffer until it is empty. This policy, denoted as exhaustive threshold
priority policy (ETPP) and introduced in [17], introduces an hysteresis in the

thresholding strategy, and thereby reduces the rapid succession of switching to and from giving priority to the real-time traffic. Although this will penalize the nonreal-time traffic, we expect less variance in the queue lengths, and therefore in the delays of the real-time traffic.

In the corresponding SPN-based polling models, extra functionality is required to distinguish between the two different possibilities that can occur when the buffer occupancy is smaller than L, i.e., either the server alternates between the two queues, or it remains at the real-time queue as this queue needs to be emptied because the buffer occupancy has been larger than L but has not been zero since then.

In Figure 13 the overall SPN is shown. Added are the two places `normal` and `strtt` (service to real-time traffic). Place `normal` initially contains a single token, indicating that the operation is normal, i.e., as before in the TPP. The immediate transition `from-normal` has an enabling function that evaluates to true whenever place `buff_rt` is occupied by more than L tokens. When it becomes enabled, it fires and puts a token in `sttrt`. This indicates that from that moment onwards all the service capacity is given to the real-time traffic, until the real-time buffer is empty again. Transition `to-normal` becomes only enabled when `buff_rt` becomes empty. Consequently, after firing, normal operation is resumed. In Table 4 we present the slightly adapted enabling functions for the ETPP.

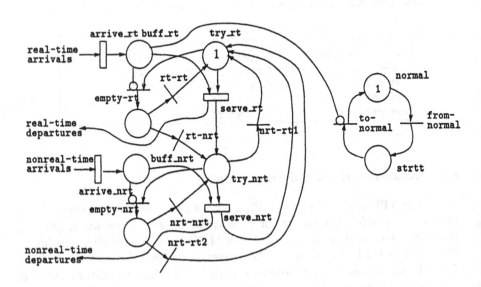

Fig. 13. The ETPP as an SPN-based polling model

Table 4. Enabling functions for the immediate transitions in the ETPP model

transition	enabling condition
rt-rt	$(\#\texttt{sttrt}= 1)$ or $((\#\texttt{buff\_rt}> 0)$ and $(\#\texttt{buff\_nrt}= 0))$
rt-nrt	$(\#\texttt{buff\_nrt}> 0)$ and $(\#\texttt{normal}= 1)$
nrt-nrt	$(\#\texttt{buff\_rt}= 0)$ and $(\#\texttt{buff\_nrt}> 0)$
nrt-rt1	$(\#\texttt{buff\_nrt}= 0)$ or $(\#\texttt{sttrt}=1)$
nrt-rt2	$(\#\texttt{buff\_rt}> 0)$
from-normal	$(\#\texttt{buff\_rt}> L)$

5.4 Some numerical results

In this section we present some numerical results. It should be noted that to the best of our knowledge, these results can not be obtained by any of the known analytical techniques.

TPP under MMPP traffic. In this section we present the analysis of the TPP under MMPP traffic. Such a combination of workload and system models can not be handled anymore with the approach presented in [29]. We use the following parameters. The server speed is 600 Mbps, or 1415094 cells/sec. The nonreal-time traffic, modelled as a Poisson process, amounts for 40% of the load. The real-time traffic is, for every source (parametrically increased from 1 to 11) described using the model of Saito and the following parameters: cells are arriving at a peak rate of 44.7 Mbps and the average bit rate is 16.8 Mbps [35]. The cell service times have an Erlang-2 distribution; using more phases to represent the deterministic length of ATM cells caused too many states in the underlying CTMC.

In Figure 14 we depict the average cell response time (delay) for increasing number of video sources. For $L = 0$, i.e., the absolute priority limiting case, the mechanism still works fine. For $L = 4$, the operation already becomes less pronounced. Moreover, for $L = 6$, the aimed-at strategy does not seem to work anymore. Indeed, under high load, the average delay for the real-time traffic is higher than for the nonreal-time traffic! This is exactly the opposite of what the mechanism aims at. The TPP does not seem to be able to cope well with the bursty character of the MMPP traffic. These effects did not show up in the study of Lee and Sengupta [29] where only Poisson sources where assumed.

TPP versus ETPP under MMPP traffic. In Figures 15 through 17 we present some performance results of the ETPP and compare it with the TPP, when the traffic is, at least for the real-time part, of MMPP type.

In Figure 15 we observe that, for a fixed threshold $L = 6$, the ETPP does indeed decrease the expected response time for real-time cells at the cost of a small increase in the expected response time for nonreal-time cells. Indeed, the earlier distinguished better performance (smaller average response time) for the nonreal-time traffic has vanished.

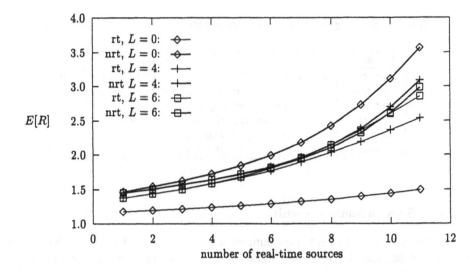

Fig. 14. $E[R]$ (in μsec) for the TPP under MMPP traffic for increasing number of real-time (video) sources

Moreover, in Figure 16 it can be observed that also the variance of the real-time buffer occupancy decreases by the proposed modification of the scheduling policy. This is a very nice property, as this also reduces the delay jitter in the real-time traffic. This property only comes at the cost of a small increase in variance of the nonreal-time buffer occupancy. A less pronounced but similar effect has been observed in case $L = 4$ (not shown here).

In Figure 17 we depict the buffer-full probability β for a fixed buffer size of 15 cells, a threshold $L = 6$, and an increasing number of supported real-time sources (note the logarithmic β-scale). The earlier mentioned video model of Saito is used. Cell service times are assumed to be Erlang-2 distributed. As to be expected, β increases with increasing load. We also observe a small improvement when comparing ETPP with TPP.

The number of states in the underlying Markov chain varies between 3153 for 1 real-time source up to 81978 states when 11 real-time sources are modelled. To solve the last model, we required 90 minutes on our SUN SParc 10 with 32 MB main memory.

6 Matrix-geometric SPNs and polling models

As the previous sections have shown, SPNs do allow us to specify with relative ease rather intricate polling models. As main disadvantage, however, the rela-

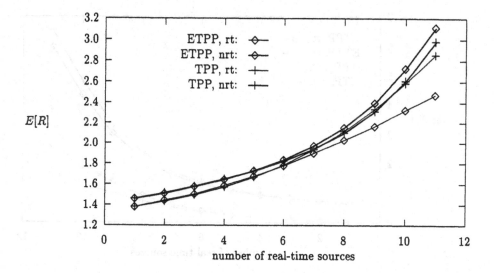

Fig. 15. The average response time $E[R]$ (in μsec), $L = 6$, for increasing number of real-time (video) sources

tive fast state space size increase should be mentioned. To overcome the latter problem, we have been investigating subclasses of SPNs for which the solution is less costly, most notably while the structure of the underlying Markov chain is exploited.

Below, we will address in Section 6.1 the background of the class of SPNs that allows for the efficient matrix-geometric solution. In Section 6.2 we will illustrate the approach by the modelling of an ATM multiplexer with the threshold-priority scheduling strategy (as discussed before).

6.1 Background

In many of the performance models appearing when studying communication systems, the state space of the underlying CTMC has the form of a two-dimensional "strip" that is finite in one dimension, and infinite but regular in the other dimension. This is, e.g., the case in models of multiplexers where there is one (unbounded) buffer, and where a single multi-state arrival process deals with job arrivals and a single multi-state service process deals with the servicing of jobs. Whenever such models occur, matrix-geometric solutions might apply to compute the steady-state distribution of the underlying CTMC. When employing such methods, the regular structure of the CTMC on the infinitely long strip of states is explicitly exploited.

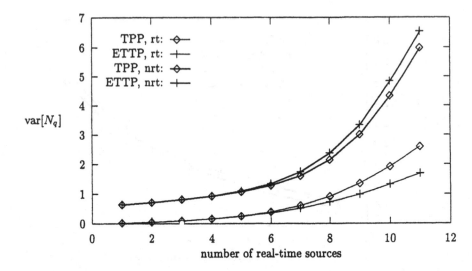

Fig. 16. The variance var[N_q] (in cell^2) of the number of queued real-time and non-real-time packets, $L = 6$, for increasing number of real-time (video) sources

The suitability of matrix-geometric solution approaches, as initiated by Neuts [34], has been known for some time, however, the ease of application of these methods has often been a problem. Until recently, these methods could only be applied when specifying the models at the state space level, i.e., by hand-coding the entries of the infinite but regular generator matrix of the CTMC. As an exception to the rule, the tool XMGM should be mentioned; with this tool a restricted class of queueing models could be specified easily, which was then subsequently solved using matrix-geometric techniques [15].

In order to facilitate the use of the powerful matrix geometric methods also to more system-oriented researchers and system developers, we have developed a class of SPNs that allows for a matrix-geometric solution [18]. This class, which is sometimes also indicated as the class of 1-place unbounded SPNs (as introduced by Florin and Natkin [11, 12]), is restricted in comparison to the general class of SPNs, however, with some experience many practical performance evaluation problems can be cast in it. A formal definition of this class of SPNs is possible, and has been given in [18]. Characteristic for this class is the fact that there is one place that can contain, in principle, an infinite number of tokens. Changes to the number of tokens in this place always take place with unit steps, and the way in which these changes can take place, i.e., the rate at which these changes occur and the resulting marking, is, starting from a certain positive threshold onwards, independent of the actual number of tokens in that place. Furthermore, also the

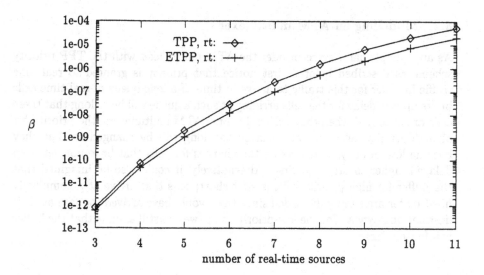

Fig. 17. The blocking probability β for increasing number of real-time (video) sources and $L = 6$

changes that take place in the marking, given a fixed number of tokens in this unbounded place, should be independent of this number for situations "above" the threshold. Extensions to this class are still possible, e.g., taking into account cases where non-unit step changes affect the buffer (thus modelling batch arrivals and/or services; see also [20]).

When the above requirements are fulfilled, the solution of the underlying infinitely large CTMC boils down to the solution of a so-called *quasi birth-death model* [18, 34]: (i) the part of the CTMC that has an unregular structure, the so-called boundary part below the above-mentioned threshold, is solved via a system of linear equations, which in general is quite small (up to a 100 unknowns); (ii) the part of the CTMC that has a regular structure, the so-called repeating part up and above the threshold, can be solved by deriving the minimal non-negative solution of a matrix-quadratic equation. Although this equation can be solved quite efficiently with the recently developed iteration scheme by Latouche and Ramaswami [28], it stil forms the most time-consuming part of the computations.

With the tool SPN2MGM we are able to specify SPNs that fulfill the above conditions. The tool then recognises the special structure of the underlying Markov chain and solves it using the matrix-geometric approach. Details on SPN2MGM can be found in [19]. The model described below has been constructed and solved with SPN2MGM.

6.2 Modelling an ATM multiplexer

As an example here, we reconsider the ATM multiplexer with the TPP priority scheme as described before. First notice that priority is granted to real-time traffic in order for this traffic to arrive in time. If a long queue of real-time cells builts up, the delay for the cells arriving at such a queue will be so long that these cells are outdated when they arrive. In some ATM multiplexers, the priority bit of high-priority cells that are delayed too long will be changed so that they become low priority, thus making the chances for cells that have not yet been delayed higher to arrive in time. Alternatively, it could also be imagined that the buffer for high-priority cells is kept short; cells that arrive at a completely filled buffer are simply discarded since they would have arrived too late at their destination anyway. For the low-priority cells, we can still assume that the buffer is infinitely large.

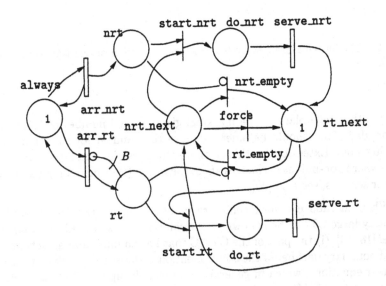

Fig. 18. The load-based cyclic polling model of a two-class ATM multiplexer under the TPP scheduling discipline represented as an SPN

Keeping the above considerations in mind, we construct an SPN model of a simple two-class TPP-scheduled ATM multiplexer as depicted in Figure 18. The two places rt and nrt model the buffers for real-time and nonreal-time traffic. Arrivals for these two classes of traffic take place via the transitions arr_rt and arr_nrt respectively. Note that transition arr_nrt is always enabled. Transition arr_rt, however, is disabled whenever there are B tokens (cells) in the buffer for real-time traffic, thus modelling the finiteness of that buffer. The fact that place always always contains a token implies that the underlying CTMC is of infinite size. Since we will use matrix-geometric techniques, this is no problem. Then, the single output line of the multiplexer should be shared by both the priority classes. Whenever there are tokens in place rt (or nrt) and the server is free and available for real-time (nonreal-time) traffic, i.e., there is a token in place rt_next (nrt_next), the corresponding starting transition start_rt (or start_nrt) fires and the transmission of the cell takes place via transition serve_rt (serve_nrt). After the service of a nonreal-time or real-time cell, the next cell to be served is, in principle, of the opposite class, represented by the depositing of a token in place rt_next respectively nrt_next. However, whenever a particular priority class is polled and the queue for that class is found empty, the other queue is tried again. This is modelled by the transitions rt_empty and nrt_empty; notice the connected inhibitor arcs. In order to avoid infinite cycling of the server between the two queues when both are empty, the transition rt_empty has as extra enabling predicate the condition (nrt > 0) so that when both queues are empty, the server waits in place rt_next. In order to give the real-time traffic class its priority when needed, an extra immediate transition force has been introduced. This transition, which has higher priority than all other immediate transitions, forces the free server to go from nrt_next to rt_next whenever its enabling predicate (rt > L) holds.

It should be noted that we have aimed at showing the load-dependent polling structure in this model. Also, as before, the model does not say anything about the actual implementation of the multiplexer.

This load-based polling SPN model does allow for a matrix-geometric solution. When the tool SPN2MGM is provided with a description of this SPN, it computes the steady-state solution of this model using matrix-geometric techniques.

As a small numerical example, consider the following parameters. We assume that both the arrival transitions have (arrival) rate 3 jobs per second and the services take on average 0.1 second. This implies a total load of 60%. The buffer size B for real-time traffic is assumed to equal 5, and the threshold L varies between 0 and 5. In Figure 19 we show the expected number of jobs waiting in places rt and nrt as a function of the threshold L. As can be observed, the larger L the less preferential treatment the real-time traffic obtains. For smaller L, the real-time traffic profits at the cost of the nonreal-time traffic. In Figure 20 we then show the probability that the real-rime buffer is occupied more than the threshold L as a function of L. This measure shows how effective the thresholding mechanism indeed keeps the real-time buffer with its desired range $[0, L]$.

7 Concluding remarks

In this paper we have presented SPN-based polling models that can be used to evaluate the performance of a number of well-known computer-communication systems such as token rings and token busses as well as ATM multiplexers. The SPN-based approach has as advantage over analytical approaches that complex scheduling strategies can often be modelled in more detail. This, however, comes

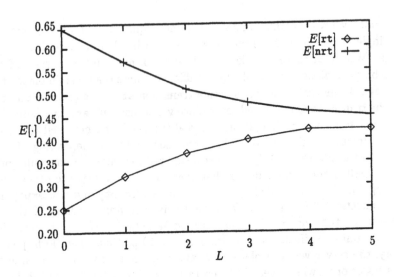

Fig. 19. The expected buffer occupancies for real-time and nonreal-time traffic as a function of L

at the cost of large CTMCs that need to be solved, and the limitation of only phase-type distributions. Depending on the application at hand, various ways can be chosen to circumvent the largeness problems. A fixed-point decomposition and a folding method have been presented that, in an approximate way, alleviate the this problem. By using DSPNs instead of (Markovian) SPNs the state-space problem can also be avoided partially, since non-exponential behaviour need not be approximated any more by the state-space enlarging phase-type distributions. The employment of matrix-geometric solutions can also help in avoiding the explicit generation of large state space, simply because the structure of the infinite state space is exploited. Of most of the above approaches, we presented examples, showing their basic operation and practical applicability. It is up to a modeller to decide on a particular approach. Which approach is best in a certain situation depends really on the modelling problem at hand as well as on the availability of tools.

As topics for further research, we see a further development of the matrix-geometric approach, e.g., by further developing the tool SPN2MGM and by speeding up the internal computations; first results in that respect can be found in [20]. The fixed-point iterative approach has not been applied yet in combination with DSPNs; this also needs further study. It is also of interest to investigate how the matrix-geometric approach can be applied within a fixed-point iteration scheme, thus allowing for multiple-place-unbounded SPNs.

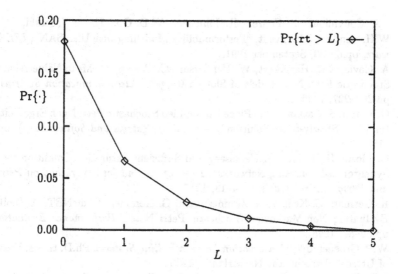

Fig. 20. The probability the buffer occupancy for real-time traffic exceeds L

Acknowledgements

The author would like to thank the anonymous reviewers for their careful comments on an earlier version of this paper.

References

1. M. Ajmone Marsan, G. Conte, G. Balbo, "A Class of Generalized Stochastic Petri Nets for the Performance Evaluation of Multiprocessor Systems", *ACM Transactions on Computer Systems* **2**(2), pp.93–122, 1984.
2. M. Ajmone Marsan, S. Donatelli, F. Neri, "GSPN Models of Markovian Multiserver Multiqueue Systems", *Performance Evaluation* **11**, pp.227–240, 1990.
3. M. Ajmone Marsan, S. Donatelli, F. Neri, U. Rubino, "On the Construction of Abstract GSPNs: An Exercise in Modelling", *Proceedings PNPM91*, IEEE Computer Society Press, pp.2–17, 1991.
4. M. Ajmone Marsan, S. Donatelli, F. Neri, U. Rubino, "GSPN Models of Random, Cyclic, and Optimal 1-Limited Multisever Multiqueue Systems", *Proceedings ACM Sigcomm*, IEEE Computer Society Press, pp.69–80, 1991.
5. W. Bux, "Token-Ring Local-Area Networks and Their Performance", *Proceedings of the IEEE* **77**(2) pp.238–256, 1989.
6. H. Choi, K.S. Trivedi, "Approximate Performance Models of Polling Systems using Stochastic Petri Nets", *Proceedings IEEE INFOCOM'92*, pp.2306-2314, 1992.
7. G. Ciardo, J. Muppala, K.S. Trivedi, "SPNP: Stochastic Petri Net Package", *Proceedings PNPM89*, IEEE Computer Society Press, pp.142–151, 1989.
8. G. Ciardo, J. K. Muppala, and K. S. Trivedi, "On the Solution of GSPN Reward Models", *Performance Evaluation* **12**(4), pp. 237-254, 1991.

9. J.A. Couvillion, R. Freire, R. Johnson, W.D. Obal II, A. Qureshi, M. Rai, W.H. Sanders, J.E. Tvedt, "Performability Modelling with UltraSAN", *IEEE Software*, pp.69–80, September 1991.

10. A. Coyle, B.R. Haverkort, W. Henderson, C. Pearce, "A Mean-Value Analysis of Stochastic Petri Net Models of Slotted Rings", *Telecommunication Systems* 6(2), pp.203–227, 1996.

11. G. Florin, S. Natkin, "One Place Unbounded Stochastic Petri Nets: Ergodicity Criteria and Steady-State Solution", *Journal of Systems and Software* 1(2), pp.103–115, 1986.

12. G. Florin, S. Natkin, "A Necessary and Sufficient Saturation Condition for Open Synchronized Queueing Networks", *Proc. of the 2nd Int'l Workshop on Petri Nets and Performance Models*, pp.4–13, 1987.

13. R. German, C. Kelling, A. Zimmermann, G. Hommel, "TimeNET: A Toolkit for Evaluating Non-Markovian Stochastic Petri Nets", *Performance Evaluation* 24, pp.69–87, 1995.

14. W.P. Groenendijk, *Conservation Laws in Polling Systems*, Ph.D. thesis, University of Utrecht, Utrecht, the Netherlands, 1990.

15. B.R. Haverkort, A.P.A. van Moorsel, D.-J. Speelman, "Xmgm: A Performance Analysis Tool Based on Matrix Geometric Methods", in: *Proceedings of the Second International Workshop on Modelling, Analysis and Simulation of Computer and Telecommunication Systems*, IEEE Computer Society Press, pp.152–157, 1994.

16. B.R. Haverkort, "Polling Models: Theory and Applications", *ÖCG-Schriftenreihe* 73, Oldenbourg Verlag, pp.237–266, 1994.

17. B.R. Haverkort, H. Idzenga, B.G. Kim, "Performance Evaluation of ATM Switch Architectures using Stochastic Petri Nets", in: *Performance Modelling and Evaluation of ATM Networks*, Editor: D. Kouvatsos, IFIP series, Chapman and Hall, London, pp.553–572, 1995.

18. B.R. Haverkort, "Efficient Solution of a Class of Infinite Stochastic Petri Nets: Theory and Applications", *Proceedings of the International Computer Performance and Dependability Symposium*, IEEE Computer Sciety Press, pp.72–81 , 1995.

19. B.R. Haverkort, "SPN2MGM: Tool Support for Matrix Geometric Stochastic Petri Nets", *Proceedings of the 1996 International Computer Performance and Dependability Symposium*, IEEE Computer Society Press, pp.219–228, 1996.

20. B.R. Haverkort, A. Ost, "Steady-State Analysis of Infinite Stochastic Petri Nets: A Comparison between the Spectral Expansion and the Matrix-Geometric Method", *Proceedings of the Seventh International Workshop on Petri Nets and Performance Models*, IEEE Computer Society Press, 1997.

21. H. Heffes, D.M. Lucantoni, "A Markov Modulated Characterization of Packetized Voice and Data Traffic and Related Statistical Multiplexer Performance", *IEEE Journal on Selected Areas in Communications* 4(6), pp.856–868, 1986.

22. ITU-T Recommendation I.371, "Integrated Services Digital Network (ISDN), Overall Network Aspects and Functions—Traffic Control and Congestion Control in B-ISDN", International Telecommunication Union, March 1993.

23. O.C. Ibe, K.S. Trivedi, "Stochastic Petri Net Models of Polling Systems", *IEEE Journal on Selected Areas in Communications* 8(9), pp.1649–1657, 1990.

24. O.C. Ibe, H. Choi, K.S. Trivedi, "Performance Evaluation of Client-Server Systems", *IEEE Transactions on Parallel and Distributed Systems* 4(11), pp.1217–1229, 1993.

25. R. Jain, "Performance Analysis of FDDI Token Ring Networks: Effects of Parameters and Guidelines for Setting TTRT", *IEEE Magazine of Lightwave Telecommunication Systems*, pp.16–22, May 1992.

26. M.J. Johnson, "Proof that the Timing Requirements of the FDDI Token Ring Protocol are Satisfied", *IEEE Transactions on Communications* 35(6), pp.620–625, 1987.

27. C. Kelling, R. German, A. Zimmermann, G. Hommel, "TimeNET: ein Werkzeug zur modellierung mit zeiterweiterten Petri Netzen", *Informationstechnik unf Technische Informatik* 37(3), pp.21–27, 1995 (in german).

28. G. Latouche, V. Ramaswami, "A Logarithmic Reduction Algorithm for Quasi Birth and Death Processes", *Journal of Applied Probability* 30, pp.650–674, 1993.

29. D.-S. Lee, B. Sengupta, "Queueing Analysis of a Threshold Based Priority Scheme for ATM Networks", *IEEE/ACM Transactions on Networking* 1(6), pp.709–717, 1993.

30. H. Levy, M. Sidi, "Polling Systems: Applications, Modeling and Optimization", *IEEE Transactions on Communications* 38(10), pp.1750–1760, 1990.

31. C. Lindemann, "An Improved Numerical Algorithm for Calculating Steady-State Solutions of Deterministic and Stochastic Petri Net Models", *Proceedings PNPM91*, IEEE Computer Society Press, pp.176–185, 1991.

32. C. Lindemann, R. German, "DSPNexpress: A Software Package for Efficiently Solving Deterministic and Stochastic Petri Nets", in: *Computer Performance Evaluation 1992: Modelling Techniques and Tools 1992*, Editors: R. Pooley, J. Hillston, Edinburgh University Press Ltd., 1993.

33. A. Lindeyer, *Phase-Type Approximations in Fixed-Point SPN Polling Models*, technical report, University of Twente, 1994 (in dutch).

34. M.F. Neuts, *Matrix Geometric Solutions in Stochastic Models—An Algorithmic Approach*, The Johns Hopkins University Press, 1981.

35. H. Saito, *Teletraffic Technologies in ATM Networks*, Artech House, Boston, 1994.

36. K.C. Sevcik, M.J. Johnson, "Cycle Time Properties of the FDDI Token Ring Protocol", *IEEE Transactions on Software Engineering* 13(3), pp.376–385, 1987.

37. E. de Souza e Silva, H.R. Gail, R.R. Muntz, "Polling Systems with Server Timeouts and Their Application to Token Passing Networks", *IEEE/ACM Transactions on Networking* 3(5), pp.560–575, 1995.

38. H. Takagi, *Analysis of Polling Models*, MIT Press, 1986.

39. H. Takagi, "Queueing Analysis of Polling Models", *ACM Computing Surveys* 20(1), pp.5–28, 1988.

40. H. Takagi, "Queueing Analysis of Polling Models: An Update", in: *Stochastic Analysis of Computer and Communication Systems*, Eds.: H. Takagi, North-Holland, 267–318, 1990.

41. A.S. Tanenbaum, *Computer Networks*, Second Edition, Prentice-Hall, 1989.

42. M. Tangemann, "Mean waiting Time Approximations for Symmetric and Asymmetric Polling Systems with Time-Limited Service", in: *Messung, Modellierung und Bewertung von Rechen- und Kommunikationssystemen*, Eds.: B. Walke, O. Spaniol, Springer-Verlag, pp.143–158, 1993.

43. J.A. Weststrate, *Analysis and Optimization of Polling Models*, Catholic University of Brabant, Tilburg, the Netherlands, 1992.

Structural Decomposition and Serial Solution of SPN Models of the ATM GAUSS Switch*

Boudewijn R. Haverkort[1] and Hessel P. Idzenga[2]

[1] Laboratory for Distributed Systems, RWTH-Aachen, D–52056 Aachen, Germany
http://www-lvs.informatik.rwth-aachen.de/
[2] Lucent Technologies, P.O. Box 18, 1270 AA Huizen, the Netherlands

Abstract. We address the performance, in particular, the cell loss ratio, of the ATM GAUSS switch under a variety of realistic video and constant bit rate traffic patterns.

We describe the operation of the GAUSS switch and derive a stochastic Petri net model for it. One problem with this model, when subjected to realistic traffic, is that it is too large (in terms of states of the underlying Markov chain) to be analysed. We circumvent this largeness problem by structurally decomposing this model in a number of smaller models that can be solved in a serial fashion, thereby using analysis results of one another. This approach not only speeds up the solution process by several orders of magnitude, it also still yields accurate results.

With respect to the GAUSS switch we show that under realistic traffic, the internal buffers need to be doubled in size, as opposed to analysis results under Poisson traffic, to yield acceptable cell-loss performance.

Concluding, this paper serves three aims: (i) it shows the suitability of stochastic Petri nets in the context of ATM system analysis; (ii) it illustrates a structural decomposition method circumventing the state space explosion problem; and (iii) it derives more detailed performance results for the GAUSS switch than has been possible previously.

1   Introduction

ATM (Asynchronous Transfer Mode) switches and multiplexers have been subjected to performance evaluations of many kinds. Analytical performance evaluation studies have mainly focussed on individual ATM multiplexers under symmetric traffic conditions. "Closed-form" analysis techniques that are often employed are matrix-geometric techniques [8, 10, 23], and generating functions [17]. When the systems to be analysed become more complex, or the traffic conditions become more complex or asymmetric, these analytical techniques fall short. In most of these cases, simulation studies are then performed. However, also simulation suffers from some drawbacks, most notably its relatively high cost, especially in obtaining accurate estimates for small quantities such as blocking probabilities, and the error-prone process of coding simulation programs.

---

* This work his been done while B.R. Haverkort was assistant professor at the University of Twente, and H.P. Idzenga was writing his M.Sc. thesis [14] there.

In between the closed-form analytical and the simulation approach, i.e., "in between" with respect to both the modelling capabilities and the evaluation costs, lies the numerical approach based on stochastic Petri nets (SPNs). SPNs allow for a very flexible construction and solution of, possibly large, continuous-time Markov models of ATM switches. These models can be solved numerically, using powerful current-day software packages and workstations. Advantage of the SPN approach over simulation is that rare-events are much less of a problem. Also, by its formally well-established semantics, SPN models are easily constructed and less error-prone than C-coded simulation programs. By the fact that a numerical solution is employed, more generality in the models can be achieved than with the closed-form analytical approaches, as will be illustrated in this paper.

It is surprising how little SPNs have been used for the performance evaluation of ATM-related systems despite its attractive properties regarding model construction and solution. We are only aware of a few other SPN-based performance studies in the ATM/B-ISDN context (Broadband Integrated Services Digital Network). In [15], Kant and Sanders analyse the Knockout switch under non-uniform and bursty traffic. The conclusions from their analyses are that the cell loss probabilities under the non-uniform and bursty traffic conditions are much higher than under more mild, i.e., Poisson, traffic conditions. In this respect, the SPN-based approach revealed more detailed characteristics than earlier performed analytical performance studies. Haverkort *et al.* [11] analyse a number of cell scheduling policies in ATM multiplexers under a variety of cell arrival patterns. This study extended some results derived by Lee and Sengupta [17]. Finally, at a recent Dagstuhl seminar (May 22-26, 1995), various presentations of SPN-usage in the context of ATM were reported [22, 25]. One problem with the use of (continuous-time) SPNs might be the lack of possibilities to include deterministic timing (see also [6] for an overview, or [3, 20, 9] for the state-of-the-art in deterministic and stochastic Petri nets (DSPNs)). In this paper, we will use phase-type distributions to model non-exponential behaviour (as also proposed in [2])

In this paper we will analyse the performance of a particular ATM switch, the GAUSS switch developed by De Vries at KPN Research [1, 28]. In our analyses, we focus on the determination of the cell loss ratio. One of the problems that comes along when employing SPNs for the modelling of such large systems, especially under realistic traffic conditions, is the state-space explosion. In this paper, we therefore present a structural decomposition of the overall model and show that by a serial solution of a number of submodels, thereby using results from previously analysed submodels in the analysis of next submodels, results in a far less time- and space-consuming solution trajectory, while the performance measures of interest are shown to be very accurate, as compared to the overall solution.

Our structural decomposition and serial solution bears resemblance with the (fixed-point) decomposition approach proposed by Ciardo and Trivedi [5]. We do, however, not require a fixed-point iteration; a "one-way" pass through the submodels is sufficient, i.e., the *import graph* (terminology from [5]) is acyclic. Secondly, the way we propose to interrelate the submodels is very much tailored towards the specific application.

In this paper we use the SPN formalism as proposed by Ciardo *et al.*, as supported by their package SPNP [4]. The employed terminology therefore also refers to the class of SPN models as supported by that package.

This paper is organised as follows. In Sect. 2 we describe realistic cell arrival processes that will be used for our analyses. Sect. 3 describes the operation of the GAUSS ATM switch. The modelling of this switch using stochastic Petri nets is presented in Sect. 4. The analyses carried out are described in Sect. 5. We conclude the paper in Sect. 6.

2 Workload Models

To obtain reliable results of a performance analysis of an ATM switch, its workload model should accurately represent the statistical properties of the real traffic. Simple traffic can be described using Poisson and deterministic arrivals, of which the SPN models are given in Sect. 2.1. MMPP workload models are capable of modelling more complex traffic like video and data. Such models, and their corresponding SPNs will be presented in Sect. 2.2.

2.1 Poisson and Deterministic Arrivals

In a Poisson arrival process, the interarrival times of the cells are exponentially distributed. In SPNs, timed transitions (thick bars) have an exponentially distributed firing time, and, consequently, a Poisson arrival process can be modelled with only one timed transition, as depicted in Fig. 1(a). If a token is present in place pop (population), then transition arr (arrival) fires after an exponentially distributed time with mean $1/\lambda$. In this arrival process, two successive arrivals can take place very quickly. In real systems, however, cell generations will be one packetizing-time apart, which means that the interarrival time has a more deterministic behaviour. A deterministic interarrival time can be approximated with an Erlang-k distribution, with squared coefficient of variation $C_v^2 = 1/k$. An SPN model for this distribution is given in Fig. 1(b), and is a sequence of exponentially distributed delays (arr.i), all with firing rate $k \times \lambda$. The token in place arridle guarantees that only one token can be present in the places queue.i ($1 \le i \le k - 1$). As for the Erlang-k distribution, also for other phase-type distributions [23] simple SPN representations can be found, such as e.g., proposed by Chen *et al.* [2].

2.2 MMPP Arrival Processes

Although requests for data services can sometimes be modelled adequately using Poisson or deterministic arrivals, telephony services are better modelled as

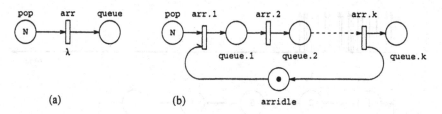

Fig. 1. SPN representation of (a) Poisson process, and (b) Erlang-k renewal process

on-off sources with deterministic interarrival times [13]. Video traffic has statistical properties that cannot be described by simple renewal processes; short- and long term correlations demand the use of nonrenewal models. Markov modulated Poisson processes (MMPPs) offer the flexibility to model these traffic characteristics. In current-day video coding algorithms like H.261 [7] and MPEG [18], the amount of data in one frame is dependent on the rate of movement in the video sequence. Video sequences in which a lot of movement is present, or a lot of scene-changes occur, require more data for a good representation of an image. After coding, the ATM Adaptation Layer (AAL) segments this data to meet the 48 byte payload requirement of ATM cells. Two possible algorithms to transmit the information are:

1. The AAL buffers the cells and transmits them to the receiver at a controlled speed, dependent on the amount of data to be transmitted. The cells are distributed uniformly over one frame-time τ (1/30st of a second);
2. The AAL buffers the cells and transmits them at the highest possible speed. In this case, cells are transmitted in a burst, in which the generation rate is constant.

Maglaris [21] models video as a bit stream with discrete levels of bit rates, describing this as a superposition of 20 identical on-off sources. This results in an MMPP with 21 different states, of which state 0 has bit rate 0, and state 20 denotes the peak rate. This reflects the properties of the first option mentioned above. Saito [26] has a very elegant model for the second option: one frame is divided into k phases, with equal exponentially distributed sojourn times τ/k. This results in the frame length having an Erlang-k distribution. In the first p phases the source generates cells at a rate λ_v, the peak rate, while during the remaining $(k-p)$ phases, no cells are generated (see Fig. 2(a) in case $k = 5$). The ratio p/k is chosen such that peak and mean rate match with those of the real (measured) video signal.

The model of Saito can easily be implemented using SPNs, as depicted in Fig. 2(b). The transitions vt.i all have firing rate k/τ, similar to the transition rates in the MMPP. A token circulates through places video.i, inhibiting the firing of transition arr if it is present in places video.2, video.3 or video.4. If transition arr is enabled, it fires with rate λ_v. Similar to the extension of a Poisson process to a deterministic arrival process, we can extend the interarrival

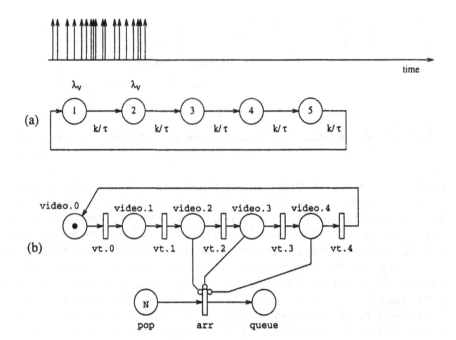

Fig. 2. (a) Saito's video model; (b) SPN representation of this video model

times of the video cells to be deterministic. This too, represents the packetizing time that occurs during generation of ATM cells. In this case, the single transition arr is replaced by a set of transitions arr.i ($0 \leq i \leq k-1$), all with firing times $k \times \lambda_v$, that represent an Erlang-k distribution.

3 The GAUSS Switch

The ATM GAUSS switch (Grab Any UnuSed Slot) has been proposed by De Vries at KPN Research [1, 28] (see also [24, Section 5.5]).

The basic structure of the GAUSS switch is comparable to the Knockout switch [24, 29]: it is a time-division switch, that has an input broadcast bus for every input port, and an output port interface connected to the bus for every output port, as depicted in Fig. 3. The main difference between the GAUSS and the Knockout Switch is the output module. The output module of the GAUSS switch is depicted in Fig. 4. Every output module has one Shift Register (SR), one Speed Adaptation Output buffer (SAO-buffer), and a number of Shift Register Units (SRU), one for each input port.

Consider output module i. Due to the use of the broadcast bus, cells from input j arrive at SRU j of output module i, even if these cells are not destined for output i. The SRU filters cells that are destined for other outputs, and stores ATM cells that have to be transmitted through output port i. At the beginning (the left side) of the SR, an Empty Slot Generator (ESG) produces empty slots,

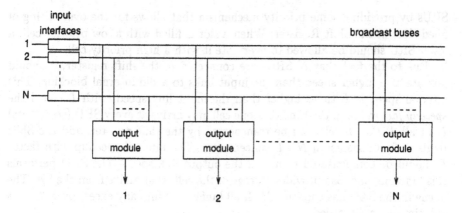

Fig. 3. Basic structure of the GAUSS switch

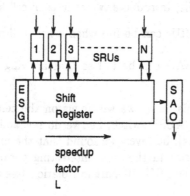

Fig. 4. An output module of the GAUSS switch

in which exactly one ATM cell can be transported, at a rate of L times the speed of the in- and output links. If the first SRU has a cell to be transmitted, it grabs the empty slot and places a cell in it, marking the empty slot as used by setting the *full* indication. If the buffer of the SRU is empty, the empty slot advances one position and can be filled by the second SRU.

The SRUs normally have a very limited storage space of C buffers. Suppose the SRU can store one cell ($C = 1$), then there is a non-zero probability that a second cell arrives at the same SRU when its buffer still contains a cell that has not been able to grab an empty slot. In this case, one of the two cells competing for the buffer has to be discarded, according to a previously determined selection scheme: either the cell that arrived later will be discarded or the cell that did not have the opportunity to grab an empty slot will be overwritten.

This mechanism is unfair for the SRUs that are connected at the end of the SR (to the right), as these SRUs have a higher probability of finding a slot filled than those near the beginning (to the left) of the SR, whereas at the first SRU the slots are always empty. It is possible to decrease the unfairness for the lower

SRUs by providing some priority mechanism that allows for the overwriting of filled slots in the Shift Register. When a slot is filled with a low priority cell, a lower SRU should be allowed to overwrite it with a high priority cell.

Due to the fact that N SRUs are connected to the shift register, it should operate at a higher speed than the input links to avoid internal blocking. This internal speed is L times higher than the input link speed, with L called the *speedup factor*. So, in the time that one cell may arrive at every SRU (so, in total N at all SRUs), L cells can be transported by the shift register, and the Shift Register functions as an $N : L$ concentrator. The internal speedup with factor L should be compensated to match the output link speed. The SAO performs this function, and also provides storage of the cells that arrive from the SR. The queue in the SAO has capacity M. If all buffers are full, any excess over M cells will simply be discarded.

One can conclude that the GAUSS switch (without low or high priority distinction between cells) introduces two sources of cell loss:

1. the buffer of the SRUs can be full when a new cell arrives; one cell is then deleted;
2. the queue in the SAO can be full so that arriving cells there have to be discarded.

In the remaining of this paper we will focus on the cell loss behaviour of the SRU/SR-part of the GAUSS switch, i.e., we do not address the SAO. We did address the SAO in [14], however, we found that the most challenging performance evaluation issues and the most interesting performance results can be found when studying the SRU/SR-part in isolation (see also [14]).

4 Stochastic Petri Net Model of the GAUSS Switch

SPNs offer a flexible method to model every desired detail of a system in theory, but in practice there are limitations with regard to the firing time distributions of transitions, and the size of the model. The size of a model of the complete GAUSS switch makes it necessary to apply a decomposition: we can focus on the SRU/SR combination and the SAO separately. Since the behaviour of the SAO does not have any influence on the performance of the SRU/SR combination, we will only address the performance of the SRU/SR.

We first address an integrated model of the SRU/SR in Sect. 4.1. This model, due to size considerations, cannot be solved in practice. We therefore propose a series of decomposed models in Sect. 4.2. How these decomposed models are used to obtain performance measures for the overall model is explained in Sect. 4.3.

4.1 Overall SPN Model of the SRU/SR

Consider a GAUSS switch with four inputs. We can model the SRU/SR part of the switch using SPNs, as depicted in Fig. 5, where we have chosen $N = 4$. A finite number of cells is stored in place pop; this number is chosen large enough so that it is as if the model is open (in a queueing theoretical sense). Arrivals of these cells at the inputs are modelled as firings of the timed transitions arr.i (the interarrival times are negative exponentially distributed for the time being), after which the just-arrived cells are stored in a buffer sru.i, with capacity C. Transition block.i is an immediate transition that is enabled if there are more than C cells in place sru.i (using the SPNP enabling() function). The immediate firing of this transition will limit the number of cells in sru.i to C. We have chosen to use this way of modelling, instead of a multiple inhibitor arc (with multiplicity C) from sru.i to arr.i in order to ease the computation of the cell loss ratio (CLR). When using inhibitor arcs, no cells would be lost since arrivals where blocked anyway when the buffer is full, yielding a CLR=0 (according to the definition of CLR given below). In our modelling approach, the number of arriving cells, i.e., the throughput of arr.i can be larger then the number of served cells (i.e., the throughput of service.i) simply because arriving cells at a full buffer are immediately returned to place pop. It should furthermore be noted that since we deal with a finite-customer population model, the throughputs of transitions arr.i and service.i are not known *a priori* since these transitions might not always be enabled. From these two computed throughput values, the CLR for SRU i can finally be computed as:

$$CLR_i = \frac{tput(\texttt{arr.i}) - tput(\texttt{service.i})}{tput(\texttt{arr.i})}. \tag{1}$$

Places slots.i and transitions genslot.i model the deterministic generation of empty slots with speedup factor L, compared to the link rates. We approximate the deterministic slot generation time by using an Erlang distribution with 10 phases (the squared coefficient of variation is then only 0.1).

If no cells are present in places sru.i, the empty slots propagate through places choice.i via transitions next.i, to return to the empty slot generator again. If cells are present, however, these empty slots will be filled, which is modelled by firing of transition grab.i, and the empty slot will not be available anymore to downstream SRUs (transitions next.i are disabled whenever sru.i is non-empty, again using the enabling() function provided by SPNP). Transitions service.i provide for the return of slots to the slot generator, and of the cells to place pop. We comment on the timing of service.i later.

In the model of Fig. 5, Poisson arrivals are used (transitions arr.i). In Sect. 2, we concluded that the Poisson process is not the most appropriate choice to model more complex arrival processes like video. It is possible to extend this model to MMPP arrivals, but the state space of the model will grow too large. Suppose we want to model video arrivals by replacing transitions arr.i with Saito's video model, as depicted in Fig. 2(b). Per input, this modification will introduce 5 extra states, so for four inputs, the number of states will increase with a (multiplicative) factor $5^4 = 625$. The number of non-zero elements of the generator matrix of this model with the video traffic will be about $1.21 \cdot 10^8$, whereas for a Poisson arrival process, this is "only" 193620. The former model is too large to analyse on a current-day workstation.

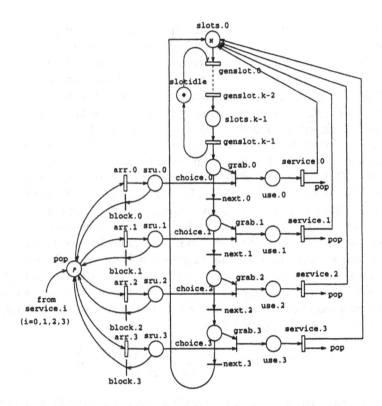

Fig. 5. SPN model of SRU/SR of the GAUSS switch with four inputs

4.2 The Decomposed Model

Size considerations urge us to apply a decomposition which results in a model for each separate SRU. The problem that arises here is the distribution of empty slots in the Shift Register. When a decomposition between SRUs is applied, the SRUs should know how many, and which, slots passing through the SR are already used by upstream SRUs. The best way to remember this is to use an aggregation of all traffic models of the higher SRUs to control the generation of empty slots. This method, however, has as disadvantage that a superposition of several models will give a model in which the number of states equals the product of the number of states of the composing models. For the second SRU, this model will only be the model of the first SRU; for the last SRU, however, all models have to be used, and an insufficient decrease of the number of states will be obtained. An approximation of the distribution of the used time slots is therefore necessary.

Suppose $L = 4$, and the rate of the output link is $\mu = 5$, then $L \times \mu = 20$ empty slots will be generated every time unit (see Fig. 6). The first SRU will find every slot empty, so no approximation is necessary there. This SRU will occupy several empty slots, dependent on the traffic characteristics for that SRU.

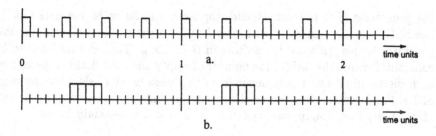

Fig. 6. (a) uniform distribution of used slots; (b) clustering of used slots per time unit

Suppose, the first SRU occupies on average 4 slots every time unit. Then the next SRU will only have 16 slots left. One extreme is that the occupied slots are uniformly distributed over the 20 generated slots, as depicted in Fig. 6(a), but it is also possible that the four occupied cells cluster, so there is a sequence of 4 occupied slots in each time unit (see Fig. 6(b)). The first distribution is very regular, and will underestimate the cell loss of the next SRUs compared to the real situation. A regular distribution is favourable, because between the filled slots, there are still 4 empty slots left that can be filled. In the second case, if the capacity of the SRU $C = 1$, and a quick succession of two arriving cells takes place during a cluster of occupied slots passing by, one of these two cells will be discarded because the first cell had no opportunity to grab an empty slot before the second one arrived. This succession would not result in a cell loss in the first (regular) case because immediately after the occupied slot, an empty slot follows.

In this new model, the empty slots for the first SRU are generated with an Erlang-distribution with 10 phases, to reflect the deterministic character of the interslot times. An extra on-off process is then used to model the different distributions of the occupied slots as perceived by downstream SRUs. If this on-off process is in the *off*-state, then no empty slots are generated, which means that the currently generated slots are occupied by cells form upstream SRUs; the sojourn time in the off-state is $1/\beta$. If the on-off process is in the *on*-state, then true empty slots are generated, which means that the currently generated slots are not yet occupied by upstream SRUs; the sojourn time in the on-state is $1/\alpha^3$. It should be noted that the residence times of the on-off process in its states *on* and *off* are taken to be exponentially distributed. This does not necessarily correspond to the actual slot-usage patterns of the SRUs. It turns out (see also Sect. 5) that this assumption does not influence the usability of our approach.

Let L be the speedup factor, μ the output link rate, and *used* the number of empty slots already occupied by upstream SRUs (per unit of time). In the case of Fig. 6(a), only one occupied slot is generated at a time. This means that

---

[3] An alternative idea would be to generate empty slots at all times, and to decide *after* they have been generated whether they can still be used (they are still empty) or whether they should be discarded (the are already in use).

the generation of empty slots should stop for the duration of one time slot. In the ESG, slots are generated with rate $L \times \mu$, so the required off-time for this generator is then $(L \times \mu)^{-1}$, resulting in $\beta = L \times \mu$. The ratio α/β has to be constant, because the ratio of the number of empty and used slots is constant for each distribution, i.e., it only depends on the *mean* usage of slots by upstream SRUs. Furthermore, since the total capacity per unit of time is constant as well ($L \times \mu$), we have: $empty + used = L \times \mu$, or $used = L \times \mu - empty$. Since

$$\frac{no\ empty\ slots}{empty\ slots} = \frac{1/\beta}{1/\alpha} = \frac{\alpha}{\beta} = \frac{used}{empty} \Rightarrow \alpha = \frac{\beta \times used}{empty}, \tag{2}$$

we can compute the following values for α and β as follows:

$$\alpha = \frac{L \times \mu \times used}{L \times \mu - used}, \quad \text{and} \quad \beta = L \times \mu. \tag{3}$$

In the case of Fig. 6(b), α and β will have different values. In this distribution, all used slots cluster together in one time-unit. If *used* slots are used by upstream SRUs, then the time it takes to generate *used* slots is $used \times (L \times \mu)^{-1}$. β will be the rate of the *off* time, and will thus be:

$$\beta = \frac{L \times \mu}{used}. \tag{4}$$

Because the ratio of α and β has to be constant again, α is calculated similarly to the procedure above. The resulting values are:

$$\alpha = \frac{L \times \mu}{L \times \mu - used}, \quad \text{and} \quad \beta = \frac{L \times \mu}{used}. \tag{5}$$

The ratio α/β will be constant for every distribution of empty slots over the generated slots, but the absolute values α and β will be different, depending on how much an actual situation reflects either Figure 6(a) or 6(b).

In case the used slots are uniformly distributed amongst the empty slots, the performance of the SRU/SR combination will be overestimated, and *we will use the uniform distribution of Fig. 6(a) as a lower bound on the actual CLR.*

In the above approach we only take into account information about the first moment of slot-usage by an SRU. Advantage of such an approach is its simplicity, disadvantage is that explicit and independent expressions for *both* α and β are not given, but only for their ratio α/β. There are basically two ways to overcome this problem:

1. one can study the slot-usage pattern in an SRU in more detail so that more information about also the second moment of slot-usage becomes available, i.e., one finds how much the actual slot-usage pattern reflects Fig. 6(a) or 6(b);

2. one can experimentally investigate which absolute values of α and β, given their ratio, do provide the best results, when comparing the decomposed model with an overall (non-decomposed) model.

In order to avoid more mathematical derivations, and by the difficulty in obtaining higher-order moments via the tool SPNP, we decided to take the second approach. Therefore, before presenting various performance evaluations of the GAUSS switch in Sect. 5, we will first have to evaluate our approach experimentally.

To conclude this section, we now present the model of a single SRU in Fig. 7. Places slots.i, slotidle and transitions genslot.i perform the empty slot generation with an Erlang-k distribution, and places sloton, slotoff and transitions onoff and offon modulate this empty slot generation to account for the already occupied slots. The firing rate of transition onoff is α, the firing rate of transition offon is β.

Timed transition arr represents the arrival of cells at the input of the SRU. For simplicity, in this figure it has an exponentially distributed firing time, but it can have an Erlang distribution to represent deterministic interarrival times or it can be modulated by a realistic traffic source. Place sru can maximally contain C tokens by the fact that transition block will be enabled only if the number of tokens in place sru exceeds C (again by using an enabling function).

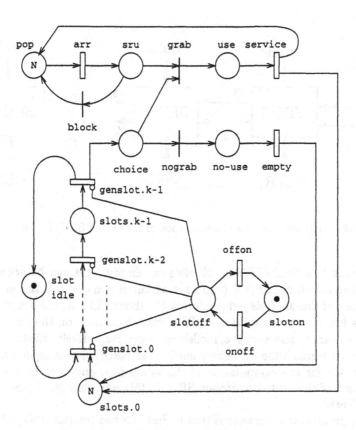

Fig. 7. SPN model of one SRU

If an empty slot arrives at place choice, transition grab fires if a token (cell) is present in place sru, otherwise nograb fires. This behaviour is enforced by enabling functions. Transitions service and empty finally take care of the return of used and empty slots to their particular places at a very high rate to guarantee a quasi-infinite population of cells and slots. Their throughput represents the number of used and empty slots per time unit by this SRU. Transitions empty and service have been chosen to be timed with a very high rate, instead of being immediate. In this way, we can directly compute their throughputs and use them to compute the CLR according to (1). A different approach would have been to keep these latter two transitions immediate and to compute their throughputs indirectly; this would have kept the state spaces slightly smaller.

4.3 Overall Model Solution: A Serial Approach

In the previous section we have presented an SPN model of a single SRU, however, this model can still be parameterized via the rates of transitions onoff (α) and offon (β). This is exactly what we need to do when we want to evaluate the performance characteristics of the overall system (see also Fig. 8 where we have indexed the rates α and β with the SRU number they are used in).

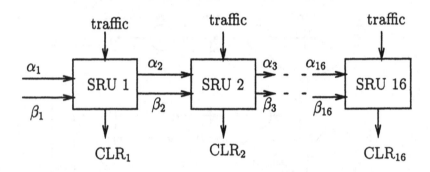

Fig. 8. Serial approach for the evaluation of the overall SRU/SR combination

For the first SRU (SRU 1), all slots are empty. This can be accomplished by setting $\alpha = 0$ and $\beta > 0$ (and have as initial marking a token in sloton). The result of this analysis yields information about CLR_1 and about the number of slots left over for downstream SRUs, via information on the throughput of transitions empty and service, modelling empty, respectively, filled slots passing by an SRU through the SR. These numbers are used, via the equations derived in Sect. 4.2, for the computation of the values of α and β to be used for the evaluation of the next downstream SRU. In this way, serially, all the SRUs can be evaluated.

One problem that remains is that in Sect. 4.2 two (extreme) ways have been proposed to compute α and β for the next downstream node from the information on the throughputs of transitions *empty* and *used*. Since we did not derive explicit and independent results for both α and β, we will resolve this single degree of freedom experimentally in Sect. 5.

5 Evaluation

In this section we will present evaluation results for the SRU/SR of the GAUSS switch. After specifying the parameters in Sect. 5.1, we start with a validation of the proposed decomposition in Sect. 5.2. We then continue with an analysis of the total SRU/SR with Poisson and more deterministic arrivals in Sect. 5.3. The speedup factor L and the number of buffers also have influence on the performance of the SRU/SR; this will be analysed in Sect. 5.4. The combination of the SRU/SR with a complex video workload is investigated in Sect. 5.5.

5.1 Numerical Parameters

The bandwidth of each outgoing link of the switch is 155.520 Mbit/s, which results in 366792 53-byte cells/s. The internal speed of the SRU/SR is L times higher. The utilization of each outgoing link for Poisson and deterministic traffic is $\rho = 0.9$, with traffic uniformly distributed over all inputs, while for video traffic we assume that the utilization $\rho = 0.9$ if all video sources produce cells at their peakreate.

The video sources are modelled according to Saito [26] as discussed in Sect. 2.2. During the first two phases cells are produced at peakrate: 44.7 Mbit/s, i.e., 105425 cells/s, and during the remaining three phases, no cells are generated. The average bit rate will thus be: $2/5 \times 44.7$ Mbit/s = 17.88 Mbit/s, almost similar to the values mentioned in [27].

When using SPNP [4], GAUSS-Seidel and the Successive Over-Relaxation procedures are employed (with a precision of 10^{-13}, if possible) for the determination of the steady-state probabilities.

5.2 Validation of the Decomposition

To validate the decomposition of the SRU/SR combination, we will compare the model of the complete SRU/SR with four inputs (see Fig. 5), and the decomposed model of all separate SRUs (Fig. 7). If the results of these two models match, results of the decomposed model with more than four input ports can be interpreted with increased confidence.

For this analysis, α and β are chosen such that the average number of series of occupied filled slots is 0.5, 1.0, and 1.5. The other parameters are given in Sect. 5.1.

Fig. 9 gives the CLR of all input ports for two values of the speedup factor: $L = 2$ and $L = 4$, for the complete model and the decomposition. First notice that the CLR increases with increasing port number. This is no surprise; the more downstream the port is, the higher the likelihood that slots are occupied

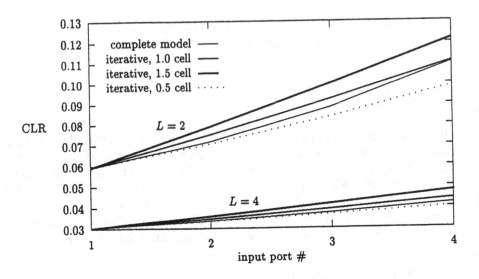

Fig. 9. Cell loss ratio of all inputs for speedup factors $L = 2$ and $L = 4$

by cells from downstream inputs. As could be expected, the combination of α and β in which on average 1.5 used slots are clustered, gives the highest CLR. The CLR of the complete model is slightly less than the CLR of the decomposed model with on average one occupied slot between empty slots. This can be explained by the fact that in the decomposed model, the modulating on-off sources have exponentially distributed firing times, which causes the generation of empty slots not being disabled for *exactly* one slot time, but for *on average* one slot time. In the complete model, the time between the generation of empty slots exactly is an integer number of slot times, because these cells are removed by transitions grab.i. As is generally known, variance in a distribution causes worse performance. This variance could be compensated for by changing the exponential on-off distribution to an Erlang on-off distribution, but this would be at the expense of a large number of extra states (we did not pursue this further).

In Fig. 10, the CLR of SRUs 2 and 4 is plotted as a function of the speedup factor L. In this plot, one can see that for $L = 4$, the graphs of the complete model and of the combination of α and β that results in on average one occupied slot between the empty slots fall together, so for Poisson arrivals, these values for α and β are a good choice. In conclusion, this evaluation shows that the decomposition into separate SRUs yields acceptable results when compared to the complete model.

The continuous-time Markov chain underlying the overall SPN has 21420 tangible markings (and 193620 nonzero elements in the generator). On our Sun

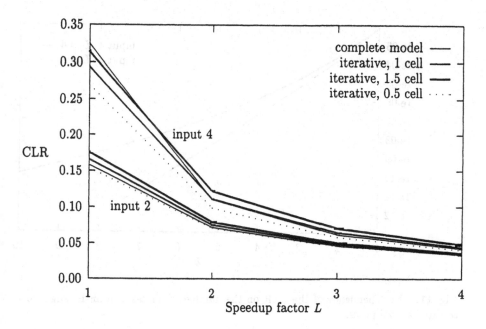

Fig. 10. Cell loss ratio of inputs 2 and 4 for varying speedup factor L

Sparc 10, it took about 1200 seconds to solve such a model. The decomposed models each have 624 tangible markings (2730 nonzero's). It takes only 2 seconds to solve one such a model. In total, that makes 8 seconds for the 4-SRU variant of the GAUSS switch; a speed-up by a factor 150.

5.3 SRU/SR with Poisson and Deterministic Arrivals

In this section, the influence of the number of phases in the Erlang-k arrival processes will be discussed. The CLR on input 1 and 4 of a GAUSS Switch with 16 inputs will be calculated, with one buffer for each SRU, and speedup factor $L = 4$.

Variation of the number of phases k of the Erlang-distribution has a large influence on the CLR, as can be seen in Fig. 11. This graph shows that the CLR of the first input reaches zero (i.e., is smaller than the required precision) for $k = 8$. This can be explained as follows: in case the arrival process is a simple Poisson process ($k = 1$), the probability that two arrivals occur in a period in which only one empty slot is created —and one of these two cells is discarded— is much higher than in case the interarrival time is Erlang-10 distributed. As the Erlang-10 distribution better models the deterministic packetizing delay (the time required to put 48 bytes into one ATM cell), from now on Erlang-10 arrivals will be used.

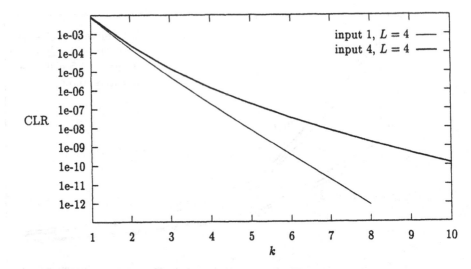

Fig. 11. The dependency of the CLR on the number of phases k in an Erlang-k distributed arrival process

5.4 Influence of the Speedup Factor L and the Buffer Size

The number of buffers per SRU will also have influence on the cell loss. In case there are two quick successive arrivals at one SRU with one buffer, one cell will be lost, whereas with two buffers, both cells can still be stored and transmitted over the SR (at a later stage). Fig. 12 shows the CLR of SRU 16 and the mean CLR of all inputs for 1 and 2 buffers per input for varying speedup factor L, and Erlang-10 distributed arrivals (deterministic arrivals). Input 16 is the last input, and will hence have the highest cell loss.

The CLR for input 16 is zero already for speedup factor $L = 2$ when two buffers are present in the SRU, whereas for one buffer, even for $L = 6$ the CLR is not lower than the precision reached. In [28], for Bernoulli arrivals and a utilization of $\rho = 0.9$, it was concluded that to have CLR=$10^{-9} \Rightarrow L \geq 9$, and that CLR=$10^{-6} \Rightarrow L \geq 6$. In our analysis, the Erlang-10 arrivals have less variance than Bernoulli arrivals, causing the CLR to be lower. The addition of one extra buffer gives a very significant improvement on the CLR. This can be useful when a video source is connected to the switch, because it can be expected that the CLR for video traffic is higher than for mild, deterministic arrivals. We are aware of the fact the the extra buffer also has its influence on the delay; we do not address this issue further.

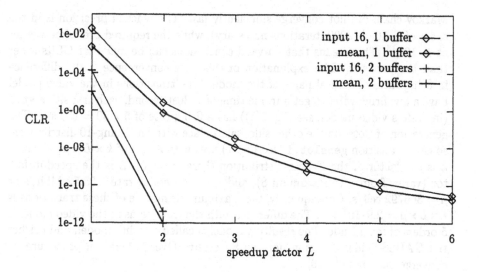

Fig. 12. CLR for input 16 and mean CLR for 1 and 2 buffers per SRU

5.5 SRU/SR with Video Workload

As pointed out in Sect. 2, the deterministic and Poisson arrival processes are not capable of representing the statistical properties of video traffic. No analysis of the GAUSS switch has been performed with realistic workload models with a high peakrate. The question to be answered in this section therefore is whether the performance of the GAUSS switch will deteriorate drastically with such sources. As usual, performance is expressed by the CLR.

We will evaluate models as just described, again using our serial decomposition approach. It should be noted that for these models, there has been no evaluation against an overall (complete) model. Instead, we took an optimistic approach, i.e., we assumed that the slot-usage pattern is fairly uniform. Thus, the computed CLRs should be interpreted as lower bounds; the actual CLR will be at least as large as the ones computed here. Another reason not to try and solve the overall model is the fact that even when using the decomposition approach some convergence problems occurred (see below); we expect them to be even bigger in an overall model.

We assume that the first three inputs of the GAUSS switch will undergo video-traffic, while the remaining 13 inputs will have (deterministic) Erlang-10 arrivals with a rate such that the utilization of the output link is $\rho = 0.9$ when, additionally, all three video-sources produce cells at their peak-rate.

This model has been analysed with one and two buffers per SRU. The addition of a video-model, however, introduced a severe problem: the solution of the

Markov chain did not converge sufficiently fast. The reached precision is in the order of $1 \cdot 10^{-8}$ (8000 iterations necessary), while the required precision was set at $1 \cdot 10^{-13}$. This means that no valid conclusions can be made for CLRs lower than the precision. The explanation of this slow convergence is the difference in firing rates of several parts of the model. The transitions in the video model have a low firing rate: to get a frame time of 1/30st second, each transition vt.i (in Saito's video model; see Fig. 2(a)) has a firing rate of $5 \times 30 = 150$ s^{-1}. The generation of slots, on the other side, takes place with an Erlang-10 distribution, so each transition genslot.i (see Fig. 7) fires with a rate of $k \times L \times \mu$, in which k is the factor of the Erlang-distribution (here: $k = 10$), L is the speedup factor (varies from 1 to maximum 8), and μ is the service rate: 155.52 Mbit/s or $\mu = 366792$ cell/s. Consequently, the maximum firing rate of these transitions is $k \times L \times \mu = 2.9 \cdot 10^7$ s^{-1}. The difference with the transitions in the video model is 5 orders of magnitude. The resulting model is called a "stiff" model, and neither the GAUSS-Seidel iteration nor the Successive Over-Relaxation procedure do converge fast enough [16].

To check whether the precision reached with 8000 iterations still gives good results, an analysis is done with a maximum of 20000 iterations. We analysed an SRU/SR combination with one buffer per SRU. The resulting CLR for the first three inputs, to which video sources are connected, is given in Table 1. The precision reached with 20000 iterations is about 10^{-9}. This table shows, that although the required precision is not reached, the cell loss rate does not change significantly for a higher number of iterations. The CLR only changes in its fourth or fifth significant digit, which means that the values obtained are fairly reliable.

input	8000 iterations	20000 iterations
1	$1.134 \cdot 10^{-10}$	$1.135 \cdot 10^{-10}$
2	$2.422811 \cdot 10^{-7}$	$2.422856 \cdot 10^{-7}$
3	$5.136237 \cdot 10^{-7}$	$5.136330 \cdot 10^{-7}$

Table 1. Influence of the number of iterations on the CLR of video-input 1, 2 and 3 for $L = 7$

As can be observed in Fig. 13, the CLR of the input ports with video workload is much higher than for those with deterministic arrivals. The CLR decreases dramatically with increasing L, resulting in cell losses of the order of 10^{-7} at $L = 8$ for the third input, which has the worst-case CLR because it is the last input with video-traffic. In this case it will be necessary to use a second buffer in each SRU to obtain the required CLR. The extra buffer of the SRU has a large influence as can be observed in the same graph; it decreases the CLR with one ($L = 1$, input 3) to six ($L = 8$, input 3) orders of magnitude. For input 3, the addition of an extra buffer results in a cell loss lower than for input 1 with one buffer per SRU, whenever the speedup factor $L > 3$. The CLR of input

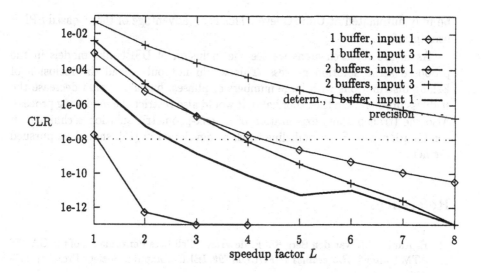

Fig. 13. Cell loss ratio of inputs 1 and 3 with videotraffic for different queue sizes compared to CLR for input 1 with deterministic arrivals

2 always lies between CLR_1 and CLR_3. As inputs 4 to 16 carry deterministic traffic, their CLR is very low ($\sim 10^{-10}$). We can conclude that the addition of one extra buffer is necessary when video traffic has to be transmitted to obtain a satisfying CLR.

6 Conclusions

As indicated in the abstract, we had three main aims with this paper: (i) to show the suitability of stochastic Petri nets in the context of ATM system analysis; (ii) to illustrate a structural decomposition method circumventing the state space explosion problem; and (iii) to derive more detailed performance results for the GAUSS switch than has been done previously.

The results we obtained regarding the third objective, see Sect. 5, already imply that the SPN approach and our decomposition are successful tools for the evaluation of ATM systems, thus satisfying the first two aims. SPNs provide a generally applicable framework; the proposed decomposition is much more application dependent. A problem that remains to be solved or reduced is to circumvent stiffness; another type of decomposition, i.e., time-scale decomposition, might help here.

The model construction and solution using SPNP has been very comfortable. The serial solution of the submodels, including their "parameter passing" can be fully automated at Unix/C level. This is an advantage of the C-based SPNP user-interface.

As future research areas we see the inclusion of DSPN-like models in the proposed serial solution process. This would not only avoid the inclusion of Erlang distributions with large numbers of phases, but also would decrease the stiffness problem. On the other hand, it would also restrict the modelling process. Also the (transparant) exploitation of matrix-geometric solution techniques in combination with SPN modelling, such as developed in [12], should be pursued further.

References

1. E. Aanen, J.L. van den Berg R.J.F. de Vries, "Cell Loss Performance of the GAUSS ATM Switch", *Proceedings INFOCOM '92*, IEEE Computer Society Press, pp.717-726, 1992.

2. P. Chen, S.C. Bruell, G. Balbo, "Alternative Methods for Incorporating Non-Exponential Distributions into Stochastic Timed Petri Nets", *Proceedings of the Third International Workshop on Petri Nets and Performance Models*, IEEE Computer Society Press, pp.187–197, 1989.

3. H. Choi, V.G. Kulkarni, K.S. Trivedi, "Markov Regenerative Stochastic Petri Nets", *Performance Evaluation* 20, pp.337–357, 1994.

4. G. Ciardo, J.K. Muppala, K.S. Trivedi, "SPNP: Stochastic Petri Net Package", *Proceedings of the Third International Workshop on Petri Nets and Performance Models*, IEEE Computer Society Press, pp.142-151, 1989.

5. G. Ciardo, K.S. Trivedi, "A Decomposition Approach for Stochastic Reward Net Models", *Performance Evaluation* 18, pp.37–59, 1993.

6. G. Ciardo, R. German, C. Lindemann, "A Characterization of the Stochastic Process Underlying a Stochastic Petri Net", *IEEE Transactions on Software Engineering* 20(7), pp.506-515, 1994.

7. CCITT Recommendation H.261, *Video codec for audiovisual services at $p \times 64$ Kbit/s*, Geneva, 1990.

8. V. Frost, B. Melamed, "Traffic Modelling for Telecommunications Networks", *IEEE Communications Magazine* 32(3), pp.70–81, 1994.

9. R. German, C. Lindemann, "Analysis of Stochastic Petri Nets by the Method of Supplementary variables", *performance Evaluation* 20, pp.317–335, 1994.

10. B.R. Haverkort, A.P.A. van Moorsel, D.-J. Speelman, "Xmgm: A Performance Analysis Tool Based on Matrix Geometric Methods", in: *Proceedings of the Second International Workshop on Modelling, Analysis and Simulation of Computer and Telecommunication Systems*, IEEE Computer Society Press, pp.152–157, 1994.

11. B.R. Haverkort, H.P. Idzenga, B.G. Kim, "Performance Evaluation of Threshold-Based ATM Cell Scheduling Policies under Markov-Modulated Poisson Traffic using Stochastic Petri Nets", in: *Performance Modelling and Evaluation of ATM Networks*, Editor: D. Kouvatsos, Chapman & Hall, pp.551–572, 1995

12. B.R. Haverkort, "SPN2MGM: Tool Support for Matrix Geometric Stochastic Petri Nets", *Proceedings of the 1996 International Computer Performance and Dependability Symposium*, IEEE Computer Society Press, pp.219–228, 1996.

13. H. Heffes, D.M. Lucantoni, "A Markov Modulated Characterization of Packetized Voice and Data Traffic and related statistical Multiplexer Performance", *IEEE Journal on Selected Areas in Communications* 4(6), pp.856-868, 1986.
14. H.P. Idzenga, *Performance Analysis of ATM Switch Architectures using Matrix Geometric Methods and Stochastic Petri Nets*, M.Sc. Thesis, Department of Electrical Engineering, University of Twente, 1994
15. L.A. Kant, W.H. Sanders, "Loss Process Analysis of the Knockout Switch using Stochastic Activity Networks", *Proceedings of the 4th International Conference on Computer Communications and Networks* (September 20–23, Las vegas, USA), pp.344–349, 1995.
16. U. Krieger, B. Müller-Clostermann, M. Sczittnick, "Modelling and Analysis of Communication Systems Based on Computational Methods for Markov Chains", *IEEE Journal on Selected Areas in Communications* 8(9), pp.1630-1648, 1990.
17. D.-S. Lee, B. Sengupta, "Queueing Analysis of a Threshold Based Priority Scheme for ATM Networks", *IEEE/ACM Transactions on Networking* 1(6), pp.709–717, 1993.
18. D.H. Le Gall, "MPEG: A Video-Compression Standard for Multimedia Applications", *Communications of the ACM* 34(4), 1991.
19. C. Lindemann, R. German, "DSPNexpress: A Software Package for Efficiently Solving Deterministic and Stochastic Petri Nets", in: *Computer Performance Evaluation 1992: Modelling Techniques and Tools*, Editors: R. Pooley, J. Hillston, Edinburgh University Press, 1993.
20. C. Lindemann, "An Improved Numerical Algorithm for Calculating Steady-State Solutions of Deterministic and Stochastic Petri Net Models", *Performance Evaluation* 18, pp.79–95, 1993.
21. B. Maglaris, D. Anastassiou, P. Sen, G. Karlsson, J.P. Robbins, "Performance Models of statistical multiplexing in packet video communications", *IEEE Transactions on Communications* 36(7), pp.834-844, 1988.
22. B. Müller-Clostermann, "DSPN Modelling of Usage Parameter Control in ATM Networks", presented at Dagstuhl seminar 9521, May 22-26, 1995.
23. M.F. Neuts, *Matrix Geometric Solutions in Stochastic Models—An Algorithmic Approach*, The Johns Hopkins University Press, 1981.
24. C. Partridge, *Gigabit Networking*, Addison-Wesley, 1993.
25. A. Puliafito, M.B. Krishnan, K.S. Trivedi, I. Viniotis, "Buffer Sizing of ABR Traffic in an ATM Switch", presented at Dagstuhl seminar 9521, May 22-26, 1995.
26. H. Saito, M. Kawarasaki and H. Yamada, "An Analysis of Statistical Multiplexing in an ATM Transport Network", *IEEE Journal on Selected Areas in Communications* 9(3), pp.359-367, 1991.
27. H. Saito, *Teletraffic Technologies in ATM Networks*, Artech House, Boston, 1994.
28. R.J.F. de Vries, *Switch Architectures for the Asynchronous Transfer Mode*, Ph.D. thesis, University of Twente, 1992.
29. Y.S. Yeh, M.G. Hluchyi, and A.S. Acampora, "The Knockout Switch: A Simple, Modular Architecture for High-Performance Packet Switching", *IEEE Journal on Selected Areas in Communications* 5, pp.1274-1283, 1987.

COSTPN for Modeling and Control of Telecommunication Systems

Hermann de Meer[1], Oliver-Rainer Düsterhöft[1], and Stefan Fischer[2]

[1] Department of Computer Science, University of Hamburg, Vogt-Kölln-Str. 30, 22527 Hamburg
[2] Praktische Informatik IV, University of Mannheim, D-68131 Mannheim

Abstract. The design of modern telecommunication systems is a complex task since many parameters, mostly of stochastic nature, have to be taken into account in order to achieve desired performance values. Stochastic Petri Nets (SPNs) are a well-known modeling and analysis tool for such systems. In addition, the ability to adapt system operations to quickly changing environment or system conditions is of great importance. Therefore, a new framework for the extension SPNs is presented in this paper which introduces elements providing means for a *dynamic optimization* of performability measures. A new type of transition is defined offering a feature for specification of controlled switching, called *reconfiguration*, from one set of markings of a SPN to another set of markings. In a numerical analysis, these optional reconfiguration transitions are evaluated in order to optimize a specified reward or cost function. The result of the analysis is a set of strategies which tell the controller of the system when to fire enabled reconfiguration transitions and when to remain in the current state. The extended SPNs are called *COSTPNs (COntrolled STochastic Petri Nets)*. For the numerical analysis, COSTPNs are mapped on EMRMs (Extended Markov Reward Models). Computational analysis is possible with algorithms adopted from Markov decision theory, including transient and stationary optimization. This paper introduces the new COSTPN model, discusses the algorithms necessary for the mapping of COSTPNs on EMRMs and shows how COSTPNs can be applied for the modeling and control of a typical telecommunications system, namely a multimedia server. Major emphasis is put on the introduction of new enabling and firing rules for reconfiguring transitions and on the illustration of the new modeling approach by means of the multimedia server example.

Keywords: stochastic Petri nets, performability, dynamic optimization, extended Markov reward models, Markov decision theory.

1 Introduction

In the early eighties Petri nets were extended with the notion of time, for example by Molloy [MOLL 81]. Marsan et al. promoted stochastic Petri Nets (SPNs) for the modeling and performance analysis of multiprocessor systems [MARS 84], and SPNs have been widely applied since then [MARS 95]. SPNs were extended to stochastic reward nets (SRNs) by Trivedi, Ciardo and Muppala [TRIV 91]. Recently, progress has been made, for example by Choi et al. [CHOI 94] and German et al. [GERM 95], extending the class of models beyond continuous time Markov chains to include deterministic and general type time distributions. Many more approaches and tools related to SPNs do exist, most of them being covered in the overview paper by Haverkort and Niemegeers [HAVE 96].

Often, it has been advocated, for instance by Kramer and Magee [KRAM 85], that large scale and distributed systems should be provided with techniques allowing dynamic reconfiguration in the presence of environmental changes that affect their running conditions. Adaptation seems to be a particularly promising approach for the management of communication systems with multimedia applications, as these impose challenging computational and timing requirements on resources [VOBO 96,DEME 95]. But strategic knowledge is required in order to execute adaptation and reconfiguration actions in a most effective manner. It has been argued that quantitative modeling support could provide useful guidelines for required control decisions in this respect [DEME 92,HAVE 94].

Thus, the extension of SPNs with features directly providing decision support for such adaptation and reconfiguration tasks seems to be a promising approach, due to the wide acceptance of SPNs for performance and performability modeling and the relatively high level of SPNs. Therefore, a new type of transition is introduced that offers a feature for specification of controlled switching, called *reconfiguration*, from one set of markings of a SPN to another set of markings. In a numerical analysis, these optional reconfiguration transitions are evaluated in order to optimize a specified reward or cost function. As the result of an application of a numerical evaluation procedure, control strategies are computed that allow the optimization of a specified performance or performability measure. The resulting strategies can be directly applied to control adaptation and reconfiguration in the modeled system. The extended SPNs are called *COSTPNs (COntrolled STochastic Petri Nets)*.

For the numerical analysis, COSTPNs have to be mapped on a model representation that allows the application of some optimization and evaluation algorithms. We propose to use EMRMs, which were originally introduced for this purpose [DEME 92,DEME 94]. Algorithms from Markov decision theory were adopted to provide techniques for transient and stationary optimization of performability measures. For the mapping of COSTPNs on EMRMs, it is necessary to modify the standard algorithm applied in the generation of extended reachability graphs (ERGs) in ordinary SPN analysis. In addition, EMRMs then have to be constructed from the modified ERGs.

To show that the new technique is useful for modeling and analyzing modern telecommunication systems, it is applied to a multimedia server in a distributed setting. Such a server receives calls of different types and has to decide whether to admit a given call or not, and how many resources are to be reserved. Accepted calls contribute to the revenue a server receives, but they also use the server's scarce resources. Overload situations may lead to penalties to be paid. The server's decisions depend on a wide variety of stochastic parameters, which the new technique takes into account in producing reward-optimal strategies for the server.

As another interesting problem of practical relevance, reservation strategies for variable-bit-rate (VBR) and bursty traffic are investigated. It is a difficult and generally unsolved task to trade hard quality-of-service (QoS) guarantees based on expensive peak-rates reservation schemes with efficient resource utilization based on unexpensive average-rate reservation schemes. Average-rate reservation assumptions can be applied in the limit if a very large number of multimedia applications share the same resources. But in practice, a limited number of multimedia sources are supported at the same time so that some intermediate-rate reservation schemes are required. It is our intention to contribute to the solution of this problem with the help of COSTPNs by investigating *dynamic rate reservation* schemes.

The rest of this paper is organized as follows: In Section 2, the concept of EMRM is briefly repeated. The new control structure of COSTPN (reconfiguring transition) is introduced in Section 3. In particular, the new enabling and firing rules for reconfiguring transitions are explained. The full mapping algorithm of COSTPN on EMRM is discussed in detail in Section 4. Section 5 then introduces a multimedia server example model, discusses several model variants, and presents a large variety of results obtained with the COSTPN modeling tool PENELOPE. The features of the tool itself, which provides a usage-friendly environment for optimization and computational experimentations, are beyond the scope of this paper. In-depth descriptions can be found in [DEME 96,DEME 97]. Section 6 finally concludes the paper.

2 Extended Markov reward models

Performability modeling [PERF 92] makes extensive use of Markov reward models (MRMs). Let $Z = \{Z(t), t \geq 0\}$ denote a continuous time Markov chain with finite state space Ω. To each state $s \in \Omega$ a real-valued *reward rate* $r(s)$, $r : \Omega \to \mathbb{R}$, is assigned, such that if the Markov chain is in state $Z(t) \in \Omega$ at time t, then the *instantaneous reward rate* of the Markov chain at time t is defined as $X(t) = r_{Z(t)}$. In the time horizon $[0, ...t)$ the *total reward* $Y(t) = \int_0^t X(\tau)d\tau$ is accumulated. Note that $X(t)$ and $Y(t)$ depend on $Z(t)$ and on an initial state $i \in \Omega$; the dependence on i is indicated by subscript notation: $X_i(t)$ and $Y_i(t)$. The probability distribution function $\Psi_i(y, t) = P(Y_i(t) \leq y)$ is called the *performability*. For ergodic models the instantaneous reward rate and the time averaged total reward converge in the limit to the same overall reward rate $E[X] = \lim_{t \to \infty} E[X_i(t)] = \lim_{t \to \infty} \frac{1}{t} E[Y_i(t)]$. The introduction of reward functions provides a framework for a formal definition of a "yield measure" or a "loss measure" being imposed on the model under investigation.

EMRMs provide a framework for the combined evaluation and optimization of reconfigurable systems by introducing some new features for MRMs [DEME 92,DEME 94]. EMRMs are the result of a marriage between Markov decision processes and performability techniques. A *reconfiguration* arc, which can originate from any Markov state of a model, specifies an optional, instantaneous state transition that can be controlled for an optimization. The resulting strategy is commonly time-dependent. The so called *branching* states provide another feature of EMRMs. No time is spent in such states, but a pulse reward may be associated with them. The introduction of branching states has motivation similar to the introduction of immediate transitions to stochastic Petri nets [MARS 84], so that branching states also are called *vanishing* states.

Reconfiguration arcs denote options to reconfigure from one state to another. At every point of time a different decision is possible for each reconfiguration arc. A strategy $S(t)$ comprises a tuple of decisions for all options in the model at a particular point of time t, $0 \leq t \leq T$. Strategies can be time dependent, $S(t)$, or time independent, $S = S(t)$.

A strategy $\hat{S}(t)$ is considered *optimal* if the performability measure under strategy $\hat{S}(t)$ is greater equal than the performability measure under any other strategy $S(t)$. With X^S, $Y_i^{S(t)}(t)$, $Y_i^S(\infty)$ denoting the overall reward rate, the conditional accumulated reward and the accumulated reward until absorption gained *under strategy S or S(t)* respectively, a strategy $\hat{S}$ or $\hat{S}(t)$ is optimal, iff

$$
\begin{cases}
E[Y_i^{\hat{S}(t)}(t)] \geq E[Y_i^{S(t)}(t)] \, \forall S(t) \, \forall i & \text{transient optimization,} \\[2mm]
E[X^{\hat{S}}] \geq E[X^S] \, \forall S & \text{stationary optimization (ergodic),} \\[2mm]
E[Y_i^{\hat{S}}(\infty)] \geq E[Y_i^S(\infty)] \, \forall S \, \forall i & \text{stationary optimization (nonergodic).}
\end{cases}
$$

The aim of applying EMRMs is to compute such optimal strategies and the corresponding performability measure, not only for the time parameter, but also for the other model parameters. These strategies could be used for adaptation control strategies in the management of system with environmental changes affecting their running conditions.

For the optimization approach we refer back to the performability framework and provide two types of methods for the computation of optimal strategies and performance functions:

- **Transient optimization**, where the expected accumulated reward $E[Y_i(t)]$ is used as an optimization criterion. The algorithm for an analysis within a finite period of time $[0, T)$ has been introduced in earlier work [DEME 92, DEME 94] and is derived from Euler's method for the numerical solution of ordinary differential equations [STEW 94]. It extends an approach of Lee and Shin [LEE 87], who had introduced and proved the correctness of transient optimization for *acyclic* CTMCs (continuous time Markov chains), to *general-type* CTMCs.
- **Stationary optimization**, which is performed for an infinite time horizon $[0, \infty)$. As optimization criteria, we distinguish between *time averaged mean total reward in steady-state*, $E[X] = E[X_i] = \lim_{t \to \infty} \frac{1}{t} E[Y_i(t)]$ for all i, where $E[X]$ is independent of initial state i for a selected strategy, and the *conditional accumulated reward until absorption*, $E[Y_i(\infty)] =$

$\lim_{t \to \infty} E[Y_i(t)]$, which is computed for non-ergodic models containing absorbing states. In the latter case, $E[Y_i(\infty)]$ can be dependent on an initial state i. The optimization itself is performed by deployment of variants of *value iteration* or *strategy iteration* type methods [DEME 92], relying on numerical algorithms such as Gaussian elimination, Gauss-Seidel iteration, successive over-relaxation (SOR), or the Power-Method. All these methods are implemented in the PENELOPE tool and can be deliberately chosen for a computation or they are by default automatically selected, thereby adapting to the actual model structure.

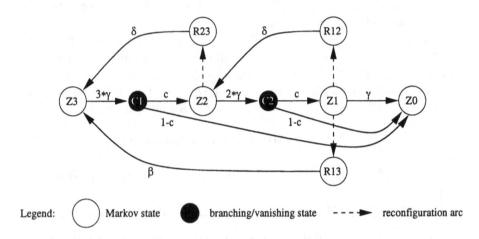

Fig. 1. EMRM of a network node with multiple protocol machines.

These basic definitions are illustrated in Figure 1. The EMRM shown there models a possible situation in a network node, where several protocol machines can process in parallel the connections via that network node. Being in state Zi means that i protocol machines are active. Each active protocol machine can fail after an exponentially distributed time, characterized by the rate γ. When a protocol machine fails, the whole system fails with probability $1 - c$ ($c < 1$) due to common effects of incomplete error coverage such as error propagation or defective failure analysis. This coverage related effect is modeled by the branching/vanishing states $C1$ and $C2$. When one or two of the protocol machines have failed, *offline* repair can be performed. This option is modeled by possible reconfigurations such that a system management can choose the optimal decision (repair or no repair) according to the current situation and the modeled performability measure. Three reconfiguration arcs are specified in the model: $Z2 \to R23$, $Z1 \to R12$ and $Z1 \to R13$. When the system is in one of the states from which a reconfiguration arc originates ($Z1$ or $Z2$), it has the option to stay in this state or to reconfigure to the destination of the reconfiguration arc. From state $Z1$, two reconfiguration arcs are originating. The system has the option to stay in this state, to reconfigure to $R12$ and to perform local repair or to reconfigure to $R13$ for global repair with longer duration. Exactly one of these possibilities has to be chosen, conditioned on being in state $Z1$ at a particular point of time. The states $R23$, $R12$ and $R13$ model the system under offline repair, i.e., the system is to be switched off and no useful work can be accomplished during repair. The mean duration of repair for one protocol machine is $1/\delta$, for two protocol machines it is $1/\beta$. One yield measure for this system is the performance as a function of the number of active protocol machines. A reward function specifying the corresponding yield measure then is:

$$r(s) = \begin{cases} i & : & if \ s = Zi, \\ 0 & : & otherwise. \end{cases}$$

A state of the system is evaluated the better – with respect to the modeled yield measure – the more protocol machines are active. On the other hand, the risk to fail and to reach the absorbing state $Z0$ increases with the number of active protocol machines, when more than one protocol machine are active. There is clearly a trade-off to be considered when evaluating the relative merit of different system states, which becomes even more complicated due to the effect of offline repair. As another important aspect, a failure in state $Z1$ leads always to the absorbing state $Z0$ with no chance of repair. Note, the non-ergodic model structure due to existence of an absorbing state suggests the application of a time-dependent (transient) evaluation and control to be most useful.

An example strategy $\mathbf{S}(t)$ at $t = 150$ would be:

$$\mathbf{S}(150) = \begin{pmatrix} Z2 \to Z2 \\ Z1 \to R12 \end{pmatrix}.$$

The interpretation of $\mathbf{S}(150)$ is as follows: if the system is in state $Z2$ at time $t = 150$, it should stay there ($Z2 \to Z2$). If it is in state $Z1$ at time $t = 150$, it should be reconfigured to $R12$ ($Z1 \to R12$). In general, six possible strategies can be applied at every point of time. The strategies are obtained by combining combinatorically every possible reconfiguration decision in state $Z1$ with every possible reconfiguration decision in state $Z2$.

A transient optimization is performed for the modeled protocol machines in a network node in order to determine which configuration (one, two or three protocol machines active) is optimal for a given parameter set. The investigated time horizon covers the interval $[0, T)$, $T = 1000$. The parameter set used in the optimization is $c = 0.9$, $\delta = 0.01$, $\beta = 0.005$ and $\gamma = [1e-06, 1.5e-01]$. With discrete time steps of 0.1 and discrete units of 1.1548 (logarithmic factor) for the parameter γ, the whole decision space is composed of $10000 * 84 = 84000$ investigated parameter combinations. For every parameter combination the optimal strategy is determined by the optimization analysis, i.e., the comparison of the respective performability measure for six possible strategies. The resulting optimal strategies are presented in Figure 2.

The strategy graphs show the investigated decision as a function of t and γ. Time is represented in a reverse pattern, namely as the *remaining time* $t' = T - t$,

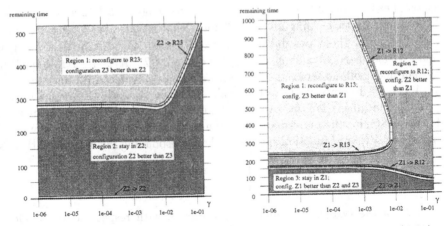

Fig. 2. Transient optimization strategies for state $Z2$ (left) and state $Z1$ (right).

where T is the time horizon and t is the elapsed time. The investigated decision space is divided into regions by *strategy curves*. These curves mark the reward-optimal time instants of strategy changes. A strategy curve, marked such as $Z1 \rightarrow R13$, is interpreted as follows: If the system is in state $Z1$ and is classified — in relation to the parameters t' and γ — to be *above* the curve, the system reconfigures from state $Z1$ to state $R13$. If the current situation is classified to be *below* the curve, an alternative decision applies. All strategy curves for one state divide the decision space into regions, as it is presented in Figure 2.

Often, it is interesting not only to apply the resulting strategy regions for control of the modeled system, but also to interpret their actual meaning and shape. This may provide important insights into the system's operations. Offline repair, adverse effects of various configurations on the reliability, and the impact of different levels of performance (degree of parallelism) have to be weighted and traded. So even for this rather simple problem, the pattern of decisions appear to be quite complex and cannot easily be anticipated.

From the results in Figure 2 strategic insights are to be concluded. If the protocol machines' failure rate γ is sufficiently small and the remaining system run time t' long enough, strategies (Region 1 for $Z1$ and Region 1 for $Z2$) which lead to $Z3$ (three protocol machines) via global repair $R13$ and repair $R23$, respectively, have to be followed. Under these circumstances, the higher the level of performance, the better it is. For a shorter remaining time, cost of (offline) repair is increasingly less compensated for by the higher degree of parallelism (Region 2 for $Z2$; first local repair according to Region 2 and then no repair according to Region 3 for $Z1$). The strategies are more complex for higher failure rates γ. Strategies become strongly time dependent due to the adverse effects of the different configurations in the more error prone case. Here, the dependability issues are dominating the control strategies and a lower degree of parallelism is considered to be of advantage.

Note the interesting shape of the strategy regions in the right part of Figure 2. These regions are bounded by curves which mark the strategy changes, each one due to its own particular reasons. Consider lower part of curve $Z1 \rightarrow R12$ first (bounding Region 3). This curve is mostly flat, but *decreases* finally for increasing values of γ. It expresses the fact that in an increasingly error prone environment, offline repair in state $Z1$ is more and more attractive to avoid system failure transition to state $Z0$. Curve $Z1 \rightarrow R13$, on the other hand, is (slightly) *increasing* for higher rates of γ. This reflects the higher risk of system failures as more components get involved (see vanishing state $C1$ of the EMRM in Figure 1). Finally, consider the upper part of curve $Z1 \rightarrow R12$ (bounding Region 1). This curve is interesting in that it marks the strategy change between global repair (Region 1) and *possibly* two sequential phases of local repair (Region 2). Note that this strategy keeps the option to re-decide on the second repair step, depending on the actual duration of the first repair step. This is reasonable, since configuration $Z3$ appears to be less attractive as γ increases, as can also be seen from the left part of Figure 2.

3   Controlled stochastic Petri nets

Extending the modeling approach of SPNs in such a way that dynamic optimization of performability measures, as with EMRMs, can be performed on a higher abstraction level has the following advantages:

- To model large systems with Markov reward models is difficult and error-prone because of the largeness and complexity of the systems.
- Often the modeler of a system is also the designer of the system, so she/he often thinks in the dimensions of the system's structures. It is much easier for her/him to use a high-level modeling tool where abstractions for system structures exist than to understand low-level unstructured Markov reward models.
- The control mechanism for optimization of systems with reconfiguration adds another level of complexity to the modeling and analyzing process. A high-level specification of control seems to be of predominant importance.

Consequently, a new feature to support dynamic optimization is introduced to SPNs. It comprises a control structure that allows to specify a controlled switching between sets of markings of an SPN. Such a controlled switching is interpreted as a *reconfiguration* in the modeled system.

A reconfiguration is modeled by the firing of a new type of transition, called a *reconfiguring* transition. The introduction of reconfiguring transitions leads to a new modeling tool, called COSTPN, and provides a way to combine the classical performability modeling of SPNs with the option to *dynamically optimize* measures. A reconfiguration on the level of EMRMs means a switching from one state to another state. A switching between states in Markov models corresponds to a switching between markings in SPNs. The Petri net standard to model state transitions is to specify a transition which is enabled in the respective markings and can fire to reach other markings. Therefore, realizing the new concept of reconfiguration with a new type of transition seems to be a natural choice. Accordingly, the *enabling rule* for reconfiguring transitions corresponds to those of timed transitions in SPNs.

New and most important for COSTPN, however, is the *firing rule* of reconfiguring transitions. This rule implements the dynamic optimization of the performability measures. By means of a reconfiguring transition the modeler specifies, when a reconfiguration *could* be performed — the enabling of reconfiguring transition is controlled by the modeler. The enabling rule of reconfiguring transitions is thus applied in the *model generation phase*, in which an EMRM is constructed from the COSTPN. In the *model evaluation phase*, in which the constructed EMRM is computationally analyzed, the firing rule of reconfiguring transitions is applied in order to determine when a reconfiguration *has to* be performed in order to optimize a given performability measure and to follow optimal strategies.

In the following, enabling and firing rules of reconfiguring transitions for COSTPNs are discussed in more detail. Then, the formal definition of COSTPNs is presented.

3.1   Enabling and firing rule of reconfiguring transitions

The markings of an SPN are partitioned into two types: *vanishing markings* and *tangible markings*[1]. Vanishing markings are characterized by the fact that at least one immediate transition is enabled and can fire. No time is spent in such markings. Immediate transitions always have higher priority than timed transitions. In tangible markings, no immediate transitions are enabled, but timed transitions can be enabled and can fire. A tangible marking is called absorbing, if no timed transition is enabled in it. We adopt these definitions to COSTPNs.

The newly introduced control mechanism can only be applied in tangible markings, but a reconfiguration itself is assumed to be instantaneously performed. The property that a reconfiguration can only be applied in tangible markings, has to be reflected by the *enabling rule* of reconfiguring transitions. The basic conditions to enable a transition – enough tokens in the input places and not too many tokens in inhibitor places – hold also for reconfiguring transitions. Because reconfiguring transitions can only be enabled in tangible markings, their priority has to be lower than the priority of immediate transitions. Their priority is the same priority as of timed transitions.

The *firing rule* of reconfiguring transitions is based on the goal to optimize a performability measure, which is defined through a reward structure. Reconfiguring transitions are enabled together with timed transitions and the conflict between the firing of enabled reconfiguring transitions and enabled timed

---

[1] In the *ERG* and in the Markov reward models derived from an SPN vanishing markings correspond to *vanishing states* and tangible markings correspond to *Markov states*.

transitions has to be solved. The solution of these conflicts with regard to the optimization leads to the computation of optimal strategies. Whenever the modeled system resides in a tangible marking, in which a reconfiguring transition is enabled, it can (i) instantaneously reconfigure to the marking which is reached through the firing of the enabled reconfiguring transition; no timed transition can fire in the current marking in this case. Or it can (ii) stay in the current marking and *not* fire the enabled reconfiguring transition, so that the enabled timed transitions can fire in the current marking in their usual manner. The decision which option to select is based on the comparison of optimization criteria as described in Section 2. The option with the highest expected reward is selected from the possibilities. For transient optimization, the expected accumulated reward $E[Y_i(t)]$ is computed. For stationary optimization, either the time averaged mean total reward $E[X]$ is computed, if the model is ergodic, or the accumulated reward until absorption $E[Y_i(\infty)]$, if the model is non-ergodic.

More than one reconfiguring transitions can be enabled in a tangible marking. In this case, the number of options corresponds to the number of enabled reconfiguring transitions plus one, since every firing of an enabled reconfiguring transition corresponds to one option, and not to fire any enabled reconfiguring transition is another option.

The configuration of the modeled system is dynamically controlled in every marking, in which a reconfiguration (firing of an enabled reconfiguring transition) can be applied, with respect to the given performability measure (reward structure).

3.2   Definition of COSTPN

Before defining COSTPNs in detail, some abbreviations are introduced. RS denotes the *reachability set* of a COSTPN, where a reachability set is the set of all markings which can be reached through the firing of transitions from an initial marking M_0. RG denotes the *reachability graph* of a COSTPN. RG is a directed graph (N,A), where the set of nodes N corresponds to the set of markings in the reachability set RS and the set of arcs A is defined by the reachability relation between the markings in RS. There is an arc between a marking M_i and a marking M_j, if M_j can be reached through the firing of an enabled transition t_k in M_i. The arc $(M_i \rightarrow M_j)$ is labeled with the name of the transition t_k. RS and RG are generated according to the enabling pattern immanent to a COSTPN. ERG denotes the *extended reachability graph* of the COSTPN. The ERG is produced from the RG in order to generate an EMRM. In this transformation the stochastic elements, like rates and probabilities, of the COSTPN are used to transform the transitions of the RG into transitions of a Markov chain.

The formal definition of COSTPNs can now be presented:

Let $\textbf{COSTPN} = (PN, T_1, T_2, T_3, W, Pr, Rew)$, where

$PN = (P, T, I, O, H, M_0)$ is the underlying Petri net.

P : Set of places, $T = T_1 \cup T_2 \cup T_3$: Set of transitions, $I \subset P \times T \times \mathbb{N}$: Set of input arcs with multiplicity, $O \subset T \times P \times \mathbb{N}$: Set of output arcs with multiplicity, $H \subset P \times T \times \mathbb{N}$: Set of inhibitor arcs with multiplicity,

M_0 : initial marking.

$T_1 \subseteq T$ with $T_1 \cap T_2 = \emptyset$ and $T_1 \cap T_3 = \emptyset$. T_1 is the set of timed transitions.

$T_2 \subseteq T$ with $T_2 \cap T_1 = \emptyset$ and $T_2 \cap T_3 = \emptyset$. T_2 is the set of immediate transitions.

$T_3 \subseteq T$ with $T_3 \cap T_1 = \emptyset$ and $T_3 \cap T_2 = \emptyset$. T_3 is the set of reconfiguring transitions.

$W : (T_1 \cup T_2) \to \mathbb{R}$. W is a function defined on the set of timed and immediate transitions, which allows the definition of the stochastic component of the COSTPN. W can be marking-dependent. The quantity $W(t_i)$ is the (marking-dependent) firing rate, if t_i is timed, and the (marking-dependent) firing weight, if t_i is immediate.

$Pr : T \to \mathbb{N}$. Pr is the priority function that maps transitions onto natural numbers representing their priority level. The priority level of timed and reconfiguring transitions is always set to 0 and the priority level of immediate transitions is always greater than 0. The enabling of the transitions in a marking of a COSTPN depends on the priority levels of the transitions.

$Rew : RS \longrightarrow \mathbb{R}$ defines the reward structure of a COSTPN, that assigns a reward rate or a pulse reward to any marking of a COSTPN. The reward structure specifies a performability measure, which is optimized in the analysis.

By means of the defined priority function Pr the formal definition of COSTPN reflects the properties of the enabling rule of reconfiguring transitions. The priority level of reconfiguring transitions is always lower than the priority level of immediate transitions such that enabled immediate transitions fire before any reconfiguring transition. The enabling rules of the transitions of a COSTPN are necessary for generation of RS and RG of the COSTPN. The set of enabled transitions in marking M is formally given by:

$$EN_{COSPTN}(M) = \{t \in T \mid t \in EN_{PN}(M) \wedge$$
$$Pr(t) = max\{Pr(t_i) \mid t_i \in EN_{PN}(M)\}\}.$$

where $EN_{PN}(M)$ denotes the set of enabled transitions in marking M in the underlying Petri net PN.

The firing rule of reconfiguring transitions is applied in the model evaluation phase of a COSTPN in order to optimize a given performability measure and to compute the optimal strategies.

3.3 Example of a COSTPN

To clarify the basic definitions, the call admission control component (CAC) of a multimedia server, later analyzed in more detail in Section 5, is presented. The CAC is basically modeled by a queue, in which all arriving calls are stored and wait for acceptance or rejection. A call is lost if not admitted during some waiting time of random mean duration. *Call acceptance* is modeled by the new control mechanism: firing one of the *reconfiguration transition* $R1, \ldots, Rn$ represents acceptance of a new call of type $1, \ldots, n$, respectively, where n is the number of different call types. Arriving calls may be of different types with different resource requirements and different source models. Therefore, there is more than one reconfiguring transition present. The described scenario is reflected by the COSTPN depicted in Figure 3. The transitions are described in Table 1.

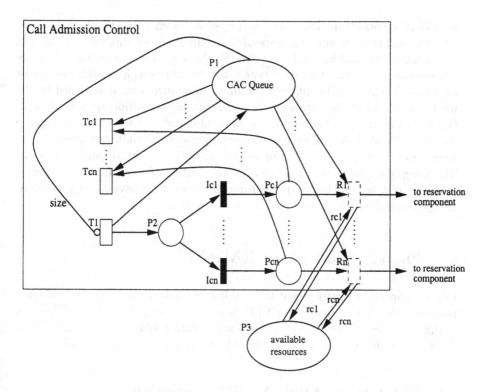

Fig. 3. COSTPN for the modeled call admission control in the multimedia server.

The calls are represented by tokens in the places $P2, Pc1, \ldots, Pcn$. Place $P1$ models the length of the CAC Queue, $\#P1 =$ current length. Place $P3$ contains tokens representing the available resources of the modeled multimedia server. An appearance of a token in $P2$ models the arrival of a call caused by the firing of timed transition $T1$. It is assumed that the arrival and also the timeout processes are described by independent, exponentially distributed random variables with parameters r_T1 and r_Tci, respectively. Calls will thus arrive with a rate of r_T1 and the mean duration of timeout is $1/r\_Tci$. When a call arrives, the length of the CAC queue is updated. The maximum length of the CAC Queue is bounded by *size*, implemented by the inhibitor arc $P1 \to T1$ with multiplicity *size*. By

transition	type	priority	firing rate/weight
$T1$	timed	0	r_T1
$Tc1, \ldots, Tcn$	timed	0	r_Tci
$Ic1, \ldots, Icn$	immediate	1	$wc_1, \ldots, wc_n$
$R1, \ldots, Rn$	reconfiguring	0	-

Table 1. Transitions of the sample COSTPN.

a probability mass function it is determined of which type the arrived call is. The probability mass function is specified by the firing weights $wc_1, \ldots, wc_n$ of the immediate transitions $Ic1, \ldots, Icn$. So the appearance of a token in place Pci models the arrival of a call of type i. If there are enough available resources for a call of type i — the different resource requirements being modeled by the multiplicities rci — the corresponding reconfiguring transition Ri is *enabled*. In this case, a reconfiguration could be performed in order to accept the call and to transfer the call to the call processing components of the multimedia server. The timed transitions $Tc1, \ldots, Tcn$ model the already mentioned timeout process. The firing of one of these transitions end the waiting of a call for acceptance — the call is rejected and leaves the multimedia server. Whenever a call is accepted or rejected the length of the CAC Queue is updated.

4 Construction of an EMRM

In the following we briefly repeat the well-known algorithm to generate a Markov reward model (MRM) from an SPN [TRIV 91]. Then, we point out the differences of that algorithm to the one proposed for generating EMRMs from COSTPNs. To complete this section, the extended algorithm is presented in detail.

4.1 Sketch of the SPN → MRM algorithm

To generate an MRM from an SPN the following algorithm can be used:

1. Generation of the RG of the SPN according to the enabling rule.
2. Transformation of the RG into an ERG by replacing the transition names labeling the arcs by the firing rate (timed transition) or by the ratio of the firing weight (immediate transition) to the firing weights of all enabled immediate transitions.
3. Association of the defined reward structure of the SPN to the states of the ERG. Pulse rewards assigned to vanishing states are only allowed if steady-state analysis is applied.
4. Elimination of vanishing states, where no absorbing loops of vanishing states may exist. The existence of such a loop implies a stochastic discontinuity.

5. Transformation of the transition matrix U into an infinitesimal generator matrix Q of the MRM.

$$Q_{i,j} = \begin{cases} U_{i,j} & if \ i \neq j \\ -\sum_{\forall k, k \neq j} U_{i,k} & if \ i = j \end{cases}$$

4.2 Extensions to the basic algorithm

The main differences between the algorithm SPN $\rightarrow$ MRM and COSTPN $\rightarrow$ EMRM are as follows:

1. COSTPN has *different enabling rules.*
2. During the transformation of a *RG* into an *ERG* the arcs labeled with reconfiguring transitions must be transformed into reconfiguration arcs.
3. Pulse rewards are generally allowed at vanishing states, even if transient analysis is performed.
4. As defined in Section 2, vanishing states are allowed in EMRMs, but in the current implementation [MAUS 90] vanishing loops can not be treated. In order to give the modeler the freedom to model everything except absorbing vanishing loops, the algorithm COSTPN $\rightarrow$ EMRM has to be extended by the elimination of vanishing loops. So the modeler need not to examine the COSTPN to avoid vanishing loops.
5. The transformation of the transition matrix U of the *ERG* into an infinitesimal generator matrix Q need not to be done, because EMRMs may deliberately contain vanishing states with pulse rewards attached to them.

The elimination of the vanishing loops is implemented by the following algorithm:

1. Identify all vanishing loops in the *ERG*. This can be accomplished by using standard graph search algorithm for identifying strongly connected structures [BAAS 88].
2. Identify all subgraphs containing a vanishing loop. Such a subgraph contains the vanishing states of the vanishing loop, entrance states z_{en}, from which the loop is entered, and exit states z_{ex}, to which the loop is left. A subgraph is described by the following transition matrix:

$$U_{subgraph \ with \ loop} = \begin{pmatrix} C & D \\ E & F \end{pmatrix}$$

(Submatrix **C** contains the probabilities of transitions between vanishing states within the loop. Submatrix **D** contains the probabilities of transitions from each vanishing state of the loop to each exit state. Submatrix **E** contains rates and probabilities of transitions from each entrance state to each vanishing state in the loop. Matrix **E** also contains the reconfiguring transitions from Markov entrance states to each vanishing state in the loop. Since transitions between entrance and exit states need not be considered in the loop elimination procedure, submatrix **F** contains only zero elements.)

3. **FOR** every subgraph, $U_{subgraph\ with\ loop}$, containing a vanishing loop **DO**:

   (a) **FOR** any reconfiguring transition through which the loop can be reached **DO**:

     Replace the reconfiguring transition so that the loop cannot be reached through it. A new vanishing state is inserted so that the reconfiguring transition leads to the new vanishing state, from which the loop is reached with probability 1.

   (b) Determine the set of all pairs of states $\{(z_{en}, z_{ex})\}$, where the loop is entered from entrance state z_{en} and left to exit state z_{ex}.

   (c) **FOR** every such pair (z_{en}, z_{ex}) **DO**:

     Compute the expected accumulated reward $R_{en,ex}$ for the passage from z_{en} to z_{ex}, only passing through states of the loop, under the condition that the loop is entered from z_{en} and left to z_{ex}:

$$R_{en,ex} = \frac{R_{en \to ex}}{1 - q_{loop}} + \frac{R_{loop} * q_{loop}}{(1 - q_{loop})^2}.$$

     – $R_{en \to ex}$: pulse reward of the direct paths from z_{ex} to z_{en}. No state is visited more than once on these paths.

     – q_{loop}: probability of passing through the loop ($q_{loop} < 1$, no absorbing loops are admissible in a valid ERG).

     – R_{loop}: pulse reward gained by passing through the loop exactly once.

Note that $R_{en,ex} = 0\ \forall\ (z_{en}, z_{ex})$, if no pulse reward is attached to any vanishing state of the loop.

(For a derivation of $R_{en,ex}$ consider the following infinite sum, which converges:

$$R_{en,ex} = R_{en \to ex} +$$
$$(R_{en \to ex} + R_{loop}) * q_{loop} +$$
$$(R_{en \to ex} + 2 * R_{loop}) * q_{loop}^2 + \cdots$$

The first term is the reward accumulated on direct paths from z_{en} to z_{ex}. The second term is the reward accumulated on paths from z_{en} to z_{ex} passing through the loop exactly once. The third term is the reward accumulated on paths from z_{en} to z_{ex} passing through the loop twice, and so on.)

   (d) Eliminate the loop from $U_{subgraph\ with\ loop}$ by applying the standard algorithm [TRIV 91]:

$$U_{subgraph\ without\ loop} = E * (I - C)^{-1} * D.$$

     (I is the identity matrix with the dimension of C.)

   (e) For every pair of states (z_{en}, z_{ex}), for which $R_{en,ex} \neq 0$, a new state z_v is generated, which is associated with computed reward $R_{en,ex}$. z_v is inserted between the pair of state (z_{en}, z_{ex}). The transition from z_{en} to z_{ex} is replaced by a transition from z_{en} to z_v. The rate/probability of this new transition is the same as of the transition from z_{en} to z_{ex}, which results from the elimination of the loop in step (d). From z_v a transition leads to z_{ex} with probability 1.

(f) Replace $\mathbf{U}_{subgraph\,with\,loop}$ in the ERG by the subgraph computed in the steps (a) through (e).

4.3 Summary of the extended algorithm

The COSTPN $\to$ EMRM generation algorithm can be summarized as follows:

1. Create the RG according to the enabling rule of COSTPN as described in Sec. 3. For an illustration see Figure 4 for the RG of the CAC component described in Section 3.3 (The parameters of the CAC COSTPN are set to the following values for the generation of the RG: length of CAC Queue = 2, number of call types = 3 and number of resource units = R.).

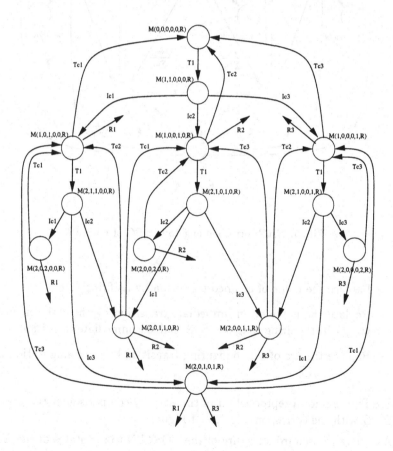

Fig. 4. RG of the CAC component which is described in Section 3.3.

2. Transform the RG into an ERG by replacing the transition names/arcs as follows:

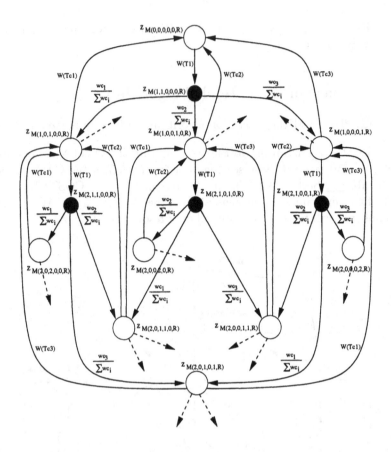

Fig. 5. *ERG* corresponding to the *RG* of Figure 4.

- Replace the name of a timed transition by its firing rate.
- Replace the name of an immediate transition by the ratio of its firing weight to the firing weights of all enabled immediate transitions.
- Replace the arc of a reconfiguring transition by a reconfiguration arc.

See Figure 5 for a reproduction of this step by comparison of the presented *ERG* with the corresponding *RG* of Figure 4.

3. Associate the reward structure of the COSTPN to the states of the *ERG*.
4. Eliminate all vanishing loops by applying the algorithm described above. Absorbing vanishing loops are not allowed in the *ERG*.

The resulting EMRMs can be directly analyzed by the implemented numerical algorithms [DEME 92].

5 Modeling and Analysis of a Multimedia Server

The goal of this section is to show how COSTPNs and the corresponding EM-RMs can be used to model and analyze telecommunication systems in order to optimize their operations with respect to certain reward measures. As an example, a multimedia server is presented that, for instance, provides clients with video or audio clips from a disk, or works as a central node in video-conferencing systems. In the first part, the server is modeled as a COSTPN, and it is discussed how interesting variants can be obtained applying different COSTPN features. The second part, then, presents a number of results of analyses of these models, carried out with the help of the PENELOPE toolset. Finally, the results and experiences gained with COSTPN, used for the design of telecommunication systems, are summarized.

5.1 A Multimedia Server Model

5.1.1 Resource Reservation based on Stream Aggregations

Continuous media streams like audio or video are special in that their transmission and processing require certain timing relations between single data units within a stream to be maintained, since otherwise, the semantics of these media are lost. In order to reach this goal, the necessary resources for transmitting and processing of data units have to be available when they are needed. One way to assure this is to make *resource reservations* within end systems, used to produce or present such a stream, and networks, used to transport it. Considering variable bit rate (VBR) streams like MPEG movies, if one wants to ensure the timely handling of every single data unit (and thus a 100 % guarantee), resources are to be reserved for the stream's peak rate. However, this policy leads to huge over-reservation in non-peak periods and thus makes sub-optimal use of usually scarce resources. Since such strict guarantees are often not necessary for multimedia systems, resources should better be reserved for some lower rate, trading higher utilization of resources for limited risks of guarantee violations. When supporting a very large number of streams, where streams are multiplexed over the reserved resources, there are usually sufficient resources available for streams sending at their peak rates since others are not using all of their reserved resources at this very moment, if reservation is based on an average rate assumption. When admitting a limited number of calls, though, which is the most relevant case, the problem remains to determine the optimal amount of resources to be reserved, trying to avoid over-reservation but keeping guarantee violations within a specified (usually very small) limit. Reservation strategies need to be determined lying somewhere between peak-rate and average rate reservation scheme. It will become obvious that COSTPNs are a suitable tool to support a system designer in finding the reward-optimal solution.

The basic operation mode of the multimedia server discussed here is as follows: an incoming call is analyzed with respect to its parameters and the current system conditions. If the call can be admitted, resources will be reserved accordingly. The call processes its data units for a certain time and finally releases the reserved resources and leaves the system. Call parameters to be taken into account include call length, call arrival rate, rewards to be gained, or the calls burstiness, while system conditions include the number of already active calls, available number of free resources, possible types of calls, etc. Since, in addition to this complexity, many parameters are of a stochastic nature, COSTPNs are a very suitable tool to model and analyze such systems. Timed and immediate transitions can be used to model the stochastic behavior of system components while *reconfiguring transitions* allow the modeling of possible *decisions* inside the managing component of the server. Analyzing the equivalent EMRM of the server model with numerical methods leads to the generation of strategies which tell the managing component what decisions to take at which point of time.

The server model developed here consists of four main components: call admission control (CAC), resource reservation, accounting unit, and source models, the latter representing the behavior of different types of calls.

The "intelligent" component in this server is the call admission control, where it is decided whether it is advantageous to admit an incoming call or not. This unit has already been described in Section 3.3, see Figure 3. The decision process is modeled by a number of reconfiguration transitions, one for each possible call type.

As soon as a call has been admitted by the CAC, it enters the reservation component (Figure 6, transition and place descriptions in Tables 2 and 3, respectively). The main task of this component is to replace the resources to be reserved for this call from the set of available resources (place $P3$) into the pool of reserved resources (place $P4$). This is modeled by transitions $Ir1$, ..., Irn, with n denoting the number of call types. The call token itself is forwarded to the corresponding source model component. Since the server multiplexes several calls over the set of reserved resources, all resources in $P4$ can be used by all active calls.

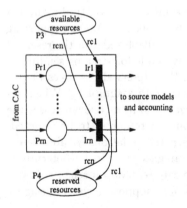

Fig. 6. Reservation component.

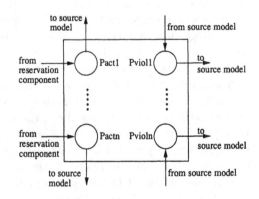

Fig. 7. Accounting component.

Henceforth, we advocate a conceptual separation between resource reservation, admission control, and resource usage. This allows for a flexible and efficient

Transition	Type	Prio.	Rate/Weight	Meaning
Iri	immed.	1	1	reserve resources for a call of type i

Table 2. Transition description of the reservation component.

Place	Meaning
Pri	calls of type i waiting for reservation
$P3$	freely available resources
$P4$	reserved but currently unused resources

Table 3. Place description of the reservation component.

control mechanism to be openly deployed and ultimately used for a dynamic optimization by service providers. Modes of operation can be taylored towards specific QoS requirements, cost constraints, and individual source traffic models in a highly dynamical manner. In particular, we are looking here at multiplexing of multimedia streams for a flexible schema of reservation and usage of resources.

The accounting component (Figure 7, Table 4) is responsible for recording the revenue the server receives for hosting calls, but also the penalties to be paid whenever a call does not receive the level of service it expects. Usually, revenue is generated by those markings containing a token in one or more of the places $Pact1, \ldots, Pactn$. The number of tokens in these places models the number of active calls of the corresponding type. Thus, the service user usually pays for the service as long as his call is active[2]. In contrast, penalties (= negative reward) have to be paid by the server as soon as a token appears in one or more of the places $Pviol1, \ldots, Pvioln$.

Place	Meaning
$Pacti$	active call of type i
$Pvioli$	token here indicates possible guarantee violation

Table 4. Place description of the accounting component.

Finally, the system components describing the different types of calls are modeled by the COSTPN of Figure 8. Transition and place descriptions can be found in Tables 5 and 6, respectively. The idea is the following: an active call of a given type i is represented by a token in Place $Pi1$. The call produces, at a given rate, data units to be processed by the server. This process is modeled by transitions $Ti1$ and $Ii1, \ldots, Iim$. $Ti1$ describes the fact that new data units are ready for processing, while the immediate transitions (which remove the token in $Pi2$ immediately) model a probability mass function over the data size. Thus, by varying the number of these immediate transitions, the firing probability

---

[2] Other models are certainly possible and make sense — an example will be given later.

assigned to each of them, and the number of tokens moved to place $Pi3$, the burstiness of call type i is modeled. In principle, it is possible to model any type of burstiness by adding or removing transitions, or by changing the firing weights. If one wants to model, for instance, an MPEG stream with the frame pattern IBBPBBPBB, then it would be useful to introduce three immediate transitions, one for each type of frame. The probability ($=$ firing weight) for the I-frame transition would be $\frac{1}{9}$, that for P-frames $\frac{2}{9}$, and the one for B-frames $\frac{6}{9}$. The I-frame transition would, for instance, move 10 tokens to $Pi3$, the B-frame transition would produce 5 tokens, and the B-frame transition 3 tokens.

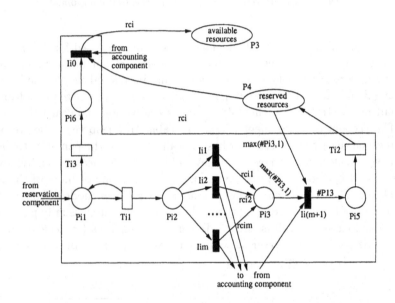

Fig. 8. Source model for call type i.

Transition	Type	Prio.	Rate/Weight	Meaning
$Ti1$	timed	0	r_Ti1	prepare data processing
$Ti2$	timed	0	r_Ti2	process data unit
$Ti3$	timed	0	r_Ti3	finish call
$Ii0$	immed.	1	1	release resources
$Iij, 1 \leq j \leq m$	immed.	1	w_Iij	decide about number of data units
$Ii(m+1)$	immed.	2	1	move data units to processor

Table 5. Transition description of the source model.

Tokens appearing in $Pi3$ will be immediately forwarded to $Pi5$ (holding the data units currently being processed) by transition $Ii(m+1)$ if and only if there

Place	Meaning
$Pi1$	call waiting for processing or finishing
$Pi2$	calls producing data units for processing
$Pi3$	data units waiting for processor
$Pi5$	data units currently being processed
$Pi6$	finished calls

Table 6. Place description of the source model.

are currently enough resources in the reserved resources pool $P4$[3]. It may well happen that resources are not available , since resources are not reserved at peak rates per stream, but calls are rather multiplexed over the reserved resources in an aggregated fashion as a function of the current state, i.e., the number of already admitted streams sharing the same reserved resources. Imagine a situation where 3 of the above MPEG calls have been admitted and each of them made a reservation of 5 resource units per frame. Assume further that one call is just processing an I-frame and has removed 10 units from the resource pool. If now another call has to process an I-frame, it has to wait until the first call has freed at least 5 resources before its 10 tokens can be moved from $Pi3$ to $Pi5$.

The processing of data units is modeled by transition $Ti2$, which removes tokens from $Pi5$ at a given rate. Whenever a data unit has been processed, its assigned resource is moved back to the pool $P4$ and made available for other active calls.

When a call has finished its task, the length of a call being described by the rate of transition $Ti3$, it releases its resources (immediate transition $Ii0$) and puts them back into the pool of free resources $P3$. Also, the token for this active call is removed from the corresponding place $Pacti$ in the accounting component.

The overall model of the multimedia server (called model type A from now on) is depicted in Figure 9 for n call types. In addition to the descriptions and pictures of the single components, this global view shows the token flow between the single components. From the CAC, if a call of type i is accepted, its token is moved to the reservation component by Ri. As soon as the resources have been reserved, the token enters both the corresponding source model and the accounting unit ($Ir1$ to Irn). A call leaving the source model also removes the active call token from the accounting unit ($Ii0$). Another connection between source model and accounting unit concerns the handling of penalties. The server will have to pay a penalty if it cannot provide enough resources when a call needs them. The start of this resource test is modeled by each of the transitions $Ii1$ to Iim by moving a token to $Pvioli$ in the accounting unit. In the error-free case, this token will be immediately removed by transition $Ii(m+1)$. The corresponding marking is then vanishing and will not lead to a negative reward. If, however, there are not enough resources available in the pool $P4$ to remove

---

[3] The arcs from and to the accounting component are explained later.

all tokens from $Pi3$, the marking becomes tangible, which leads to a negative reward (cost).

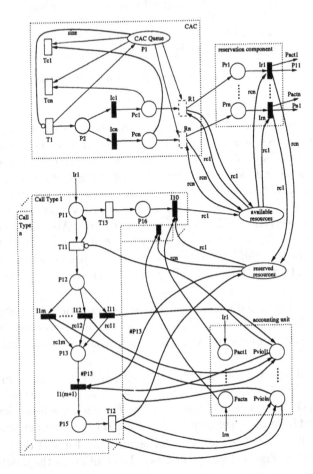

Fig. 9. Complete model of the multimedia server.

5.1.2 Strategies for an Adaptive Reservation Mechanism

A variant B of the server model is shown in Figure 10 (transition and place descriptions in Tables 7 and 8) for a one-call-type system. The goal of this model is to always reserve the optimal amount of resources. Therefore, no reservations are made a priori when a new call becomes active; rather, the system can dynamically decide when it should add resources to the pool of reserved resources or remove them from there. Resources should be reserved when they are necessary to provide the requested service guarantees; otherwise, they should be released. In addition to $R1$, the model therefore has two further reconfiguring transition $Rres\_in$ and $Rres\_out$ which add or remove a single resource unit. Preparing a resource unit for use by active calls requires some time; this is modeled by place P_res_prep and transition T_res_prep. Releasing a resource can be done in zero time. A maximum of 3 resource units can simultaneously be in preparation for usage. To keep the model bounded, an inhibitor edge has to be introduced

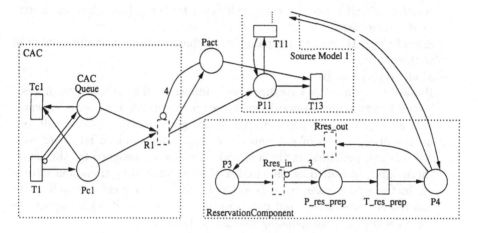

Fig. 10. Variant of the multimedia server.

Transition	Type	Prio.	Rate/Weight	Meaning
T_res_prep	timed	0	$r\_T\_res\_prep$	prepare resources for processing
$Rres\_in$	reconf.	0	—	reserve resource
$Rres\_out$	reconf.	0	—	release resource

Table 7. Transitions of model type B.

Place	Meaning
P_res_prep	resources being prepared for processing

Table 8. Places of model type B.

between $Pact$ and $R1$, since there is no longer a direct relation between a call and a certain number of resources. A maximum of 4 calls is allowed to be active in this model. Finally, place $P16$ and transition $I10$ in the original source model (of type A) as well as place $P2$ and transition $Ic1$ have been removed, since they are no longer necessary[4]. The rest of the model remains the same as in model type A of Figure 9 and is therefore left out in Figure 10.

---

[4] These places were introduced in model type A for modeling reasons.

With these basic model definitions, we now build a set of server configurations for the experiments. These models and their characteristics are listed in Table 9. The names assigned to each model will be used again in the experiment description. The models are basically distinguished along four characteristics:

– Number of call types: for every call type, the model contains one source model component.
– Size of the CAC queue: this parameter influences the maximum number of waiting calls.
– Model type (A or B)
– Reward structure: this structure mainly influences the optimization process. Two different types of reward structures are investigated: a positive-reward-only structure, where positive rewards are given for a supported call, but no penalties in case of guarantee violations, and a mixed structure with rewards and penalties. In the first case, rewards are assigned for the actual throughput, measured by the number of tokens moved by transition $Ii(m + 1)$. In the second case, rewards are assigned for active calls as well as for non-reserved resources[5] and penalties have to be paid if tokens appear in $Pvioli$ while the corresponding marking is tangible.

Name	#call types	queue size	model type	reward structure
Model 1.1	1	1	A	0 if tangible marking #P13 if vanishing marking and $I13$ is the only enabled transition
Model 1.2	1	1	A	$\#Pact1 * reward1 - \#Pviol1 * penalty1$ $+\#P3 * resource\_reward$ (tang.) 0 (vanish.)
Model 1.3	1	3	A	$\#Pact1 * reward1 - \#Pviol1 * penalty1$ $+\#P3 * resource\_reward$ (tang.) 0 (vanish.)
Model 2.1	2	2	A	$\#Pact1 * reward1 - \#Pviol1 * penalty1$ $+\#Pact2 * reward2 - \#Pviol2 * penalty2$ $+\#P3 * resource\_reward$ (tang.) 0 (vanish.)
Model 3.1	1	1	B	$\#Pact1 * reward1 - \#Pviol1 * penalty1$ $+\#P3 * resource\_reward$ (tang.) 0 (vanish.)

Table 9. Server configurations to be used in the experiments.

The overall task of the CAC is now to take all these parameters into account and take reward-optimal decisions (i) in admitting or rejecting calls in type A

---

[5] This assignment models the fact that resources could be used for other applications running on the server.

models and (ii) additionally in reserving the optimal amount of resources in type B models. Translation of the multimedia server COSTPN into an EMRM allows for a numerical analysis of the system and the generation of reward-optimal strategies. In the remainder of this section, we will present a number of results obtained for several multimedia server configurations.

5.2 Results and Strategies

With the server configurations described in Table 9, several experiments have been carried out, using the already mentioned toolkit PENELOPE. Basically, the experiments and their analysis can be classified as follows:

- Experiments with model type A and one type of calls. The experiments try to answer the question, when a waiting call of this type should be admitted and when it should be rejected. These experiments are based on model group 1.
- Experiments with model type A and two type of calls. Here, situations are investigated in which calls of both types are waiting. The question is whether one (or more) of them should be admitted at all, and if so, which type it (they) should be of. Model group 2 builds the base for these experiments.
- Experiments with model type B. Since resource reservation and call admittance are decoupled in this model, the question rather is, how many resources should be reserved if a certain number of calls is active. These experiments are based on model group 3.

First, the COSTPN has to be translated into the corresponding EMRM for every model. Table 10 lists the sizes of the COSTPN and the corresponding EMRM[6]. The initial marking of every COSTPN is given by 8 tokens in Place $P3$ and no token in any other place. This models the idle system with 8 available resource units. The toolkit allows both transient and stationary optimizations.

Model Name	Places	COSTPN			EMRM				
		Timed Trans.	Immed. Trans.	Rec. Trans.	Markov States	Vanish. States	Timed Arcs	Prob. Arcs	Rec. Arcs
1.1	12	5	5	1	202	334	614	406	64
1.2	12	5	5	1	202	334	614	406	64
1.3	12	5	5	1	404	784	1430	928	192
2.1	20	9	10	2	12600	19422	58320	31194	3003
3.1	11	6	3	3	2380	2344	8844	3384	4032

Table 10. Sizes of COSTPN and corresponding EMRM for different models.

---

[6] For simplification and state space reduction reasons, CAC and reservation unit are integrated in model type A by removing places $Pr1$ to Prn as well as transitions $Ir1$ to Irn, and assigning the latters' functionalities to the reconfiguration transitions $R1$ to Rn. In other words, the reconfiguration transitions do not only test for the availability of resources, but also make the reservations itself.

In the latter case, the numerical analysis and optimization is carried out with an infinite time horizon which means that if a system is in a given state, then irrespective of time the same strategy will be applied (depending only on the system's parameter space, but not on the remaining time). In the transient optimization process, in contrast, it is assumed that the modeled system will complete its operation at a certain point in time (finite time horizon). For the server example, a transient optimization makes sense when the server eventually stops its operations, let's say, every day at midnight. Stationary optimization could be applied if the server runs continuously without ever stopping. The experiments presented here mostly deal with transient optimization since it is the more interesting and realistic case.

The presentation of the experiments always follows these guidelines: first, the experiment setting is described, followed by the presentation of the results and finally an interpretation and analysis of certain strategy curves. The latter is a difficult task sometimes, since the results depend on a huge number of parameters which cannot all be easily integrated into an interpretation. Therefore, usually only certain aspects can be described here. Basically, the following three questions are always answered:

(i) What does the curve/region express?

(ii) Why has the curve/region its actual shape?

(iii) How are curves/regions in the same figure related to each other?

The results are presented in form of diagrams like the one in Figure 2. Whenever it makes sense, the parameter space will be divided into regions in which a certain strategy has to be applied. This is the case when strategies are displayed for a set of markings which belong together. Such a set of markings is characterized by predicates which are fulfilled by the markings. If the diagram contains strategies for a set of unrelated markings, strategy switching curves are displayed and described rather than regions.

Analysis of model 1.1 The basic model for the first series of experiments is model 1.1. As source model, a call type with two different possible resource requirements of $rc11 = 1$ and $rc12 = 3$ units is selected. The reservation unit always makes a reservation of $rc1 = 2$ units whenever such a call is admitted. The server possesses $\#P3 = 8$ resource units to assign to calls, leading to a maximum of 4 calls to be admitted. The only decision the CAC has to take in such a system is whether a waiting call should be admitted or not. Later, it will be shown that in more complex systems, there will also be more complex decisions.

For the analysis of model group 1, the parameter values given in Table 11 are used. The unit of time is one second. Thus, the mean inter-arrival time of calls $1/r\_T1$ is 10 seconds, the mean duration of calls $1/r\_T13$ is 60 seconds, etc. For every experiment, one of the parameters is varied which is then stated explicitly. The reward structure used is the positive-only one with rewards being assigned for vanishing markings where only $I1(m + 1)$ can fire.

r_T1	r_Tc1	r_T11	r_T12	r_T13	w_I11	w_I12
1/10	1/10	1/5	1/5	1/60	2	1

Table 11. Basic parameter set for model group one experiments.

First, we are interested in the influence certain call type parameters have on the optimal strategies. Therefore, we vary the ratio between the weights of transitions $I12$ and $I11$. The bigger this ratio is, the more often transition $I12$ fires, which in turn means that the data rate and thus the resource requirements increase. Figure 11 presents results for model 1.1 within this setting. Curves three and four in this figure show the strategies to apply in the completely idle system state and in the system state with one active call, which is not producing or processing any data. In both cases, a new call is waiting. Both curves are identical with the x-axis which means that these strategies are always applied. Both express that whenever the system is in one of these two states, the waiting call should *always* be admitted. For the third curve, this is intuitively clear, since admitting a call can only lead to additional rewards. Also, admitting a second call is always advantageous. This is not immediately clear, since a second call could lead to overload situations which may be less likely in situations with only one active call. However, with the given parameter settings, this is not the case here. More interesting, in this respect, are curves one and two. Both strategies described by the curves are to be applied when the system has three active calls and one waiting call. In curve one, all of the reserved resources (six) are currently used for data processing, while in the second case, only five of six are in use. Both curves state that only until a certain amount of remaining runtime a new call should be added. If the remaining time of the server operations drops below this limit, the waiting call will not be admitted anymore. The reason for this is, the new call probably would lead to an overload situation which means that no further reward can be generated. This is the case here, since no resource units are currently free to serve the new call. If there is more time left, a new call should be admitted, since then, there is enough time to free the resources eventually and serve the new call then.

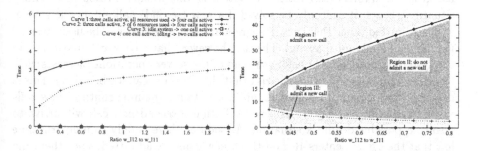

Fig. 11. Several results for model 1.1.

Fig. 12. Strategy regions when the system is completely idle and a new call asks to be admitted (Model 1.2).

Both curves reflect a shape that would have probably been anticipated. If transition $I12$ fires much less often than $I11$ (low-load case, depicted in Figure 11), then a new call can be admitted nearly to the end of the server operations. When the ratio and thus the data rate increases, more resources will be used for this call and it is necessary to stop admitting calls earlier.

Finally, to compare curves one and two, call admitting is stopped earlier in the markings with all resources currently being used (curve one). This is also obvious, since in the second situation one resource has already been released and it is more likely that a new call will find the necessary resources to process at least parts of its data.

Analysis of model 1.2 In this model, the mixed reward structure builds the main difference to model 1.1. If not stated otherwise, the reward values for the experiments are set as given in Table 12.

reward1	resource_reward	penalty1
1	0	8

Table 12. Reward values for one-call-type-experiments.

Strategies in the idle system state. Figure 12 shows results comparable to curves three and four in Figure 11. It displays the strategies to be applied in the idle system state, with no calls being active, but one call requesting admittance. The most obvious differences with respect to the positive-only reward function is that it is not always advantageous anymore to admit a call to the idle system. Why does this situation occur? The reason is that only two resource units are reserved for a call, but it may happen that three units are necessary, exactly when transition $I12$ fires. In this case, the system cannot process the data (fire transition $I13$) since it cannot get enough tokens out of place $P4$. Thus, no reward is generated, but penalties have to be paid due to the token in $Pviol1$ and the fact that the current marking is tangible. Therefore, it is sometimes useful not to admit a call when the system is idle. If such a resource deadlock situation occurs, it can only be solved if a new (second) call enters the system and is admitted, since then, new resources are available that can be used to serve the first call (call multiplexing). The risk, however, that no new call may arrive in the system (transition $T1$) before the end of the server operations increases with less time remaining, and thus the risk that the situation cannot be solved. This is described by Region I in Figure 12: up to a certain remaining runtime, new calls are always admitted, since it is very likely that a second new call will arrive to resolve a possible deadlock situation. As soon as there is only so much runtime left that the system enters Region II, no new calls are admitted, since then, the risk is too big that transition $I12$ fires and no second call arrives. Finally, very shortly before the end of the server's operations, the system enters Region III and may therefore admit a new call again. Now, it is even more unlikely that a second call arrives, but it is also less likely that $I12$ fires. All in all, the tradeoff between rewards and penalties is positive, so it is advantageous to admit a new call in this short time horizon.

The influence of the call's data rate, increasing from left to right on the x-axis, is obvious: the more data there are to be processed, the more likely it is that a deadlock situation will appear as described above. Thus, it is better to stop admitting calls earlier which is expressed by the monotonously increasing curve between Regions I and II in the diagram. Comparably, the curve between Regions II and III is monotonously decreasing, since with increasing data rate it is more and more likely that $I12$ fires shortly before the end. It becomes thus less and less interesting to admit a call again.

An interesting curve, not shown in this diagram but related to it, is to be applied in situations where there is already one active call, and a new call asks to be admitted. This curve is identical with the x-axis, which means that in these situations the call will always be admitted. This is due to the fact that multiplexing gains can only be achieved if at least two streams are simultaneously active. Based on the given resource-aggregation reservation schema, situation does improve if additional streams, accompanied by additional resources, are admitted.

Strategies in active system states. Figure 13 shows how the behavior of the system changes in different system states (non-idle states as opposed to the idle state as referred to in the previous diagram). This figure contains three curves, which indicate strategy switches in three different but comparable states: curve i applies in a state where i calls have just completed their operations and are about to leave the system ($\#P16 = i$, $\#P11 = 0$), while one call is asking for admission.

Most of the time, the waiting call will be admitted. Shortly before the end of the server operations, the systems is advised to stop admitting calls. This is again due to a possible deadlock situation, which may occur because in only a few seconds, only one call will still be active (the newly admitted one). With increasing data rate, the system stops admitting calls earlier, since a deadlock situation becomes more likely.

In model 1.2, resources are only released when a call finally leaves the system (by firing transition $I10$). This is also the reason why the three curves are not identical. The more calls are currently in the system (even if they are in the process of leaving), the more resources are still available for multiplexing. Thus, with three leaving calls, a new call will be admitted for a longer time, since it takes more time for the three calls to leave the system and to release their resources than it takes for one or two calls. This again means that more resources are available for multiplexing for a longer time, thereby reducing the deadlock risk.

Varying the reward. Figure 14 shows how different reward values influence the system behavior in the idle state: the diagram includes curves for reward1 values

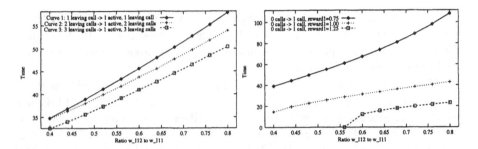

Fig. 13. Several strategies for an active system.

Fig. 14. Influence of different rewards assigned to active calls (reward1).

of 0.75, 1 and 1.25. Each curve separates the decision space into regions where it is advantageous or disadvantageous to admit a new call. The general behavior of the system is as in the previous case: until only a certain runtime remains, the waiting call will be admitted. Afterwards, the system is advised to remain in the idle state. Investigating a single curve, the same influence of an increasing data rate as above becomes obvious. A comparison of the three curves reveals that with increasing reward it becomes more attractive, even until shortly before the end, to admit the waiting call, since the collectible rewards compensate for a possible penalty.

Varying the call duration. The last two diagrams for model 1.2 show results when other parameters are varied, with the parameter w_l12 set back to its value shown in Table 11. In Figure 15, the duration of a call has been varied between 1000 and 10 seconds, which is equivalent to a rate r_t13 between 0.001 and 0.1 (logarithmic scale, *duration* = $1/rate$). Again, the decision space is divided into the three regions, similarly as already observed in Figure 12. Thus, usually a new call will be admitted (Region I), but there is a region (Region II) where it is better not to admit the call. In the call duration region between 1000 and 50 seconds (corresponds to a rate between 0.001 and 0.02), the system behavior does not change very much. However, when the call becomes too short, as depicted in the rightmost part of the diagram, then it becomes less interesting to admit a new call, so admittance is stopped earlier. The reason is that such a call may not be able, during its short lifetime, to accumulate enough rewards in order to compensate a possible penalty.

Varying the call inter-arrival time. Figure 16 shows the influence of a variation in the call inter-arrival time on the idle state, expressed by parameter r_t1. The diagram covers inter-arrival times between 10 and 6.6 seconds. Obviously, even such a minor variation has a strong influence on the system's strategies. The longer it takes for new calls to arrive (the further left part of the diagram), the better it becomes to stop admitting calls early. With $r\_t1 = 0.1$ (inter-arrival time of 10 seconds), the server stops accepting calls 45 seconds before the end, while with $r\_t1 = 0.15$ (inter-arrival time of 6.6 seconds), calls will be accepted until 10 seconds before the end of the operation time. The reason for this behavior

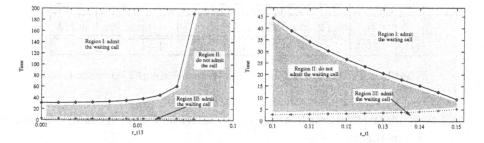

Fig. 15. Influence of varied call duration in the idle state (model 1.2).   **Fig. 16.** Influence of call inter-arrival time in the idle state (model 1.2).

obviously is that with longer inter-arrival times the risk increases that resource deadlocks for one call cannot be resolved, since no new call arrives for a long rime.

Also, there is a short period before the end where calls are admitted again due to the low probability of deadlock situations.

Analysis of model 1.3 For model 1.3, where the CAC maximum queue size is 3, only one interesting detail should be mentioned here: in this model, it is often useful to admit more than one call at the same time. Consider a situation where 3 calls are waiting in the queue and the system is idle otherwise (has no active calls). With the given parameter setting, the strategy then always is to simultaneously admit all 3 calls. Thus, deadlock situations as described above can be avoided. Comparably to model 1.2, however, if only one call is waiting, then there are situations where the system does not immediately admit the call.

Analysis of model 2 Adding a second call type to a system increases the complexity. For the third set of experiments, we introduce a new type with an average resource usage of 4 units, and two possible burst sizes of 2 and 5 units. This time, we select a CAC queue size of 2. Obviously, the complexity has increased a lot: we have more than 32000 states compared to 536 in the one-type-one-place model and 1188 in the one-type-three-places model. Also, there are a lot more states where decisions have to be taken: we have 3003 reconfiguration arcs compared to 64 and 192, respectively. Accordingly, the computation time for the numerical analysis increases dramatically. This, however, is only a minor problem, since all the computations can be done before the system is entering production phase. All results can simply be stored in tables which may then be accessed by the system's decision unit — the CAC — during runtime.

The parameter values used for this experiment are shown in Table 13. Both call types are very similar, with the difference of the resource consumption mentioned above and the weights of the corresponding immediate transitions. For call type 1, the occurrence of one data unit is twice as likely as that of three data units (w_I11 compared to w_I12), while for call type 2, the occurrence of two and five units is equally likely (w_I21 equals w_I22).

r_T1	$r\_Tc1 =$ r_Tc2	$r\_T11 = r\_T12 =$ $r\_T21 = r\_T22$	$r\_T13 =$ r_T23	w_I11	w_I12	$w\_I21 =$ w_I22	$w\_Ic1 =$ w_Ic2
1/10	1/10	1/5	1/60	2	1	1	1

Table 13. Parameter set and values for model group 2 experiments.

Table 14 introduces the basic parameter values for the reward structure used in these experiments.

reward1	penalty1	reward2	penalty2	resource_reward
1	10	2	10	0.1

Table 14. Reward values for two-call-type-experiments.

Strategies when one call of type 1 is active. Figure 17 shows the strategy regions for situations where one call of type 1 is currently active and one call of each type is requesting admission. In the corresponding experiment, the reward to be paid for calls of type 2 has been varied between 0.5 and 7.5. The diagram shows that with low rewards for call type 2, only the call of type 1 is admitted. In such a situation, it obviously makes more sense to wait for another call of type 1. To the contrary, the system is advised to admit only the call of type 2 if the reward is very high (right part of the diagram). If, in both cases, there is only a short run-time remaining, both calls are admitted, since it is quite unlikely that a new call of the type with the higher reward will arrive. Also, there is a region with *reward2* values between 3 and 4, where it is always advised to admit both calls. This is the region where there is an equilibrium between risks and chances for both types of calls.

Strategies when one call of each type is active. Now, what happens, if the call of type 2 has been admitted in the situation above. The new situation is displayed in the diagram of Figure 18 where one call of each type is active and one call of type 1 is waiting. The same situation may, by the way, also occur if the active calls where running for a while and then a new call of type 1 requests admission. The diagram is quite similar to the one before, with the difference that Region I there is no longer present in the new situation. This is no surprise, since there is no need to make a difference between admitting one or two calls anymore. As one would have anticipated, the shape and position of Region III in Figure 17 and Region II in Figure 18 are the same, since a reconfiguration does not take any time.

Strategies when two calls of type 1 are active. Figure 19 shows the strategy regions for situations with two active calls of type 1 and one call of each type waiting. With the given resolution of *reward2*, the decision space is divided into

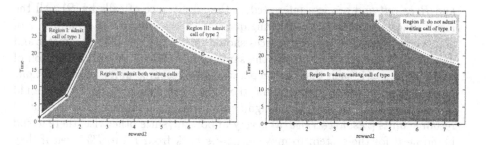

Fig. 17. Strategies when one call of type 1 is active with one call of each type waiting.

Fig. 18. Strategies when one call of each type is active with one call of type 1 waiting.

two regions, and most of the time and for most values of *reward2* displayed here, the call of type 2 is accepted. This is the case since there are already two calls with four reserved resources running, so the risk of a guarantee violation is much lower than in the situation of Figure 17. Therefore, the higher reward of call type 2 gains more weight than the risk introduced by the high peak rate. It should also be noted that there is no region where two calls can be admitted — the reason simply being that there are not enough resources left for two calls.

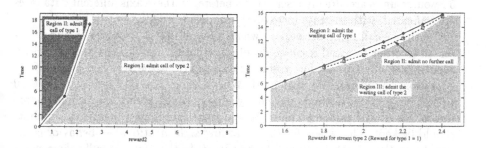

Fig. 19. Strategies when two calls of type 1 are active with one call of each type waiting.

Fig. 20. Strategies when two calls of type 1 are active with one call of each type waiting (detailed).

Finally, Figure 20 reveals an interesting detail which can only be seen with a higher resolution for *reward2*. There is a small zone between the call admission regions where no calls are admitted. In this region, it obviously is not advantageous to admit any one call. There seems to be an equilibrium similar to the one detected in Figure 17, but this time in a way that not two, but no call is admitted.

Analysis of model 3 Model type B has been introduced to optimize the resource reservation process by not providing a fixed relation between the number of calls and the resource units to be reserved, as it was the case in model type A. The experiments thus concentrate on the reconfiguration transitions $Rres\_in$

and *Rres_out* and are less interested in the admittance of calls by $R1$. All the result diagrams answer the following question: if the system has a number x of active calls, then how many resources should be in pool $P4$? The parameter and value set for the experiments are the same as in Table 11, with the addition of $r\_T\_res\_prep = 1$, such that the preparation of a resource units takes about one second. For the reward parameters, the values are shown Table 15. It should be noted that in this case, the assignment of a positive reward value to unused resource units (*resource_reward*) is very important, since otherwise, there would be no need for the system to move resources back from $P4$ to $P3$, even if they are not necessary to support active calls.

reward1	resource_reward	penalty1
1	0.1	8

Table 15. Reward values for model B experiments.

Resource usage for one active call. Figure 21 shows the strategy regions when the system currently serves one active call ($\#Pact = \#P11 = 1$), no further calls are waiting ($\#Pc1 = 0$), the active call is currently not processing any data, and 5 resource units are already reserved. As before, on the x-axis different data rates are indicated, with the rate increasing from left to right.

With one call being active, one would assume that 3 resources should be reserved in order to avoid resource deadlock situations, as described above. However, this is only true for low average data rates or at the very end of the server operation, as indicated by region III. With a value of w_I12 between 4 and 10 ($w\_I11 = 8$), the system would nearly always (until about 2.5 seconds before the end) reserve 4 resource units. If the average rate is even bigger, then 5 units should be kept. The reason for this — at first sight irrational — behavior is the possibility that another call will be admitted. In that case, it will take a while (around 1 second) to reserve and prepare new resources to support the new call. With high data rates, the risk of the occurrence of the call's peak rate is also high, and with not enough resources available also the risk of a penalty increases. Thus, with the given values for unreserved resources and penalty, it is advantageous for the system to keep one or two additional units ready for immediate use if necessary. Increasing the value for unreserved resources (*res_rew*) would change this picture in that it would become more advantageous to keep resources unreserved and only move them to $P4$ when a new call really is admitted.

Resource usage for two active calls. In Figure 22, the regions are shown for two active calls, with the other parameters being the same as above. In this situation, currently 5 resource units are reserved. For low and medium average rates (Region II, w_I12 between 2 and 8), the typical reservation will then be 5 units, so the strategy would be to remain in the current state and not to add or release a unit. With lower rates (not shown in the diagram) or with only a

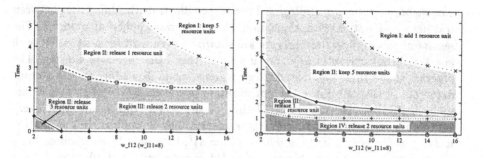

Fig. 21. Strategy regions for 1 active call when 5 resources are currently reserved. **Fig. 22.** Strategy regions for 2 active calls when 5 resources are currently reserved.

few seconds being left, even a reservation of 4 (Region III) or 3 units (Region IV) would be enough. With higher rates, it would be better for the system to add another unit, in order to be prepared for frequent peak rates and new calls coming in.

Resource usage for three active calls. The strategy regions for three active calls are shown in Figure 23. In the diagram, the dominance of Region I for nearly all displayed data rates is obvious. So the optimal reservation for three calls is 6 units for nearly the complete system runtime. Only with the relation $w\_l12 : w\_l11$ falling below 0.5, 5 resource units would be sufficient, while very close to the end, even 4 or 3 units will be enough to serve the calls.

Resource usage for four active calls. Finally, Figure 24 shows the results when 4 calls are active. Even in this situation, 6 resource units are usually enough for the data rates covered in the diagram. With lower rates, even 5 units would suffice. Compared to the models of group 1, where 2 units have been statistically reserved per call, the success of the resource usage optimization becomes apparent. Model type A used for that group is not able to realize the additional multiplexing gain when the number of calls becomes larger.

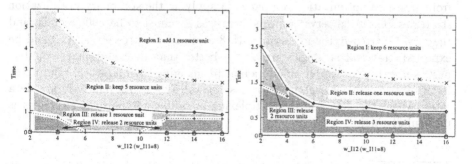

Fig. 23. Strategy regions for 3 active calls when 5 resources are currently reserved. **Fig. 24.** Strategy regions for 4 active calls when 6 resources are currently reserved.

Transient vs. stationary strategies. In order too illustrate the differences between transient and stationary optimization, the latter was applied at model 3. The results of this optimization process are shown in Figure 25 for a situation with one active but currently idle call and no further calls currently waiting. The diagram shows three different strategy regions, again depending on the average data rate of the call. Since the optimization is stationary, these strategies are to be applied during the whole lifetime of the server; this means that every time the system is in this state, the same strategy is applied. If the relation $w\_I12 : w\_I11$ is smaller than 0.5 $(w\_I12 = 4)^7$, then only three resource units are necessary to support this call. If w_I11 is between 4 and 10, then 4 units should be reserved, while beyond $w\_I12 = 10$, the system should keep its current 5 reserved resources. As in the situation described in Figure 21, this increase in necessary resource units is due to possibly arriving new calls which may have to be served at their peak rate very quickly.

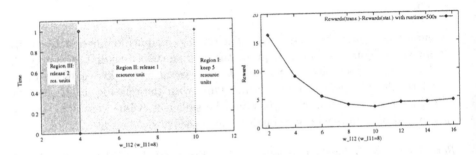

Fig. 25. Stationary optimization results for a system with one active calls and 5 reserved resources.

Fig. 26. Performance comparison graph between transient and stationary optimization.

Finally, Figure 26 displays a performance comparison between strategies obtained with transient and stationary optimization. The line in the diagram indicates the additional reward which can be expected when applying the strategies from transient optimization compared to applying the stationary optimization strategies. The total server runtime here was assumed to be 500 seconds, and the server started in the idle state with 8 available resource units. As could be expected, transient strategies are always better than their stationary counterparts since they are better able to adapt to changing environmental conditions. In this setting, this is especially the case for call types with relatively low data rates, where the gain due to multiplexing according to transient strategies is particularly high.

---

[7] It should be noted that 0.5 is probably not the exact value. This is due to the restricted resolution of w_I12 in this experiment. For real-world systems, one would probably conduct the analysis with a much finer resolution.

5.3 Summary of the results

In this section, some of the results and experiences gained with the multimedia server model are summarized. The summary is divided into four parts of assessments:

(i) multimedia server control problems which were discussed in depth,

(ii) COSTPN as a tool for integrated modeling and control of telecommunication systems,

(iii) a more specialized part in which the results for the server model are shortly summarized, and

(iv) a more global view on strategies for the multimedia server control model.

Assessment of the investigated control problems. Two different important classes of multimedia control problems have been comprehensively studied:

(i) A rich set of *admission control regions* were identified allowing to optimize revenue by trading risk based cost and reward functions. These studies were performed on the basis of a static (fixed rate) reservation scheme.

(ii) The important question of *reservation strategies for bursty traffic* was tackled. Reservation regions were identified allowing for dynamic control in the case of supporting a medium number of traffic streams. The results complement traditional average-rate reservation schemes based on the "law of large numbers". The applicability of average-rate reservation schemes is limited in practice, since the underlying law does not hold in general. COSTPN, in contrast, allows to compute strategy regions for an optimal multiplexing control based on a *dynamic rate reservation* scheme.

General assessment of COSTPNs. The descriptions in this paper provide evidence that COSTPNs are a suitable tool for modeling and controlling real-world telecommunication systems. Among the advantages the following seem to be the most important ones:

- COSTPN is a high-level stochastic modeling tool that supports a system's specification in a structured way. The underlying SPN model is a well-known formalism which will contribute to the acceptance of COSTPN. Large *Markovian decision models* can be easily specified and evaluated.

- Due to the high level of abstraction, changes in system design can be incorporated quickly. This is important, since results of analyses will often lead to system redesign. Large and relatively low-level Markov models would otherwise be much more difficult to understand and to modify.

- On top of the ordinary SPN formalism, an added value is provided by the introduction of *reconfiguring transitions* allowing for a *dynamic optimization*. Resulting strategies are presented in relation to the Petri net specification in a rather high level. Performability modeling and optimization techniques are integrated into a single framework.

Assessment of the server model results. Looking especially at the multimedia server model, the following are the most important results:

– Both the qualitative and quantitative aspects of call admission depend on a huge number of parameters. For some of them, even small parameter modifications lead to vastly different results (call inter-arrival time in Figure 16, for instance).
– The EMRM model size increases quickly with an increasing number of call types. This shows the superiority of COSTPN over EMRM. Large EMRM models also mean long computation times in the model evaluation phase. Since these computations are done offline and not during system runtime, this is a minor problem. Note, we are explicitly interested in the interesting cases were "the law of large number" does not apply, i.e., where the number of simultaneously admitted multimedia streams is relatively small. Due to this fact, the size of the state space is naturally limited so that state-space based methods are most suitable and accurate.
– With respect to overall optimization of server operations, model B seems to be better suited than A, since it better adapts to resource needs of a given number of calls. Obviously, certain decisions are better taken by the system itself rather than by the designer. Usually, a designer cannot take all relevant parameters into account. The task of the designer should be to describe the system with as few restrictions as possible and then let the system find the best configurations, based on the current parameter values.

Strategy assessment. Section 5.2 presented a large number of strategies and explanations for them. Therefore, a short summary may be helpful.

– *Model group 1:* For these models and the given parameter values, the decision whether to accept a call or not is almost always "to accept". The only situation where "not to accept" can be the better decision is when currently no call is active. One can easily imagine, however, that this would be different for different parameter sets. Assigning a higher reward value to unused resources (due to their possible use for other applications) is one such option. If, in addition, increasing scarceness of resources is expressed by higher reward values, this effect can be amplified and used as a way to control one application in a system where other applications run concurrently. To put it differently, the interesting question is how many multimedia streams at least need to be simultaneously supported with a given resource-aggregation reservation schema in order to trade for multiplexing risks. The underlying assumption thereby is, the total amount of reserved resources is proportional to the number of admitted streams.
– *Model group 2:* Here, the question is which kind of call should be accepted if there is more than one waiting. Again, the results strongly depend on the parameter values, and on the number and type of currently active calls. The more calls are already active, the more interesting it becomes to admit the high-reward, but also high-risk call of type 2, since with more calls being

active the multiplexing effect sets in and decreases the risk of a guarantee violation for single calls.

- *Model group 3:* In this model group, adaptive resource reservation strategies were investigated for a given number of active calls. As could be expected, the advantage of multiplexing comes into effect only with a higher number of calls. For only one or two calls, a relatively high number of resource units was necessary. The diagrams show that as soon as the system detects that too much resources are used it releases them. If the usage of additional units can be anticipated for the near future, the system is advised to reserve resources in advance, due to the reservation delay introduced by the new reservation component.

6 Conclusion

We have introduced controlled stochastic Petri nets that can be applied for simultaneous evaluation and optimization of performance or performability models. A new type of transition was defined, providing the means for a specification of *optional reconfigurations* on the level of stochastic Petri nets. Different algorithms are adopted under the unifying high-level paradigm of reconfigurability, as the basis for a computation of *adaptation and optimization strategies*. Both transient and stationary strategies can be computed and used as a rich basis for control decisions. The algorithm for a mapping of COSTPNs on EMRMs, on which the optimization algorithms can be directly applied, was discussed in detail. The applicability of COSTPNs to an optimization problem was demonstrated by means of a comprehensive case study.

Further studies are necessary to fully incorporate COSTPNs into our tool environment, which provides an easily usable interface for generation, execution, and evaluation of a series of optimization experiments [DEME 97].

References

[BAAS 88] Baase S.: *Computer algorithms, introduction to design and analysis*; Addison-Wesley, 2nd Edition, chapter 4 (1988).

[VOBO 96] von Bochmann G. and A. Hafid "Some Principles for Quality of Service Management", *Technical Report, Universite de Montreal*, (1996).

[CHOI 94] Choi H., V. G. Kulkarni, and K. S. Trivedi. : "Markov Regenerative Stochastic Petri Nets". *Performance Evaluation* 20, pp. 335–357, (1994).

[DEME 92] De Meer H. : "Transiente Leistungsbewertung und Optimierung rekonfigurierbarer fehlertoleranter Rechensysteme"; *Arbeitsberichte des IMMD der Universität Erlangen-Nürnberg* 25 (10) (October 1992).

[DEME 94] De Meer H., K. S. Trivedi and M. Dal Cin : "Guarded Repair of Dependable Systems"; *Theoretical Computer Science*, Special Issue on Dependable Parallel Computing 129, pp. 179–210 (July 1994).

[DEME 95] De Meer H. "Modeling and management of responsive systems: the quality of service for multimedia applications in high speed networks as an example", *2. ISSAT Intern. Conf. on Reliability and Quality in Design*, Orlando, Fl, (March 1995).

[DeMe 96] De Meer H. and H. Ševčíková : "XPenelope User Guide for XPenelope Version 3.1"; Technical Report FBI-HH-M-265/96, CS-Department, University of Hamburg (October 1996).

[DeMe 97] De Meer H. and H. Ševčíková: "PENELOPE - dependability evaluation and the optimization of performability"; *9th Intern. Conf. on Computer Performance Evaluation*, St. Malo, France. Lectures Notes on Computer Science (LNCS 1245), pp. 19-31 (June 1997).

[Germ 95] German R., Ch. Kelling, A. Zimmermann and G. Hommel : "TimeNET: A Toolkit for evaluating non-Markovian Petri nets"; *Performance 24*, pp. 69-87 (1995).

[Have 94] Haverkort B.R. and L.J.N. Franken : "The performability manager, dynamically reconfiguring distributed systems"; *IEEE Network 8 (1)*, pp. 24-32 (1994).

[Have 96] Haverkort B.R., I.G. Niemegeers : "Performability Modeling Tools and Techniques"; *Performance Evaluation 25 (1)*, pp. 17-40 (March 96).

[Kram 85] Kramer J. and J. Magee : "Dynamic configuration of distributed systems"; *IEEE Transactions on Software Engineering 11 (4)*, pp 424-436 (April 1985).

[Maus 90] Mauser H. : "Implementierung eines Optimierungsverfahrens für rekonfigurierbare Systeme"; Master thesis, University of Erlangen-Nürnberg (1990).

[Mars 84] Marsan M.A., G. Conte and G. Balbo : "A class of generalized stochastic Petri nets for the performance evaluation of multiprocessor systems"; *ACM Transactions on Computer Systems 2 (2)*, pp. 93-122 (May 1984).

[Mars 95] Marsan M.A., G. Balbo, G. Conte, S. Donatelli and G. Franceschinis : *Modeling with generalized stochastic Petri nets*; John Wiley & Sons, Series in parallel computing (1995).

[Moll 81] Molloy M.K. : "On the integration of delay and throughput measures in distributed processing models"; Ph.D. thesis, University of California (1981).

[Lee 87] Lee Y.H., K.G. Shin : "Optimal Reconfiguration Strategy for a Degradable Multimodule Computing System"; *Journal of the ACM 34 (2)*, pp. 326-348 (1987).

[Perf 92] "Special Issue on Performability"; *Performance Evaluation* (Feb. 1992).

[Stew 94] Stewart W.J. : *Introduction to the Numerical Solution of Markov Chains*; Princeton University Press (1994).

[Triv 91] Trivedi K.S., G. Ciardo and J.K. Muppala : "On the solution of GSPN reward models"; *Performance Evaluation 12*, pp. 237-253 (1991).

Stochastic Colored Petri Net Models for Rainbow Optical Networks

G. Franceschinis °, A. Fumagalli* and A. Silinguelli * *

° Università degli Studi di Torino
Dipartimento di Informatica
E-mail: giuliana@di.unito.it

* Politecnico di Torino
Dipartimento di Elettronica
E-mail: andreaf@hp0tlc.polito.it

Abstract. In this paper, Stochastic Well-formed Nets (SWN) are used to model an optical multi-receiver system and study its performance. A novel approach to resolving packet contention (loss), occurring in optical multi-receiver nodes when the number of packets concurrently reaching the node exceeds the number of available receivers, is presented: it is based on the combined use of fiber delay lines and photonic switches, to temporarily store (delay) optical packets prior to reception. In order to effectively delay the contending packets and improve the multi-receiver throughput three novel control strategies are presented and their performance analyzed using SWN models. Techniques specific to this formalism are applied to cope with the state space explosion problem. A further optimization is achieved by altering the models in such a way that exact performance results can be obtained with reduced computational complexity.

1 Introduction

In this paper we present the performance analysis of an optical system that uses a novel approach to resolving packet contention (loss) arising when the number of simultaneous arriving packets directed to the same node exceeds the number of available receivers. The proposed solution is all-optical in the sense that it does not make use of electronic hardware to achieve packets buffering, thus overcoming the so called electronic bottlenecks in optical networks. The basic idea is to implement an *optical buffer* by using fiber Delay Lines (DL) and photonic space SWitches (SW). The basic approach, termed Quadro (QUeueing of Arrivals for a Delayed Reception Operation), was first introduced in [19].

The system studied in this paper is an extended version of Quadro, called Multi-Receivers Quadro (*MR-Quadro*), in which r receivers work in parallel to

* A. Fumagalli worked under the support of the Italian CNR, through Progetto Finalizzato Trasporti 2 and of the Italian MURST. G. Franceschinis was supported by the Italian MURST under the 40% project. A. Silinguelli was supported by CSELT.

allow the simultaneous reception of up to r packets per slot. The r receivers share the same optical buffer, so that the destination contention penalties can be greatly alleviated without requiring additional receivers, which would increase the overall electronics complexity and speed at the node.

The performance study of MR-Quadro is based on the construction and analysis of Stochastic Well-formed Net (SWN) models. The models are used as a high level description of the stochastic process (Markov chain) capturing the system behavior. The reason for chosing the SWN formalism instead of other possible languages for the specification of the underlying MC, is due to the fact that SWNs allow to mitigate the so called "state space explosion problem" in an automatic way. This is achieved by using techniques that automatically take advantage of the symmetries to group states into 'macro states" (equivalence classes of states), and then perform the analysis at the macro states level[15, 28, 29, 37]. The analysis techniques used in the study presented in this paper automatically exploit this opportunity.

The rest of this section comprises two parts: the first part is an introduction to the networking background and problem, the second part introduces the SWN formalism and the associated analysis techniques. Section 2 presents the proposed approach to contention resolution in optical multi-receiver nodes. Section 3 describes a number of contention resolution strategies that can be used in the considered multi-receiver system. Section 4 presents the SWN models, discusses their complexity and proposes how to reduce the SWN state space size still obtaining exact results. Section 5 shows performance results and Section 6 summarizes the work done and discusses the suitability of the SWN formalism for modeling the proposed optical multi-receiver systems.

Networking background. On the basis of different underlying technologies three network generations can be identified [27]. The first generation uses copper wire and microwave technologies, the second uses fibers as the transmission medium, but maintains traditional electronic switching, which creates a fundamental performance bottleneck. Interest in much higher capacity third generation networks is driven by technological developments and a variety of expected applications [3]. Proponents of the third generation networks advocate "all-optical", in which data, once sent into the network, is not converted back into electronic form until it reaches its destination. All-optical networks are targeted to yield a nearly four orders of magnitude capacity increase by circumventing the switching limitations of electronics [3, 4, 5, 18, 27, 33, 34].

With this large potential optical bandwidth, a particularly challenging problem is the integration of many different applications, spanning from very low bandwidth consumers, e.g., remote monitoring, to bandwidth-intensive applications, such as video (see [1, 2, 35]). To realize this goal, the efficient resource sharing provided by optical packet switching is mandatory. To obtain a third generation, optical packet switched network that does not suffer from electronic bottlenecks, it is of critical importance to find a way for dealing efficiently and *optically* with the fundamental resource contentions.

Different approaches have been investigated so far addressing several of the key problems related to optical networks (see [33, 34] for a complete survey). Among these, a very promising basic technological direction investigated to provide a way to multiplex the huge optical bandwidth is WDM (Wavelength Division Multiplexing). WDM allows the large bandwidth to be multiplexed into multiple wavelength channels, each operating at a speed which is compatible with the available electronics speeds [12].

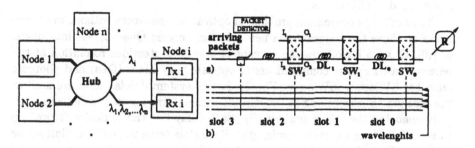

Fig. 1. WDM optical star system with n nodes.

Fig. 2. (a) Architectural and (b) logical representation of QUADRO with 2 DLs.

In this paper we shall focus on one particular WDM topology, the passive optical star, that is a good potential solution for Gigabit/sec communication in the local area [23, 26, 39]; Fig. 1 shows a generic passive optical star network with n nodes. In particular, one testbed realization of such a network topology, termed *Rainbow*, has been realized by IBM [13]. The concepts and solutions described for this topology, suitable for contention resolution in the receiver, can be generalized to other topologies, such as ring and regular mesh networks, as shown in the survey paper [20], in fact the Quadro solution can be generalized to deal with other types of resource contentions (e.g., contentions arising when several packets arriving to a switch have to be routed to the same output channel).

Control of transmissions in a WDM passive optical star must deal with two issues, collisions and (destination) contentions. Collisions occur when two or more packets are being transmitted on the same wavelength: this is a phenomenon known from traditional multi-access channels and is therefore well understood. Contentions occur when several packets are arriving at a (single) receiver on different wavelengths: they are related to multi-channel (wavelength) systems and need to be addressed by new solutions. When a destination contention occurs only one of the arriving packets can be received. Rejected packets must be retransmitted, leading to wasted bandwidth, additional delay, and the need for additional buffers to store packets awaiting retransmission. For example, it has been shown that, due to contentions, at most $\simeq 0.63$ of channel capacity can be utilized in a WDM star network with single receiver nodes such as Rainbow [13].

A seemingly straightforward approach to resolving receiver contentions is obtained by using w receivers per node, where w is the number of wavelength channels [25]. The practicality of this approach is, however, compromised for two basic reasons. First, the cost of a multi-receiver node grows with the number of channels. Second, with random arrivals, the peak to average packet arrival rate is high. While at the receiver the packet processing speed has to equal the average rate, the buffering speed has to match peak arrival rate, which in this case is w times the single channel speed. The buffering process speed thus has to be several time higher than the packet processing speed, exacerbating one of the dominant problems of optical systems, the bottleneck created by the relatively lower speed of electronics.

To handle the contention problem in optical star networks without the drawbacks of above the approaches a new solution, termed Quadro, was introduced in [19]. Quadro introduces the concept of a local contentions resolution at the receiver based on the combined use of optical Delay Lines (DLs) and photonic cross-bar SWitches (SWs) (see Fig. 2). In this system d delay lines are used at the receiver to receive one of the several packets arriving in parallel and optically queueing the others sequentially. Each delay line can optically "store" up to w packets (one for each wavelength), DL-s thus serve to create a "finite-time optical buffer", from which with the help of additional mechanisms a queued packet can later be received in a finite number of time instances. The switches can be set to the cross or the bar state for each wavelength (channel) separately, to route a packet towards the next delay line (i.e., to the next position in the buffer) or to the receiver. Observe that a packet that is not received within d slot times from its arrival is lost [2]. The proposed approach entails new control problems, namely the selection of the time instant to receive a packet queued in the receiver's delay lines [19].

The system presented in this paper is an extension of the Quadro approach, originally developed for a single receiver node, to provide for the resolution of contentions in a multi-receiver node, required to support an incoming traffic that on average is larger than what a single receiver can process. This is the case of a server or a gateway node that generates and receives a relatively large fraction of the LAN traffic. Two approaches to dealing with this problem are considered. The former, requires the duplication of the Quadro device (i.e., of the optical buffer) for each additional receiver in the node. Although straightforward, this approach duplicates the cost of the single receiver solution. The latter approach consists of multiple receivers sharing one single Quadro device. We denote this approach as *MR-Quadro*.

The MR-Quadro solution has several advantages. Using $r < w$ receivers, a MR-Quadro node can receive multiple arriving packets as long as the average number of packets per slot is smaller than r, virtually eliminating the destination contention penalties. The MR-Quadro solution does not require increased

---

[2] The use of "feedback" fibers only partially resolves the packet deadline problem, in fact optical packet can circulate in the Quadro only for a limited number of times to prevent that its signal power level results too weak for correct reception. Therefore, in this paper we considered only "feedforward" Quadro designs, being any feedback design an extension still characterized by the presence of packet deadline.

electronic processing speeds, it is *transparent* to data format (e.g., modulation, transmission speed), and its DLs and SWs are shared by all the node receivers. MR-Quadro leads to a capacity significantly higher than other known schemes [13], as shown by analysis and simulation, providing an almost total utilization of channels. Finally, as shown in this paper, by means of original strategies to control the MR-Quadro, its achievable throughput becomes virtually the same throughput achieved by the more costly solution that makes use of r distinct Quadro devices. In summary, the proposed approach yields a unique, potentially optical, solution providing at the same time high throughput, low cost and easy upgradability to a larger number of receivers per node.

The SWN formalism and related analysis techniques. The aim of the study presented in this paper, is to assess the performance of the MR-Quadro solution, and to compare the performance results when different contention resolution strategies are used.

We perform an exact analysis of the system, based on the steady state distribution of a Markov chain representing its behavior. Since the number of states of the system (i.e., the number of possible distributions of packets in the optical buffer) is large even for very small values of the number of wavelengths and delay lines, it is necessary to generate the Markov chain automatically: to this purpose we have constructed a high level model using the Stochastic Well-formed Net (SWN) formalism[15], and then we have used a software tool, *GreatSPN*[17], to automatically generate and solve the underlying Markov chain and to obtain performance indices of interest defined at the level of the high level model.

The Petri Nets (PN) formalism[36] is well suited for the description (and qualitative analysis) of concurrent systems: it features a distributed state representation (a net comprises a set of *places* that may contain a varying number of *tokens*) and the possibility of defining the activities that can modify the system state (the *transitions*) in such a way that their occurrence depends on (and influences) only a local part of the state (arcs connecting each transition with a subset of places define the conditions under which an activity may occur and the state change caused by its completion). The possibility of representing PN models in a graphical form and of visualizing their dynamics by "playing the token game", makes this formalism appealing and intuitive.

The modeling power of PNs has been enhanced by introducing the so called High Level Petri Nets[29, 30], that extend the original formalism to allow a more compact and parametric model description and a more efficient analysis. The parametricity is introduced by allowing tokens of different "colors" and by adding inscriptions (expressions) to the arcs that define the color of the tokens that are withdrawn/put into places when a given activity completes (i.e., when a transition *fires*).

In order to allow the performance evaluation of timed systems by Petri net models, in the last decade several alternatives for extending the formalism in this sense were proposed (see [9]). One possible approach is to associate exponentially distributed random firing delays with transitions[24, 32, 38]. The resulting class of models, termed Stochastic Petri Nets (SPN), has been shown to be isomorphic to continuous time Markov chains. A generalization of the SPN can be obtained by allowing immediate transitions whose delay is deterministically equal to zero, leading to the GSPN formalism[7, 8]; immediate transitions thus represent instantaneous events, and are used to model "logic actions" as opposite to time consuming actions, they have an associated weight that is used to compute the probability of choosing one among several conflicting immediate transitions hence allowing to model systems with probabilistic evolution. Immediate transitions have priority over timed transitions, moreover it is possible to define a priority relation among immediate transitions.

SWNs allow the description of the timing aspects of a system in the same way as GSPNs; being an high level formalism SWNs allow to describe complex systems with rather compact models. The drawback is that the number of possible states of the CTMC associated with a SWN model may grow extraordinarily, even when the model size is small. In some cases, it has been shown that intrinsic symmetries of the modeled system allow to mitigate this "state space explosion problem"[15, 28, 29, 37]; the peculiarity of the SWN formalism is the possibility of exploiting the symmetries in an automatic way, completely transparent to the modeler. Some efficient analysis algorithms for SWNs that exploit the model symmetries have been implemented in version 1.7 of *GreatSPN* [17], a package for the design and analysis of GSPN and SWN models, that we have used to produce the results presented in this paper. In particular, the state space reduction techniques that have been used in this work are the Symbolic Reachability Graph (SRG) [15] and the decolorization technique [16]. Both methods automatically achieve exact lumping in the underlying Markov chain, with the former based on the concept of *symbolic marking*, an equivalence class grouping several *ordinary markings*, and the latter based on the structural detection and elimination of redundant (color) information in the model. In addition, the lumped Markov chain can be obtained directly without first generating the complete chain. The lumped chain has all the required information to compute the same set of performance results that can be obtained from the complete chain.

2 Contention Resolution in Optical Multi-Receiver Nodes

Fig. 1 shows a generic passive optical star n-node network representing the class of systems under consideration. Each node is connected to the hub by two fibers, carrying optical signals from the node (input fiber of the hub), and to the node (output fiber of the hub). The passive hub dispenses arriving signals from the input to the output fibers, uniformly distributing their power. For simplicity we assume that each node has one transmitter and r receivers for transmission of data packets. In Fig. 1 node i is logically represented by two separate blocks: the transmitting part Tx_i and the receiving part Rx_i. When node i wants to transmit a packet to node j, transmitter Tx_i sends data using its locked laser on a fixed wavelength, say λ_i, while receiver Rx_j tunes one of its receivers to the same wavelength. In this paper we study a star with $w = n$ WDM data channels, where channel i is dedicated to transmissions of node i. Collisions associated with multiple transmissions occurring on the same wavelength thus do not occur in this system. It is important to observe, however, that the utility of the Quadro approach remains valid also when $w < n$ [40].

We consider systems with a common clock in which time is divided into fixed slots, whose duration equals the packet transmission time plus receiver's tuning time. With $r < w$ tunable receivers each node has to know which wavelengths its receivers must be tuned on in each time slot for receiving packets which can arrive over $n - 1$ channels (excluding its own). Without restriction, in this paper we assume that this information is provided to the node by a *Packet Detector* that monitors the traffic in the fiber before the receivers. A testbed realization of such a detector is reported in [21]. In this way the destination node learns in advance about all packets destined to arrive in the next slot. Using a deterministic system-wide process the receiver will accordingly tune to one of the wavelengths for reception and, executing the same process, the transmitter can learn whether its packet has been received or must be retransmitted. In summary, this approach can guarantee the reception of up to r packets arriving at the node and loses the exceeding contending packets. As shown in [13] for $r = 1$, this loss significantly reduces performance.

The Quadro mechanism was originally introduced to virtually eliminate this contention penalty when $r = 1$ [19]. The need for $r > 1$ originates in those nodes that are more often targeted by packets, such as servers or gateways. In such nodes the average rate of received packets is $a > 1$ packets per slot, thus the number of receivers must be at least equal to $\lceil a \rceil$. As described next, the MR-Quadro generalizes the Quadro approach to any $r < n - 1$.

2.1 Quadro

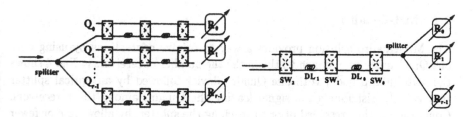

Fig. 3. Structure obtained using one Quadro with 2 DLs per receiver.

Fig. 4. Architecture of a MR-Quadro with 2 DLs.

The original Quadro implementation for a *single* receiver node is shown in Fig. 2(a). The Quadro receiver consists of d DLs, $d + 1$ 2×2 SWs (Fig. 2 shows the case $d = 2$). A SW has two inputs I_1, I_2, and two outputs O_1, O_2. Independently for each wavelength, it can be set up in two possible states: the *bar* state where the optical signal from input I_i reaches output O_i, $i = 1, 2$, and the *cross* state where an optical signal from input I_i reaches output $O_{1+(i) \bmod 2}$, $i = 1, 2$. The default state for the SW is bar for all the wavelengths. The state of the SW can be controlled and changed to cross for one or more wavelengths at any given time. Each DL delays the propagating optical signal (a packet) for one slot time. If switch SW_k is in the bar state for wavelength λ_i at a given time, then the packet possibly stored into wavelength λ_i of DL_k will be stored into DL_{k-1} (still on λ_i) at the next time slot. If instead SW_k is in the cross state for wavelength λ_i and all SW_h, with $h < k$ are in the bar state for the same wavelength, the packet "stored" in DL_k will be received in the next time slot. The mechanism just described allows to realize a "finite-time" optical buffer, with a *reception window* for each packet, whose size is function of the number of DL-s in the Quadro. The *reception process* consists of choosing a specific slot inside the window (in Fig. 2(b) slots 0,1 and 2) for the reception of each packet. The node controls the switches in such a way that only one wavelength of only one switch will be deflected to output O_1 at a time. This guarantees that only one packet will be routed to the tunable receiver at each time slot, thus avoiding contentions at the output of the Quadro. Intuitively, the DLs thus *smooth* the burstiness of the packet arrival process by redistributing packet reception instances over the reception window and consequently overcoming loss of packets due to contention.

Fig. 3 shows how r Quadro receivers can be used in a multi-receiver node. The power of arriving optical signals is equally distributed among the r Quadro receivers, Q_i, $i = 0, 1, \cdots, r-1$. Q_i's will resolve contentions and receive packets in such a way that each packet will be received only once (i.e., by only one of the Q_i's). The Quadro solution shown in Fig. 3 is a straightforward generalization of the single receiver Quadro solution. Thus, the same *optimal* strategy used to control the single receiver Quadro [19] can be still adopted in this multi-receiver node.

2.2 MR-Quadro

The MR-Quadro solution provides a way to resolve contentions by using only one Quadro device, independently on the the number of receivers r. Fig. 4 shows the MR-Quadro node, with one Quadro device followed by an optical splitter that equally distributes the signal leaving the Quadro among the r receivers. Contention is thus resolved prior to reaching the splitter, by allowing r or fewer packets to leave the Quadro device in the same slot. This "compact" solution introduces a new constraint on the control strategy that selects the packets for reception: the set of packets that leave the MR-Quadro at each slot must not contain two packets on the same wavelength because in this case a collision would arise in the output fiber leading to the receivers. As a consequence, the optimal strategy that maximizes the throughput of the MR-Quadro is not trivial to find, as the following example points out.

Consider the initial packet distribution at time $t = 0$ shown in Fig. 5(a).

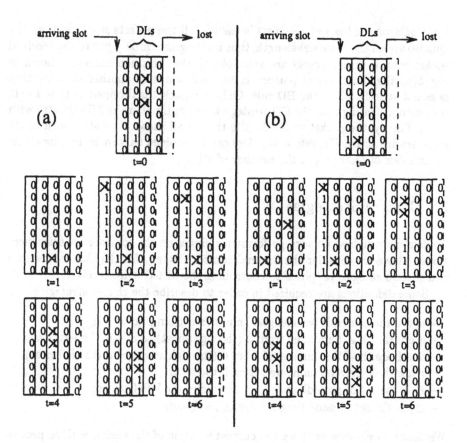

Fig. 5. (a) ED strategy and (b) alternative strategy used on a MR-Quadro with $w = 8$, $d = 3$, $r = 2$.

Each column represents a slot, and each row represents one wavelength channel. Value "1" ("0") indicates the presence (absence) of a packet destined to the node. The figure represents the distribution of packets in a $w = 8$, $d = 3$ MR-Quadro, with $r = 2$ receivers. At each time slot, packets selected for reception are marked with a ×, while the other packets will remain in the next time slot ($t = 1$) shifted by one position to the right. Arriving packets are shown in the leftmost column and lost packets are shown in the rightmost slot delimited by a dotted line. Therefore, at time $t = 0$ two packets are stored in DL_1, one packet is stored in DL_2, one packet is arriving, and no packets are lost. Let us first consider the reception sequence shown in Fig. 5(a). In this sequence the set of received packets is determined using the Earliest Deadline (ED) strategy, proven to be optimal for the single receiver Quadro [6]. The ED strategy selects for reception the rightmost packet(s) in the optical buffer, i.e., the packet with the least remaining time in the MR-Quadro. For the arrival pattern shown in this example, the ED strategy drops 2 packets at time $t = 6$. Note that at time

$t = 1$ only one of the two receivers is used as all the packets stored in the MR-Quadro are on the same wavelength, thus limiting the throughput to one received packet even if three packets are available. In the reception sequence shown in Fig. 5(b) the same arrival pattern is handled adopting another strategy that does not strictly follow the ED rule. Only one packet is dropped at time $t = 6$, thus demonstrating that the ED strategy is not optimal in the MR-Quadro with $r > 1$. Finally, the packet is lost despite that at all shown time slots both of the receivers are active. Therefore, this loss can be avoided only by increasing either the number of receivers, or the number of DLs.

3   Control Strategies

To demonstrate how Quadro performance hinges on the control strategies mentioned above we consider the optimal control strategy for the Quadro receiver and three strategies proposed to controlling the MR-Quadro receiver.

Some definitions are required in order to describe the chosen strategies.

- w are the available wavelengths, numbered from 0 to $w - 1$;
- d are the DLs, numbered from 0 (the rightmost) to $d - 1$ (the leftmost);
- $d + 1$ are the SWs, numbered from 0 (the rightmost) to d (the leftmost);
- r are the receivers, numbered from 0 to $r - 1$;
- Δ is the set of non-idling control strategies for MR-Quadro;
- Δ_1 is the set of non-idling strategies for Quadro.

We define as *position* or slot p the current location of the packet with respect to the receiver. Fig. 2(b) shows positions $p = 0, 1, 2, 3$ for a two DL Quadro receiver. Function $packet\_in^{(t)}(p, i)$ returns 1 if at time t there is a packet destined to the node in position p on wavelength i, 0 otherwise. A packet is *available* (for reception) when it is in any position $0 \leq p \leq d$. An available packet is *received* when it is routed to the receiver(s) by one of the switches. Function $fp^{(t)}(i)$, $0 \leq i < w$ returns the position of the rightmost packet on wavelength i. $FP^{(t)}$ is the set of rightmost available packets, i.e., the set of packets in the rightmost non-empty position among $0 \leq p \leq d$. Set $FP^{(t)}$ is empty when there are not available packets. The receiver's *look ahead*, l_r, is the range of additional positions whose packets' distribution is known by the receiver prior to their arrival to the (MR-)Quadro. For example, $l_r = 0$ indicates that the receiver knows the distribution of available packets only (positions $0 \leq p \leq d$), and $l_r = 1$ indicates that the receiver also knows the packet distribution in position[3] $d + 1$. Therefore, the receiver's *observation window* is the set of positions from 0 to $d + l_r$, whose size is $f = d + 1 + l_r$ slots. Let $N_{l_r}^{(t)}(i) = \sum_{k=0}^{f-1} packet\_in^{(t)}(k, i)$, be the total number of packets on wavelength i, in the observation window with look ahead l_r. Define

---

[3] Note that the look ahead is determined by the length of the fiber connecting the Packet Detector splitter to input I_2 of switch SW_2 shown in Fig. 2(a).

function $\Phi_i(p) : I\!N \to I\!N$ as

$$\Phi_i(p) = p \cdot r + i \qquad \forall i = 0, 1, \ldots, r-1 \quad \forall p = 0, 1, \ldots, d$$

and function $\nu^{(t)}(k) : I\!N \to I\!N$ as the total number of packets in positions $0 \le p \le k$ at time t

$$\nu^{(t)}(k) = \sum_{p=0}^{k} \sum_{i=0}^{w-1} packet\_in^{(t)}(p, i)$$

The *optimal reception range* for receiver j is the set of positions $0 < p < range^{(t)}(j)$, with function $range^{(t)}(j)$ defined as

$$range^{(t)}(i) = \min(\{k : \nu^{(t)}(k) > \Phi_i(k)\}, d).$$

The reason for calling this range *optimal* is that in case a receiver chooses a packet out of this range at a given time t, this choice shall surely lead to one packet loss within d time slots from t independently of the next arrivals and reception choices.

3.1   Quadro Strategy

The optimal strategy for the single receiver Quadro selects for reception one of the rightmost available packets, i.e., one packet in $FP^{(t)}$. The single receiver strategy is generalized to the r receiver node design shown in Fig. 3 as follows.

Definition 1. Quadro strategy $\delta_1 \in \Delta_1$, at time t, first unmarks all available packets, then, iteratively, for each receiver evaluates $FP^{(t)}$ considering unmarked packets only, selects for reception one packet in $FP^{(t)}$ and marks the selected packet.

Strategy δ_1 minimizes the probability of packet loss by receiving the packet(s) with the least time left to spend in the Quadro [6]. At each time t, δ_1 receives $min(r, \nu^{(t)}(d))$ packets.

3.2   MR-Quadro Strategies

Due to its architectural constraints, at each time slot t the MR-Quadro design shown in Fig. 4 cannot receive more than one packet on the same wavelength. We, therefore, define $\rho^{(t)}$, the maximum number of *receivable* packets at time t, as the number of wavelengths carrying at least one available packet. The class of strictly *non-idling strategies* Δ is the set of strategies that at each time t receive $min(r, \rho^{(t)})$ packets.

ED strategy The Earliest Deadline (ED) strategy is derived from strategy δ_1.

Definition 2. ED strategy, $\delta \in \Delta$, at time t first unmarks all wavelengths, then, iteratively, for each receiver selects the rightmost packet among those on unmarked wavelengths, and marks the selected packet's wavelength.

ED objective is to select packets with the least time still to spend in the MR-Quadro, given the constraint that there are not two selected packets on the same wavelength. An example of how ED selects packets for reception is shown in Fig. 5(a) for a specific arrival sequence.

MC and MC+1 strategies These strategies base their selection on the count of the packets in the observation window, i.e., positions 0 - $f-1$. The two considered strategies are the Most Congested Earliest Deadline with no look ahead ($l_r = 0$) (MC) and the Most Congested Earliest Deadline with one slot of look ahead ($l_r = 1$) (MC+1). Other strategies are possible and can be found in [6]. The objective of both MC and MC+1 is to give priority to packets on the most congested wavelengths, i.e., wavelengths that have at least one packet in the optimal reception range and have the largest number of packets in the observation window. By so doing, the spreading of packets (with respect to wavelengths) in the optical buffer is maximized, hence achieving a high utilization of the receivers. In addition, due to its increased size of the observation window, MC+1 aims at a more effective spreading than the one achieved by MC.

Definition 3. MC strategy, $\sigma \in \Delta$, at time t first unmarks all wavelengths, then, iteratively, for each receiver $j = 0, 1, \ldots, r$, selects the rightmost packet on wavelength i s.t.

$$fp^{(t)}(i) \leq range^{(t)}(j) \qquad \wedge$$

$$N_0^{(t)}(i) \geq N_0^{(t)}(k) \;\; \forall \text{ unmarked wavelength } k: fp^{(t)}(k) \leq range^{(t)}(j)$$

and marks the wavelength.

An example of how MC selects packets for reception is shown in Fig. 5(b) for a specific arrival sequence.

Definition 4. MC+1 strategy, $\sigma_{+1} \in \Delta$, at time t first unmarks all wavelengths, then, iteratively, for each receiver $j = 0, 1, \ldots, r$, selects the rightmost packet on wavelength i s.t.

$$fp^{(t)}(i) \leq range^{(t)}(j) \qquad \wedge$$

$$N_1^{(t)}(i) \geq N_1^{(t)}(k) \;\; \forall \text{ unmarked wavelength } k: fp^{(t)}(k) \leq range^{(t)}(j)$$

and marks the wavelength.

4 SWN Models of Quadro and MR-Quadro Control Strategies

This section is organized in three parts.

In Section 4.1 we shall present some Stochastic Well-formed Net (SWN) models of the different control strategies described in Section 3: the SWN models are used as a high level description of a stochastic process (Continuous Time Markov Chain – CTMC) describing the system behavior.

In Section 4.2 we shall present the results achieved by applying to the models the SWN analysis techniques that automatically exploit the intrinsic symmetries of the system under study. The techniques used are the Symbolic Reachability Graph [15] and the decolorization technique [16]. Both methods automatically induce a lumping in the underlying Markov chain. In the same section we also show that by taking advantage of the modeler knowledge about the system behavior it is possible to cut the state space in a even more drastic way.

In Section 4.3 we shall discuss the suitability of the SWN formalism to this application.

4.1 Construction of the SWN Models

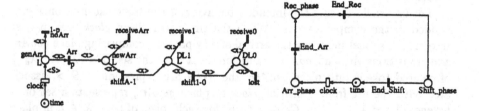

Fig. 6. The basic skeleton model of the receiver system.

Fig. 7. Sequentialization of the events occurring at each time slot.

In this section we describe the SWN models used to compute the desired performance figures. First, a basic submodel is shown that represents the hardware structure of the receiver and the events that can cause its state change. All other models in this paper are constructed by superposing additional structures on this basic submodel in order to (1) force a specific ordering in the occurrence of the events that allow the system to evolve from the state (marking) at time t to the new state at time $t + 1$ (2) model a specific reception control strategy. The basic submodel is depicted in Fig. 6: it comprises two places (DL_0 and DL_1) whose marking represents the state of the two delay lines, place *newArr* whose tokens represent the packets arrived in the current time slot, timed transition *clock* whose firing represents the end of the time slot, and several immediate transitions[4] modeling the events that cause the state change in each time slot: packets arrival (transitions *Arr* and *noArr*), packets reception (transitions *receiveArr, receive1* and *receive0*) and packets shift, with possible losses (transitions *lost, shift1-0* and *shiftA-1*). The tokens flowing in this model are either uncolored (as in plain PNs) or associated with a color taken from *color class* $L = \{\lambda_1, \ldots, \lambda_w\}$, a set containing one distinct color associated with each wavelength. A λ_i colored token in this context represents a packet on wavelength λ_i. Uncolored tokens instead are used to sequentialize different phases in the system evolution as dictated by the subnet in Fig. 7 that will be explained later. In all our models, places containing colored tokens have a color label L next to the place name, while places containing uncolored tokens do not have this color label.

Observe that this model is parametric in the number of wavelengths (that can be changed by just redefining the color class L), while a change in the number of delay lines requires a change in the net structure (the model in Fig. 6 refers to the system with two delay lines). The places representing the delay lines are not "folded" into each other in order to simplify the model description, however it is easy to modify the presented model to make it parametric also in d. It is important to observe that the parameterization on the wavelengths allows to exploit the intrinsic symmetries of the system when generating the state space (under the assumption of uniform arrivals over the set of wavelengths), while the parameterization on the delay lines has the only advantage of reducing the model size.

Let us briefly discuss the intended behavior of the basic skeleton once embedded in the complete model. When the timed transition *clock* fires, it first triggers the subnet modeling the arrival of new packets, comprising place *genArr* and transitions *Arr* and *noArr*: transition *clock* puts the set $L = \{\lambda_1, \ldots, \lambda_w\}$ of colored tokens into place *genArr* (the SWN arc expression $< S >$, associated with the arc from transition *clock* to place *genArr*, represents a set of w tokens colored $\lambda_1, \ldots, \lambda_w$). Consequently, for each token of color $\lambda_i \in L$ in place *genArr* a probabilistic choice is made to decide whether in the current time slot on wavelength λ_i a packet has arrived (firing of *Arr*) or not (firing of *noArr*). A weight is assigned to transitions *Arr* and *noArr* according to the probability (p) of a new packet arrival. Note that under symmetric workload assumption, this probability is the same for all wavelengths. The expression $< x >$ associated with the arc going from place *genArr* to transition *Arr*, means that when this transition fires, it withdraws one token of any color (x is a variable that can be bound to any color in L) from *genArr*, and puts it into place *newArr* (the same expression $< x >$ is associated with the arc from *Arr* to *newArr*). When transition *Arr* fires with x bound to a given color λ_i, we say that *transition instance*

---

[4] The presence of two types of transitions, timed and immediate with priority over timed, leads to the presence of two types of states (markings) in SWNs: tangible and vanishing. Tangible markings are states in which the system spends time (these are the states of the underlying CTMC), while vanishing markings are states that are visited by the system in zero time.

$Arr(\lambda_i)$ fires.

Next we describe the reception phase modeled by the firing of transitions $receive_i$. At most r packets (tokens) can be received (withdrawn) from the set of *available* packets (places DL_0, DL_1 and *newArr*) as determined by the chosen control strategy.

After the reception phase the shift phase follows (firing of transitions *lost*, *shift1-0* and *shiftA-1*) which causes the loss of packets left in DL_0, the shift of packets from DL_1 to DL_0 and the "storage" of the newly arrived packets into DL_1.

The first complete model we are going to derive is that shown in Fig. 8. It represents the behavior of the Quadro strategy for a r receiver node (see Fig. 3). We recall that this solution is optimal, and its packet loss probability represents a lower bound for the packet loss probability of any MR-Quadro strategy $\delta \in \Delta$.

The initial marking (state) of this model (and any other model considered in this paper) represents all the DLs empty, hence only one place, *time*, is marked with one token.

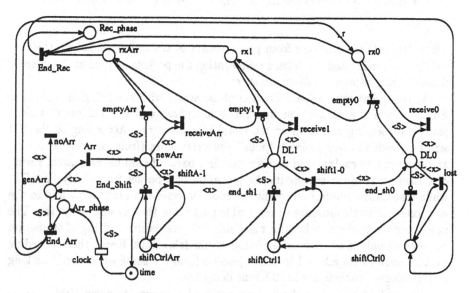

Fig. 8. SWN model of strategy δ_1 for a system with two delay lines.

Let us discuss how the "basic skeleton" and the "phases sequentializer" submodels are combined into this model together with an additional subnet that represents the ED reception strategy.

The arrival phase starts immediately after the firing of transition *clock* representing the end of the time slot (indeed, *Arr_phase* is an output place of transition *clock*). When place *genArr* becomes empty (condition checked by means of the

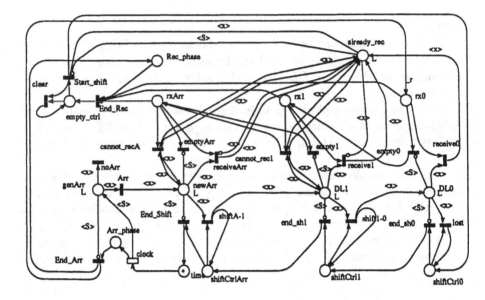

Fig. 9. SWN model of the ED strategy for a system with two delay lines.

(circle headed) *inhibitor arc* from place *genArr* to transition *End_Arr*), the "arrival phase" ends[5] and the tokens representing the packets arrived at the current time slot are "stored" in place *newArr*.

At this point the packet reception phase starts with the firing of transition *End_Arr*, which puts a token into place *Rec_phase* and r "uncolored" tokens (each representing one of the r receivers) into place *rx0*. According to the ED strategy each receiver (token) "polls" the delay lines (from place *rx0* to *rx1*, and from *rx1* to *rxArr*) and receives the first packet found in the search. Firing of transition *receive_i* represents the reception of a packet from delay line DL_i while transitions *empty_i* represent the receiver move from one delay line to the next one due to the absence of receivable packets in the current delay line. The newly arrived packets (tokens in place *newArr*) are received (firing of transition *receiveArr*) only if no packets are found in the delay lines. Recall that according to ED receivers are allowed to select packets from the same wavelength, as long as the selected packets are in different delay lines.

When all the receivers have been assigned a packet or when there are no more available packets (condition corresponding to places *rx0*, *rx1*, *rxArr* empty

---

[5] The expression $< S >$ associated with the inhibitor arc exiting from place *genArr*, represents the whole set of w tokens colored $\lambda_1, \ldots, \lambda_w$, signifying that transition *End_Arr* may fire only if no colored token from set $\lambda_1, \ldots, \lambda_w$ is left in place *genArr*, i.e., it may fire only if place *genArr* is empty.

and place *Rec_phase* with one token), transition *End_rec* fires thus starting the "shift" phase: in this model the *Shift_phase* place is actually split into three places (*shiftCtrl0*, *shiftCtrl1* and *shiftCtrlArr*), i.e., the shift phase is subdivided into three phases: 1) packet losses, 2) shift from DL_0 to DL_1 and 3) shift from *newArr* to DL_1. First, packets still present in the first delay line exit the system (firing of transition *lost*), modeling packet losses, then tokens in place DL_1 are shifted to DL_0 by firing transition *shift(i)-(i-1)*. Lastly, tokens representing packets just arrived and not received in the current time slot are moved into place *DL1* (transition *shiftA-(d-1)*). Immediate transitions *end_sh_i* moving tokens from place *shiftCtrl_i* to place *shiftCtrl_{i+1}* are necessary to guarantee the correct sequence of the previously described actions.

At the end of the shift phase, transition *End_Shift*, fires thus causing the next time slot to start (a tangible state is reached). Observe that the only non empty places after the firing of *End_Shift* are *time* (that contains one token), and DL_i, $i = 0, \ldots, d-1$ that contain one colored token for each stored packet. Thus the total number of (tangible) states of this model is $2^{(d \cdot w)}$, corresponding to the number of possible ways to distribute packets in the d delay lines.

The second SWN model we developed, represents the ED strategy when used to control the MR-Quadro system and is depicted in Fig. 9. Although similar to the previous SWN model, a major difference characterizes this second model: an additional subnet avoids the possibility of receiving more than one packet on the same wavelength at the same time slot. This additional subnet comprises place *already_rec*, that has the function to keep track of the wavelengths that have already been selected for reception, a number of inhibitor arcs going from place *already_rec* to transitions *receive_i* that prevent the reception of a packet from an already selected wavelength (this is achieved by associating the same expression $< x >$ with both the inhibitor and the input arc of transition *receive_i*), and a number of immediate transitions (*cannot_rec_i*) required to force any receiver to skip a delay line when it contains packets only on the wavelengths already selected by other receivers (this is achieved by assigning a *priority* to transition[6] *cannot_rec_i* lower than the priorities assigned to transitions *empty_i* and *receive_i*).

At the end of the reception phase, before the shift phase starts, it is necessary to "clear" place *already_rec*: this is the objective of place *empty_ctrl* and transition *clear*. This last operation would be easier if marking dependent arc weights were allowed in the formalism[22]. The number of tangible markings of this model is $2^{(d \cdot w)}$.

We now discuss the SWN model for the "Most Congested" strategy. Recall that MC selects for reception packets in "range" and on the most congested

---

[6] Both the GSPN and the SWN formalisms allow to assign a priority (positive natural number) to immediate transitions, and the firing rule prevents the firing of a transition if there is an enabled transition with higher priority. Timed transition are assigned a 0 priority. The notation $\pi(t)$ is used to indicate the priority of transition t.

wavelength, i.e., the wavelength with the largest number of available packets. In addition, MC is subjected to the constraint that any receiver pair cannot select the same wavelength for reception in the same time slot. The SWN model developed for this strategy is not shown as, due to its complexity, it would result extremely difficult to read. In alternative, a brief textual description is given that indicates how the MC model can be derived using the ED model in Fig. 9. Each transition $receive_i$ must be replaced by a set of d transitions $receive_{i,h}$, $h = 1, \ldots, d$. The firing of transition instance $receive_{i,h}(l)$ represents the reception of a packet in delay line DL_i on wavelength l, when there are h available packets on the same wavelength l. Priorities are assigned to these transitions so that $\forall h, k,\ h > k\quad \pi(receive_{i,h}) > \pi(receive_{i,k})$ (highest priority is given to the most congested wavelength) and $\forall i, j,\ i < j\quad \pi(receive_{i,h}) > \pi(receive_{j,h})$ (on a given wavelength, highest priority is given to rightmost packet). The reception phase consists of r steps, with the j-th step associated with receiver $r - j + 1$. The enabling conditions of transition instance $receive_{i,h}(l)$ at step j are: (a) a token with color l is contained in place DL_i, (b) the total number of tokens with color l in places $DL_f, f = 0, \ldots, d - 1$ and place $newArr$ is h, (c) delay line DL_i is in the reception range of receiver $r - j + 1$ and (d) there is no token with color l in place $already\_rec$. If none of the transitions $receive_{i,h}$ is enabled at one step, the model proceeds onto the next step. The SWN characterization of such a complex set of enabling conditions is quite awkward, as it requires a number of auxiliary places whose marking is associated with conditions (b) and (c). In addition, the marking update of such places requires a even larger number of immediate transitions.

The complexity of the MC model could be significantly reduced if the SWN formalism had the possibility to embed complex enabling and state change functions into the transitions which model the strategy behavior, an approach already used in the SAN formalism [31]. In addition, a more powerful state change mechanism would reduce the number of vanishing states generated during the state space construction process.

To obtain the performance figures for the MC strategy we adopted an alternative solution that allowed to overcome the complexity of the SWN model: we implemented an *ad hoc* program in which the complex sequence of vanishing states necessary in the SWN to determine the received packets in each time slot is replaced by a single C function. The program can automatically generate the aggregate state space by applying the SRG construction algorithm. This allows to resolve both the model complexity problem and the prohibitive computational cost deriving from the generation of a huge number of vanishing markings, that although not influencing the size of the Markov chain, can however prevent its construction.

The number of (tangible) states of the MC model *without* aggregation is $2^{(d \cdot w)}$.

The model devised for the MC+1 strategy is an extension of the MC model in which packet arrivals must be taken into account one time slot ahead of time. In other words, new arrivals must be generated one slot ahead of time, taking into account that these packets cannot be received immediately upon generation, but only after one slot. As a consequence, the number of tangible states necessary to model this strategy grows to $2^{(d+1)\cdot w}$.

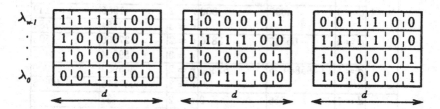

Fig. 10. Example: three equivalent states obtained through permutations of wavelengths.

Under the assumption of independent and identically distributed packet arrivals on every wavelength (i.e., uniform load), all the models presented so far are characterized by a symmetry in the wavelength dimension. This symmetry allows the reduction of the number of states of the underlying stochastic process by "lumping" markings representing "equivalent" states: two markings are equivalent if they become equal after a number of wavelength permutations. Fig. 10 shows an example of three equivalent markings. State lumping is obtained automatically by the SRG algorithm. The first two columns of Table 1 show the number of non aggregate and aggregate (i.e., symbolic) markings for the models of the MC and ED strategies, as a function of relevant system parameters. For these models the number of aggregate states is

$$\binom{w + 2^d - 1}{2^d - 1}.$$

The degree of reduction that can be achieved is bounded from above by $w!$.

4.2 Reduction of the SWN Models

The already substantial reduction of the state space automatically obtained by applying the SRG construction algorithm for SWNs shown in the previous section can be further pushed by means of two additional techniques.

The former is a reduction technique based on the elimination of redundant colors, known as decolorization [16]. This technique is applicable to the Quadro model only, as the other models, relative to the MR-Quadro, do not have the required symmetries to allow its use.

The latter technique is based on a closer observation of the system behavior that reveals an interesting property. Given the number of available receivers in the system, the number of packets that are received in a given number of time slots is upper bounded by the product r times the number of slots. As a

	n_s	$n_s^{(a)}$		η_s	$\eta_s^{(a)}$
$d=1$	16	5	$r=2$	11	3
			$r=3$	15	4
$d=2$	256	35	$r=2$	142	18
			$r=3$	236	29
w=4 $d=3$	4096	330	$r=2$	1955	149
			$r=3$	3754	284
$d=4$	65536	3876	$r=2$	27866	1595
			$r=3$	59924	3425
$d=5$	1.04858e+06	52360	$r=2$	405902	19900
			$r=3$	957815	46978
$d=1$	256	9	$r=2$	37	3
			$r=3$	93	4
$d=2$	65536	165	$r=2$	1943	18
			$r=3$	11771	40
w=8 $d=3$	1.67772e+07	6435	$r=2$	118357	194
			$r=3$	1.68629e+06	914
$d=4$	4.29497e+09	490314	$r=2$	7.8337e+06	3486
			$r=3$	2.586e+08	40067
$d=5$	1.09951e+12	6.15237e+07	$r=2$	5.46899e+08	94186
			$r=3$	4.14225e+10	2.91684e+06
$d=1$	65536	17	$r=2$	137	3
			$r=3$	697	4
$d=2$	4.29497e+09	969	$r=2$	30109	18
			$r=3$	817413	40
w=16 $d=3$	2.81475e+14	245157	$r=2$	8.00011e+06	194
			$r=3$	1.17096e+09	936
$d=4$	1.84467e+19	3.0054e+08	$r=2$	2.35491e+09	3486
			$r=3$	1.86513e+12	47872
$d=5$	1.20893e+24	1.50323e+12	$r=2$	7.39091e+11	96162
			$r=3$	3.17282e+15	4.79299e+06

Table 1. States space size as a function of some system parameters, without aggregation (n_s), with aggregation ($n_s^{(a)}$), with anticipated losses and no aggregation (η_s) and with anticipated losses and aggregation ($\eta_s^{(a)}$).

consequence, any packet available in (MR-)Quadro in excess of $r \times (d+1)$ packets will be surely lost. By removing exceeding packets upon arrival to the (MR-)Quadro, system performance remains therefore unchanged, while the number of tangible states of the system is reduced. This reduction becomes relevant when the number of receivers is considerably smaller than the number of wavelengths - the case of interest in this study.

Decolorization of the model for the Quadro system. The SWN model in Fig. 8 has some special property that can be used to transform it into an uncolored

(GSPN) model that still allows to compute the same performance results of interest, with a reduced computational cost. The model properties that are relevant to this purpose are: (1) the color of the tokens is never used to perform a synchronization, (2) the color of the tokens does not influence the probability of transition between states. Let us further elaborate on these two properties.

The firing of a transition that models a synchronization among several tokens is influenced by the color of the involved tokens when the same "variable" appears in the expression of more than one input/inhibitor arc of that transition: indeed in this case, in order for the transition to fire we require two (or more) tokens with matching color to be present/absent in two (or more) places, so that we cannot obtain the same behavior if the tokens are "decolorized" (i.e., neutral). This condition can be easily checked by analyzing the net structure (more than one input/inhibitor arc for some transition and same variable appearing in more than one expression associated with these arcs). There is one more special case of synchronization based on the token colors that corresponds to the use of expression $< S >$: if an input arc from place p to transition t has an associated expression $< S >$, then the transition is enabled only if there is at least one token of each color in place p. Clearly, in general this condition cannot be checked if the tokens are "decolorized", however if the "synchronizing" place cannot contain multiple tokens of the same color (i.e., if the place is *color safe*), the same enabling test can be obtained by simply checking the global number of tokens in that place (independently on color). Observe that the use of expression $< S >$ associated with an inhibitor arc is not a problem, because it is equivalent to checking the presence of zero tokens in the corresponding place.

The probabilistic choices performed in this model are: (1) the choice corresponding to the arrival of one packet on each wavelength, (2) the choice of one among several packets when a strategy defines a set of packets (rather than a unique packet) that can be chosen by a given receiver. Since the probability of a new arrival is identical on each wavelength and independent on the arrival of packets on the other wavelengths, the color (wavelength identity) does not influence this probability. Moreover, since the policy we have modeled lets the receiver choose only when the different possibilities do not influence the future performance of the system (in terms of packet losses), and since the receiver choice is in any case completely random and with uniform probability over the set of receivable packets, the color does not determine the probability of the conflict resolution. The automatic verification of this property is not easy in general and should be performed on a case by case basis, however our experience is that it is often the case that a restricted set of special cases easy to deal with (automatically) are very frequent in practice.

As a consequence of the above properties, it is possible to make all the colored tokens in the model "neutral", and obtain a GSPN model whose behavior is equivalent to that of the SWN we started with. The structure of the GSPN model is the same as that of the SWN in Fig. 8, the arcs with associated function $< x >$ become uncolored arcs of weight 1, the inhibitor arcs with associated function $< S >$ become uncolored inhibitor arcs of weight 1, the arc from transition *clock* to place *genArr* is changed to an uncolored arc whose weight is $|L| = w$ It is interesting to observe that the number of (tangible) states of the new GSPN model is the same as the number of symbolic states in the SRG of the SWN model. However, the computational cost for the generation of the RG is lower than that required for generating the SRG.

Observe that the model of the ED strategy in the MR-Quadro system cannot be decolorized, because it does not satisfy the first property above: each transition *receive_i* has an input and an inhibitor arc with the same variable x appearing in the corresponding expressions. The semantic of this structural property is that the MR-Quadro ED strategy needs to distinguish among packets placed onto different wavelengths, because it is not allowed to receive more than one packet per wavelength in the same time slot.

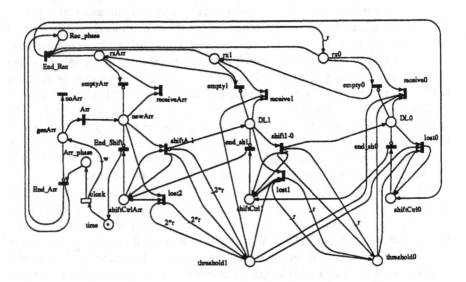

Fig. 11. GSPN model of strategy δ_1 with state reduction for a system with two delay lines.

Cutting the state space by dropping arriving packets in excess. Once we have a decolorized Quadro strategy model, we can further cut its number of states by defining an upper bound on the number of the packets in the delay lines: we do not allow a packet that will be certainly lost, to enter the DLs. If r receivers are available, we know that at most $r \cdot d$ packets can be received within d time slots. Hence, if the total number of packets in the system (after the new packets have arrived and the packets received/lost in the current time slot have already exited the system) is greater than $r \cdot d$, we can "throw away" the exceeding packets. The model in Fig. 11 implements this idea: the number of tokens in place *threshold1* is equal to the sum of the number of tokens in DL_0 and DL_1, this place is used to

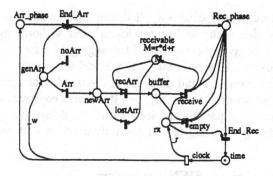

Fig. 12. GSPN model of strategy δ_1 with state reduction, "folded" model.

let at most $r \cdot d$ tokens (packets) in the two delay lines, indeed it conditions the enabling of transition *shiftA-1*; the exceeding packets are thrown away through transition *lost2* and become lost packets. The subnet of place *threshold0* works as that of *threshold1*, with an upper bound $r \cdot (d-1)$; however, since we know that all the packets that have passed through *shiftA-1* will be certainly received, the place *threshold0* is redundant: indeed, transition *lost0* has always throughput 0. For the same reason also transition *lost* never fires.

Then we can eliminate the useless place and transitions, and perform a further simplification step by folding the subnet comprising places DL_1 and all the connected transitions and the subnet including place DL_0 and all the connected transitions. This makes the packets contained in the two delay lines indistinguishable, however this doesn't change the behavior of the system since in the new model we keep only the packets that will be certainly received by the Quadro strategy, hence it is enough to know how many packets are in the system, regardless of their position. The model obtained is depicted in Fig. 12, in which place *receivable* contains as many tokens as the packets that are still allowed to enter the DLs, while place *buffer* represents the number of (certainly receivable) packets in the system[7]. The number of states of this very compact model is $r \cdot d + 1$.

The idea of throwing away the packets *in advance* can be applied to the SWN model of the MR-Quadro ED and MC strategies as well, but in this case the token colors must be retained and the places representing the delay lines cannot be folded.

A summary of the number of states in the models developed so far is reported in Table 1 (the number of states for both the ED and the MC strategy is the same): the first column reports the state space size without any type of aggregation ($2^{(d \cdot w)}$), the second column gives the number of states when the SRG

---

[7] Place *receivable* has initial marking $r \cdot (d+1)$ because the model accepts the newly arrived packets (tokens) before dismissing the packets (tokens) to be received in the same time slot.

(or decolorization) aggregation is applied, the third column gives the number of states of the models with "anticipated losses", but without aggregation, finally the fourth column gives the number of the aggregate states for the model with "anticipated losses". Observe that in the model with anticipated losses, if we fix the number of receivers and delay lines, the number of aggregate states as a function of w, grows until $w = r \cdot d$, then it remains constant, this is due to the fact that at most $r \cdot d$ packets are accepted in the delay lines, as a consequence $w - r \cdot d$ (modeled) wavelengths will certainly never contain any packet.

4.3 Suitability of the SWN Formalism

Our experience in using the SWN formalism to model the system under study indicates that SWNs are adequate to model the system (hardware) structure, and the workload. In addition, models for the simplest strategies (namely Quadro with ED) are easy to build. On the contrary, in its current form the SWN formalism does not have the necessary flexibility to easily model complex strategies such as MC and MC+1. As mentioned in a previous section in our opinion this problem could be overcome by extending the formalism to include marking dependent arc weights (of restricted type) as in [22], and more elaborated enabling and state change rules as in SAN [31]. Preferably, the extensions should not prevent the applicability of the SRG construction algorithm.

From the point of view of the performance analysis, the advantage we experienced in using a high level formalism (instead of writing an *ad hoc* program) to generate the underlying Markov chain, is that the automatic state reduction techniques could be exploited in a transparent way, moreover the results obtained were already given in terms of the high level model, i.e., of the system, so that no further work was needed to compute the performance indices of interest from the steady state solution of the underlying Markov chain.

A drawback that we have faced in the analysis phase is the large number of vanishing states generated in an intermediate step of the algorithm prior to building the underlying Markov chain[11]. In some cases the time and memory requirements in this intermediate step prevented the generation of the (relatively small) Markov chain. This problem in part is due to the non optimized SRG generation algorithm[8] implemented in the package used to solve the models, but most relevant is the impossibility of firing sets of non conflicting immediate transitions in a single step (that in this specific application would also represent more faithfully the synchronous behavior of the system). The problem is intrinsic in the model nature, and could be partially avoided by applying some well known immediate transitions reduction techniques [14, 8]: we have not used this option because we did not have a tool providing an implementation of the above

---

[8] The SRG algorithm implementation in *GreatSPN* does not yet include some optimizations (that are instead implemented in the algorithm for the generation of the RG of GSPN models) that avoid the construction of all possible interleavings, when several independent immediate transitions are concurrently enabled.

mentioned algorithms, and performing the transformation manually in this complex models would have been extremely error prone. Another possible approach to overcome this problem is to eliminate vanishing markings "on the fly" as the state space is being constructed. In our study, this problem was circumvented by means of a "modified" SRG generation algorithm tailored to our models that avoids the vanishing markings generation.

One last important remark concerns the apparent mismatch between the negative exponential stochastic timing provided by the SWN formalism and the deterministic duration of the slot in the studied system. Actually the models that we have developed use extensively the *immediate transitions* feature of the formalism and the possibility of solving in a probabilistic way the conflicts between alternative choices. In all our models there is only one timed transition (activity) whose firing (completion) represents the start of a new time slot. The state change due to arrival/reception/loss of packets takes place "instantaneously" after a new time slot starts: the model allows to compute the probability of reaching any possible "next" state. Then, after a randomly generated time period, a new state change takes place. As a matter of fact the stochastic process we need, to obtain the performance indices of interest, is the so called "embedded Markov chain" [8] associated with the CTMC derived from the SWN model, that takes into account only the probability of moving from one state to another when a state transition takes place, disregarding the time actually spent in each state. For this reason the stochastic nature of the slot duration in our model is irrelevant (in our models we have assigned an average delay of 1 to the unique timed transition, because by so doing, the steady state solutions of the CTMC and of the embedded MC coincide). A similar approach has been used in [10] for modeling ATM networks using GSPNs.

5   Performance Results

As previously mentioned, the solution requiring one Quadro device per receiver can be shown to be optimal and, under control strategy δ_1, it achieves the best obtainable performance (minimum packet loss probability) for any given node configuration (i.e., for any triple w, d, r). Numerical results are first shown for nodes making use of Quadros under varying workload, defined as the average number of arriving packets per slot per receiver ($l \in [0, 1]$). Quadro nodes are assumed to have $r \in [1, 4]$, $d \in [0, 10]$ and $w = 8$. The penalties introduced by the use of the MR-Quadro in place of the r Quadros are then evaluated for each of the three control strategies: ED, MC and MC+1.

Throughput is defined as the average number of received packets per slot, normalized to r. Packet loss probability is the average packet loss per slot, normalized to r. Packet delay is the average time spent by a packet in the Quadro receiver prior to reception. These performance indices are numerically evaluated from the SWN models presented in Section 4 with probability p defined as $p = \frac{l \cdot r}{w}$; follows the definition of the performance indices as functions of transition throughputs. Packet loss probability is given by

$$Prob_{packet\_loss} = X_{lost}/X_{Arr} = X_{lost}/(r*l)$$

where X_{lost} is the throughput of transition *lost*, that represents the average packet loss per slot. System throughput is given by

$$throughput_{sys} = X_{Arr} - X_{lost} = l*r - X_{lost}$$

and the average packet delay is given by

$$avg\_delay = \sum_{i=0}^{d} i * Prob(delay = i)$$

where $Prob(delay = i)$ is the probability for a packet to be received i time slots after its arrival, defined as[9]:

$$Prob(delay = i) = X_{received_{d-i}}/(X_{Arr} - X_{lost})$$

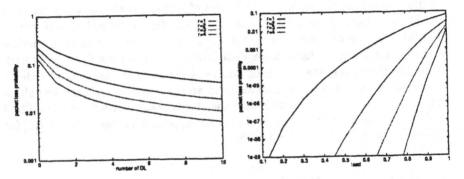

Fig. 13. Quadro strategy: packet loss probability versus d, for $l = 1$, $w = 8$, and varying values of $r = 1, 2, 3, 4$.

Fig. 14. Quadro strategy: packet loss probability versus l, for $w = 8$, $d = 5$, and varying values of $r = 1, 2, 3, 4$.

Fig. 13 shows packet loss probability versus d, for different values of r and for network load $l = 1$. For all shown values of r, packet loss probability nearly decreases of one order of magnitude in the $d = 10$ node when compared to the $d = 0$ node (Quadro is not used). Note that the number of lost packets per receiver decreases as r increases.

Fig. 14 shows packet loss probability versus l, for different values of r, in a $d = 5$ Quadro node. By limiting the offered load allowed in the network, packet loss probability can be kept below required values. For example, in a two receiver node with offered load not exceeding $l = 0.5$, packet loss probability can be maintained below 10^{-8}.

---

[9] The transition denoted $received_d$ in the formula actually corresponds to transition *receiveArr* in the nets.

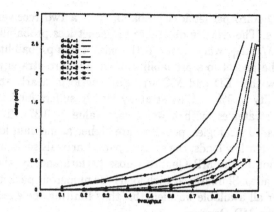

Fig. 15. Quadro strategy: packet delay versus throughput, for $w = 8$, and varying values of $d = 1, 5$ and $r = 1, 2, 3, 4$.

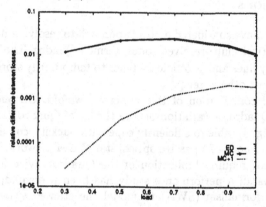

Fig. 16. Packet loss penalty versus l, for the three MR-Quadro control strategies ED, MC and MC+1 when compared to the Quadro strategy, in a node with $w = 8$, $d = 2$ and $r = 2$.

Fig. 15 shows packet delay versus throughput in eight node configurations, obtained by varying the number of DLs ($d = 1, 5$) and the number of receivers ($r = 1, 2, 3, 4$). The resulting curves show that at any given throughput, packet delay decreases if the number of receivers is increased, while it grows if the number of DLs is increased. The latter result can be intuitively explained by considering that in order to decrease packet loss the rightmost packets are always received first, thus moving the average reception position deep into the Quadro. As a consequence a $d = 5$ Quadro will receive more packets than a $d = 1$ Quadro, but on the average the received packets will have been delayed by few more DLs.

Consider now the MR-Quadro and strategies ED, MC and MC+1. We expect some performance penalties introduced by the MR-Quadro receiver when compared to the Quadro receiver node. Fig. 16 shows the relative difference of packet loss probability for each of the MR-Quadro strategies, when compared to the Quadro packet loss probability, in a two receiver ($r = 2$) node with $d = 2$ DLs. The relative difference of packet loss probability is defined as $(lost_X - lost_{\delta_1})/lost_{\delta_1}$, where $lost_X$ is the packet loss probability of strategy X and $lost_{\delta_1}$ is the packet loss probability of the Quadro strategy. The curves in the graph show that ED and MC strategies perform nearly the same, with a maximum penalty of 3%. MC+1 strategy visibly suffers for less penalties that the other two strategies, with a worst case value of 0.3%. In general, for all shown MR-Quadro strategies, penalties are higher at medium loads than at low and high loads. For low loads, the sparing packet arrivals seldom originate wavelength contention in the MR-Quadro whose performance is, therefore, close to the optimal Quadro. Similarly, at high loads the numerous packet arrivals tend to spread out over all available wavelengths, thus limiting the effect of wavelength contention in the MR-Quadro.

6 Conclusions

In this paper we have considered a novel approach to resolving packet contention occurring in optical multi-receiver nodes, termed Quadro. The approach makes use of fiber delay lines and photonic switches to temporarily store optical packets that would be lost otherwise.

The original contribution of this paper is twofold. On one hand we have proposed MR-Quadro, a variation on the theme of Quadro comprising several receivers (see Fig. 4), able to efficiently cope with packets' contentions arising at the multi-receiver node of a passive optical star WDM network. The MR-Quadro solution does not require duplication of the Quadro device for each receiver. On the other hand, a performance study has been presented, that makes use of a high level formalism (SWN) to model the studied system, and exploits the performance analysis techniques recently developed for this formalism to obtain exact performance results despite the large state space of the system when realistic configurations are considered. Moreover we have shown that simple manipulations of the models, based on intuitive considerations on the system behavior, allow to further reduce the state space size of the underlying Markov chains.

Three strategies with increasing complexity were considered for controlling the MR-Quadro: the Earliest Deadline (ED), the Most Congested (MC) and the Most Congested with look ahead (MC+1). It was shown that the MR-Quadro with the MC+1 strategy achieves packet loss probability that at most is 2% worse than the multi-receiver node making use of one Quadro device for each of its receivers.

The experience derived from using the SWN formalism in this application domain led us to the conclusion that the expressive power of the formalism is adequate to describe the system (hardware) structure, the workload and the simplest reception strategies (Quadro,ED), while it is awkward when more complex strategies, such as the MC and MC+1, need to be modeled. In conclusion, some extensions similar to those already available in other formalisms (e.g., see [22] and [31]) would be useful in the SWN, under the condition that the addition of these new features must preserve the efficient analysis techniques available in the extant SWN formalism.

References

1. *IEEE Network*, 6(2), March 1992. Special issue on Report on Gigabit Networking.

2. *IEEE Communications Magazine*, 30(4), March 1992. Special issue on Gigabit Networks.

3. *Proceedings of Gigabit Networks: Emerging Commercial Applications & Opportunities*, Washington, DC, July 1992.

4. *IEEE/OSA Journal of Lightwave Technology, jointly with IEEE Journal on Selected Areas in Communications*, 14(6), June 1996.

5. *IEEE Journal on Selected Areas in Communications, jointly with IEEE/OSA Journal of Lightwave Technology*, 14(5), June 1996.

6. Silinguelli A. Tecniche di ricezione basate su linee di ritardo ottiche commutate. Master's thesis, Politecnico di Torino, February 1996.

7. M. Ajmone Marsan, G. Balbo, and G. Conte. A class of generalized stochastic Petri nets for the performance analysis of multiprocessor systems. *ACM Transactions on Computer Systems*, 2(1), May 1984.

8. M. Ajmone Marsan, G. Balbo, G. Conte, S. Donatelli, and G. Franceschinis. *Modelling with Generalized Stochastic Petri Nets*. J. Wiley, 1995.

9. M. Ajmone Marsan, G. Balbo, and K.S. Trivedi, editors. *Proc. Intern. Workshop on Timed Petri Nets*, Torino, Italy, July 1985. IEEE-CS Press.

10. M Ajmone Marsan and R. Gaeta. GSPN models for ATM networks. In *Proc. 7th Int. Workshop on Petri Nets and Performance Models*, St. Malo', France, June 1997. IEEE-CS Press.

11. A. Blakemore. The cost of eliminating vanishing markings from generalized stochastic Petri nets. In *Proc. Int. Workshop on Petri Nets and Performance Models*, Kyoto, Japan, December 1989. IEEE-CS Press.

12. C. A. Brackett. Dense wavelength division multiplexing networks: Principles and applications. *IEEE Journal on Selected Areas in Communications*, 8(6):948–964, August 1990.

13. M.-S Chen, N. R. Dono, and R. Ramaswami. A media-access protocol for packet-switched wavelength division multiaccess metropolitan area networks. *IEEE Journal on Selected Areas in Communications*, 8(6):1048–1057, August 1990.

14. G. Chiola, S. Donatelli, and G. Franceschinis. GSPN versus SPN: what is the actual role of immediate transitions? In *Proc. 4th Intern. Workshop on Petri Nets and Performance Models*, pages 20–31, Melbourne, Australia, December 1991. IEEE-CS Press.

15. G. Chiola, C. Dutheillet, G. Franceschinis, and S. Haddad. Stochastic well-formed coloured nets for symmetric modelling applications. *IEEE Transactions on Computers*, 42(11), November 1993.

16. G. Chiola and G. Franceschinis. Colored GSPN models and automatic symmetry detection. In *Proc. 3rd Intern. Workshop on Petri Nets and Performance Models*, Kyoto, Japan, December 1989. IEEE-CS Press.

17. G. Chiola, G. Franceschinis, R. Gaeta, and M. Ribaudo. GreatSPN 1.7: Graphical Editor and Analyzer for Timed and Stochastic Petri Nets. *Performance Evaluation, special issue on Performance Modeling Tools*, 24(1&2):47–68, November 1995.

18. I. Chlamtac. Rational, directions and issues surrounding high speed computer networks. *IEEE Proceedings*, 78(1):94–120, January 1989.

19. I. Chlamtac and A. Fumagalli. QUADRO-Star: High performance optical WDM star networks. In *IEEE Globecom'91*, Phoenix, Arizona, 1991.

20. I. Chlamtac and A. Fumagalli. Quadro: An all-optical solution for resource contentions in packet switching networks. *Journal on Computer Networks and ISDN Systems*, 1992.

21. I. Chlamtac, A. Fumagalli, L. G. Kazovsky, P. Melman, W. H. Nelson, P. Poggiolini, M. Cerisola, A. N. M. Choudhury, T. K. Fong, R. T. Hofmeister, C.-L. Lu, A. Mekkittikul, D. J. M. Sabido IX, C.-J. Suh, and E. W. M. Wong. CORD: COntention Resolution by Delay lines. *IEEE Journal on Selected Areas in Communications and IEEE Journal of Lightwave Technology*, 14(5), June 1996.

22. G. Ciardo. Petri nets with marking dependent arc cardinality: Properties and analysis. In *Proc. 15th Intern. Conference on Applications and Theory of Petri Nets*, number 815 in LNCS, Zaragoza, Spain, 1994. Springer-Verlag.

23. N. R. Dono, P. E. Green, K. Liu, R. Ramaswami, and F. F.-K Tong. A wavelength division multiple access network for computer communication. *IEEE Journal on Selected Areas in Communications*, 8(6):983–993, August 1990.

24. G. Florin and S. Natkin. Les reseaux de Petri stochastiques. *Technique et Science Informatiques*, 4(1), February 1985.

25. A. Ganz and I. Chlamtac. Path allocation access control in fiber optic communication systems. *IEEE Transactions on Computers*, c-38(10):1372–1382, October 1989.

26. B. S. Glance and O. Scaramucci. High-performance dense FDM coherent optical network. *IEEE Journal on Selected Areas in Communications*, 8(6):1043–1047, August 1990.

27. P. E. Green. The future of fiber-optic computer networks. *IEEE Computer*, 24(9):78–87, September 1991.

28. P. Huber, A.M. Jensen, L.O. Jepsen, and K. Jensen. Towards reachability trees for high-level Petri nets. In G. Rozenberg, editor, *Advances on Petri Nets '84*, volume 188 of *LNCS*, pages 215–233. Springer Verlag, 1984.

29. K. Jensen. *Coloured Petri Nets, Basic Concepts, Analysis Methods and Practical Use. Volume 1 & 2*. Springer Verlag, 1995.

30. K. Jensen and G. Rozenberg, editors. *High-Level Petri Nets. Theory and Application*. Springer Verlag, 1991.

31. J.F. Meyer, A. Movaghar, and W.H. Sanders. Stochastic activity networks: Structure, behavior, and application. In *Proc. Intern. Workshop on Timed Petri Nets*, pages 106–115, Torino,Italy, July 1985.

32. M.K. Molloy. *On the Integration of Delay and Throughput Measures in Distributed Processing Models*. PhD thesis, UCLA, Los Angeles, CA, 1981. Ph.D. Thesis.

33. B. Mukherjee. WDM-based local lightwave networks part i: Single-hop systems. *IEEE Network*, 6(3):12–27, May 1992.

34. B. Mukherjee. WDM-based local lightwave networks part ii: Multi-hop systems. *IEEE Network*, 6(4):20–32, July 1992.

35. E. Nussbaum. Communication network needs and technologies - a place for photonic switching. *IEEE Journal on Selected Areas in Communications*, 6(7):1036–1043, August 1988.

36. J.L. Peterson. *Petri Net Theory and the Modeling of Systems*. Prentice-Hall, Englewood Cliffs, NJ, 1981.

37. W.H. Sanders and J.F. Meyer. Reduced base model construction methods for stochastic activity networks. In *Proc. 3rd Intern. Workshop on Petri Nets and Performance Models*, Kyoto, Japan, December 1989. IEEE-CS Press.

38. F.J.W. Symons. *Modeling and Analysis of Communication Protocols Using Numerical Petri Nets*. PhD thesis, University of Essex, May 1978.

39. H. Toba, K. Oda, K. Nakanishi, N. Shibata, K. Nosu, N. Takato, and M. Fukuda. 100-channel optical FDM transmission/distribution at 622 Mb/s over 50 Km. In *Proc. OFC'90*, San Francisco, CA, 1990.

40. E.W.M. Wong, A. Fumagalli, and I. Chlamtac. Performance evaluation of CROWNs: WDM multi-ring topologies. In *Proc. ICC'95*, Seattle, WA, June 1995.

Springer
and the
environment

At Springer we firmly believe that an
international science publisher has a
special obligation to the environment,
and our corporate policies consistently
reflect this conviction.
We also expect our business partners –
paper mills, printers, packaging
manufacturers, etc. – to commit
themselves to using materials and
production processes that do not harm
the environment. The paper in this
book is made from low- or no-chlorine
pulp and is acid free, in conformance
with international standards for paper
permanency.

 Springer

Lecture Notes in Computer Science

For information about Vols. 1–1505
please contact your bookseller or Springer-Verlag